Innovation Policy

Innovation Policy
A Guide for Developing Countries

THE WORLD BANK
Washington, D.C.

2010 The International Bank for Reconstruction and Development / The World Bank
1818 H Street, NW
Washington, DC 20433
Telephone: 202-473-1000
Internet: www.worldbank.org
E-mail: feedback@worldbank.org

All rights reserved

1 2 3 4 13 12 11 10

This volume is a product of the staff of the International Bank for Reconstruction and Development / The World Bank. The findings, interpretations, and conclusions expressed in this volume do not necessarily reflect the views of the Executive Directors of The World Bank or the governments they represent.

The World Bank does not guarantee the accuracy of the data included in this work. The boundaries, colors, denominations, and other information shown on any map in this work do not imply any judgement on the part of The World Bank concerning the legal status of any territory or the endorsement or acceptance of such boundaries.

Rights and Permissions

The material in this publication is copyrighted. Copying and/or transmitting portions or all of this work without permission may be a violation of applicable law. The International Bank for Reconstruction and Development / The World Bank encourages dissemination of its work and will normally grant permission to reproduce portions of the work promptly.

For permission to photocopy or reprint any part of this work, please send a request with complete information to the Copyright Clearance Center Inc., 222 Rosewood Drive, Danvers, MA 01923, USA; telephone: 978-750-8400; fax: 978-750-4470; Internet: www.copyright.com.

All other queries on rights and licenses, including subsidiary rights, should be addressed to the Office of the Publisher, The World Bank, 1818 H Street, NW, Washington, DC 20433, USA; fax: 202-522-2422; e-mail: pubrights@worldbank.org.

ISBN: 978-0-8213-8269-1
eISBN: 978-0-8213-8301-8
DOI: 10.1596/978-0-8213-8269-1

Cover images (clockwise from far left): © Monty Rakusen/cultura/Corbis; © E.O. Hoppé/Corbis; © Tim Pannell/Corbis; © Juice Images/Corbis; © Tony Metaxas/Asia Images/Corbis; © Tim Pannell/Corbis.

Library of Congress Cataloging-in-Publication Data
Innovation policy : a guide for developing countries.
   p. cm.
   Includes bibliographical references and index.
   ISBN 978-0-8213-8269-1 — ISBN 978-0-8213-8301-8 (electronic)
1. Technological innovations—Developing countries. 2. Technology—Economic aspects—Developing countries. I. World Bank.
 HC79.T4I5472 2010
 338'.064091724—dc22
                              2009054248

Contents

Foreword	xv
Preface	xvii
Abbreviations	xix
Overview	1
Why? The Innovation Imperative	1
What? The Government as a Gardener	2
How? A Pragmatic Agenda	3
What Is Innovation?	4
Policy Concept	5
Policy Functions	11
Policy Implementation	17
Conclusion	22
Notes	24
References	24
Introduction	25
Innovation, Did You Say?	25
What Is This Book About?	27

Part I	**Policy Concept**	29
1	**Why Promote Innovation? The Key to Economic, Social, and Environmental Progress**	31
	Innovation and Societies: The Long-Term View	32
	Technology and Economic Growth	38
	Innovation and Emerging Economies	43
	Dissemination of Technology	46
	New Global Perspectives	48

v

	Notes	50
	References and Other Resources	51
2	**How to Promote Innovation: Policy Principles**	**53**
	Take a Broad View of Innovation	54
	Adopt a "Whole-of-Government" Approach	54
	Create a Receptive and Mobilizing Environment	60
	Put Efficient Institutions and Instruments in Place	65
	Adapt to Societal Specificities	67
	Policy Conclusions for Developing Countries	68
	Notes	69
	References	70

Part II	**Policy Functions**	**71**
3	**Supporting Innovators**	**73**
	Provision of Business Services	74
	Entrepreneurship and New Innovating Firms	83
	Finance for New and Innovative Firms	89
	Bridging Institutions: Clusters and Networks	93
	Conclusions	102
	Notes	103
	References and Other Resources	104
4	**Improving the Regulatory Framework for Innovation**	**107**
	International Trade and Investment Framework	108
	Domestic Institutional and Regulatory Framework	116
	Procurement Policies for Innovation	125
	Conclusions	130
	Notes	130
	References and Other Resources	132
5	**Strengthening the Research and Development Base**	**135**
	Global Overview of R&D	135
	R&D in Developing Countries	139
	Public Sector R&D in Developing Countries	148
	Private Sector R&D in Developing Countries	150
	International R&D Cooperation and Research Programs	157
	Summary and Conclusions	159
	Notes	160
	References and Other Resources	161

6 Fostering Innovation through Education and Training — 165
Skills for a Knowledge-Based and Innovation-Driven Economy — 165
Lessons from Developed and Developing Countries — 173
Adapting the Way Learners Learn to the Knowledge Economy — 175
Beyond Formal General Education — 183
From Brain Drain to Brain Circulation — 190
Conclusion — 194
Notes — 195
References and Other Resources — 195

7 Policy Evaluation: Assessing Innovation Systems and Programs — 199
Benchmarking Innovation at the Country Level — 200
Microlevel Innovation Surveys — 206
Program Evaluation — 213
Innovation Policy Reviews — 224
Conclusions — 230
Notes — 232
References and Other Resources — 233

Part III Policy Implementation — 235

8 Policy Implementation: The Art and Craft of Innovation Policy Making — 237
Adapting Best Practices to the Local Context: The Pragmatic Innovation Agenda — 237
How to Create a Conducive Institutional Framework: The Virtuous Cycle — 255
Creating Frameworks for Change: Strategic Incrementalism — 262
Summary of Policy Principles — 267
Note — 268
References and Other Resources — 268

9 Promoting Competitive and Innovative Industries — 271
Innovation, a Global Phenomenon — 272
Agriculture — 275
Manufacturing — 283
Services — 288
Policy Conclusions — 295

	Notes	298
	References and Other Resources	299
10	**Building Innovative Sites**	**303**
	Special Economic Zones	304
	Science Parks	310
	Clusters	316
	Fostering Innovation in a City or Region	323
	Conclusion	329
	Notes	330
	References and Other Resources	331
11	**Stimulating Pro-Poor Innovations**	**335**
	How to Define Inclusive Innovations, Pro-Poor Innovations	335
	Harnessing Formal Innovation Efforts for the Poor	338
	Promoting Grassroots Innovation and Knowledge Initiatives	356
	Enabling the Informal Sector to Absorb Knowledge and Technology	362
	Notes	369
	References and Other Resources	370

Index   **375**

Boxes

O.1	A Few Examples of Innovations in Developing and Emerging Economies	5
O.2	Innovation Policies in OECD Countries—50 Years of Experience	10
O.3	Business Services for Innovators	12
1.1	Innovation Is Essential to Tackling Climate Change	37
1.2	India, an Early Innovator	39
2.1	A Brief History of Innovation Policy in OECD Countries	56
3.1	Priorities for Business Services Support Schemes	77
3.2	Knowledge Vouchers	85
3.3	Types of Incubators	86
3.4	Good Practices for Business Incubators	88
3.5	Singapore: Incubators Underpinning a Relationship Hub	89
3.6	SME Clusters in India	94
3.7	The Role of Trade Associations in Italy	96
3.8	Sector Associations in Senegal and Cameroon	97

3.9	Denmark's Network Program: Brokers and Scouts	99
3.10	Networking Programs: The International Experience	101
4.1	Brazil's Policy on HIV/AIDS	113
4.2	From Duplicative Imitation to Creative Imitation and Innovation	115
4.3	Kenya's Radical Licensing Reform, 2005–07	119
4.4	Railways and Competition	121
4.5	Variable Message Signs for British Highways	127
4.6	The Swedish Energy Agency's Procurement Procedures	129
5.1	Becoming a More Internationally Competitive, Market-Driven R&D Organization	149
6.1	What Does a Learner-Centered Classroom Look Like?	176
6.2	Learner-Centered Teaching for the Knowledge Economy	177
6.3	Entrepreneurship Program at Walhalla High School, South Carolina, United States	182
6.4	Mexico's Proactive Training Programs for SMEs	188
7.1	Examples of Indicators from Innovation Surveys	207
7.2	Additional Sources of Innovation-Related Indicators	209
7.3	Effective Policy and Program Evaluation Challenges for Designing Evaluation Schemes	214
7.4	Emerging Cross-Cutting Issues in the Evaluation of Publicly Funded Research	220
7.5	National Innovation Strategies: Lessons from OECD Country Reviews	226
7.6	Beyond GDP: Alternative Measures and Indicators of Economic and Social Progress	231
8.1	Private and Public Sector Entrepreneurs Come Together: An Irish Experience	239
8.2	Turning Scientists into Entrepreneurs: Moscow University's Science Park	248
8.3	Diaspora Member Creates First-Mover Institution in Tertiary Education	250
8.4	Members of the Diaspora Trigger Changes in Innovation Systems	257
8.5	A Framework Program to Promote Experimentation in a Rural Setting: The Spark Program	263

9.1	China, an Emerging Leader in Wind Power	273
9.2	Mauritius, Reinventing for Survival in the Global Economy	274
9.3	Main Messages from *Agribusiness and Innovation Systems in Africa*	279
9.4	Malaysia's Palm Oil Industry	281
9.5	Gold Jewelry in Turkey	284
9.6	Producing Jeans in Toritama, Brazil	285
9.7	The Software Industry in India	293
10.1	The Development of Backward Links: A Successful and a Less Successful Example	307
10.2	Attracting High-Technology Investments in an SEZ in Costa Rica	309
10.3	SEZs and Labor Circulation—A "Domestic Skilled Diaspora"?	309
10.4	A Tale of Two Countries—Investment Climate Reform	310
10.5	Zhongguancun Science Park in Beijing, China	312
10.6	Greater Sfax Development Strategy and the Science Park	313
10.7	TEKEL	314
10.8	Different Cluster Initiators	318
10.9	Demanding Local Consumers	321
10.10	The Creative Class?	325
10.11	CUORE in Naples	328
10.12	The Vancouver Agreement	329
11.1	Adapting Public Research Systems for Development Needs in India	341
11.2	Pro-Poor Innovations at University Research Centers in Africa	342
11.3	The 4 Billion at the Base of the Economic Pyramid	345
11.4	The Private Sector as Provider of Products and Services for the Poor	346
11.5	Pro-Poor Public-Private Partnerships	349
11.6	Examples of Socially Driven Pro-Poor Initiatives	351
11.7	Using Indigenous Knowledge to Improve Health and Raise Agricultural Productivity	358
11.8	Benefit-Sharing Arrangements and Intellectual Property Protection for Indigenous and Traditional Knowledge	360
11.9	Participatory Development of Improved Groundnut Varieties in Ghana	364

11.10	Public-Private Technology Information Service in Rwanda	367
11.11	Financial Institutions That Serve the Poor	368

Figures

O.1	How Innovation Contributes to Growth: A Comparison of Ghana and the Republic of Korea, 1960–2005	6
O.2	Major Technical Systems from the Middle Ages through the Present	7
O.3	Determinants of Technology Upgrading in Developing Countries: Domestic Absorptive Capacity Both Conditions and Attracts External Flows	8
O.4	Gardening Innovation	9
O.5	Innovation Policy in a Broad Perspective	10
O.6	Model for a Strong Innovation Policy	11
O.7	Scaling Up Institutional Change, from Microreforms to National Reforms	20
1.1	World Population Growth and Major Technological Events, 9000 BC to Present	32
1.2	Average Life Expectancy over the Past Two Millennia	33
1.3	Growth in Population and GDP per Capita in the Past 2,000 Years	33
1.4	Structure of the Global Technical System	34
1.5	The Industrial Revolution in Europe, 1750–1970	35
1.6	The Cognitive Revolution, 1980–2180	36
1.7	Per Capita GDP for Selected Countries or Regions, 1480–1998	39
2.1	Process Components of the Wine Industry in South Africa	55
2.2	Organizational and Marketing Elements of the Wine Industry in South Africa	55
2.3	Creating Favorable Conditions for Innovation	58
2.4	Schematic of the Innovation System in a Developing Country	59
2.5	Government Roles in Encouraging Innovation	60
2.6	Traditional Layout of Innovation Policy	66
2.7	Comprehensive Layout of Innovation Policy	66
3.1	Shanghai R&D Public Service Platform	78
3.2	Financing Cycle for New Technology-Based Firms	90
4.1	Example of a Phased Precommercial Procurement Process	128

5.1	Gross Domestic Expenditures on R&D by Area, 1996–2006	136
5.2	Relative R&D Expenditures and Number of Scientists and Engineers in G5 Countries and BRICs, 2006 PPP	140
5.3	R&D Expenditures as a Percentage of GDP for Selected Economies, 2005 PPP	141
5.4	R&D Expenditures by Sector as a Percentage of National Total in Selected Economies, 2005	143
6.1	Long-Run Trends in Skilled Emigration in Developing Countries, 1975–2000	192
6.2	Hierarchy of Diaspora Impact on Institutional Reform in Developing Countries	194
8.1	Elements of Strategic Incrementalism	264
9.1	Conceptual Diagram of an Agricultural Innovation System	278
10.1	"Island" versus "Catalyst" Special Economic Zones	305
10.2	Evolution of Science Parks over Time	312

Tables

O.1	Direct Support for Business Sector Research and Development	15
O.2	Country Contexts and Strategic Focal Points	19
1.1	Industries and Infrastructure of Each Technological Revolution, 1770–1970	36
1.2	Conventional Breakdown of Sources of Growth, 1970–2000	42
1.3	Level of Productivity in Countries of Various Incomes, 1970–2000	42
1.4	Productivity Dispersion in Brazil's Industrial Sectors, Mid-2000s	45
1.5	Percentage of Children Worldwide Who Received Basic Vaccines and Ratio to High-Income Countries, 1993 and 2003	47
1.6	Percentage of Rural and Urban Population with Access to Clean Water, 1990 and 2004	47
1.7	Percentage of Rural and Urban Population with Access to Sanitation, 1990 and 2004	48
1.8	Rate of Dissemination of Major Technologies, 1748–2000	49
2.1	East-West Contrasts in Socioeconomic Systems	67
3.1	Private Equity Fund-Raising in Emerging Markets, 2003–05	92

4.1	Example of Transaction Costs Related to the Legal and Regulatory Environment	117
5.1	R&D Performed in Government and Universities as a Percentage of GDP, 1996–2006	137
5.2	R&D Expenditure by Source of Financing: Main OECD and 10 Developing and Emerging Economies, 2005	142
5.3	Number of Patents Granted by the U.S. Patent and Trademark Office, 2008	144
5.4	Advantages and Disadvantages of Instruments for Encouraging Innovation and R&D	146
5.5	Top-10 R&D Companies from Developing and Emerging Economies, 2007	151
5.6	Number of Developing Economy Companies among the Global 1,000, 2007	152
5.7	Utility Patents Granted by the U.S. Patent and Trademark Office to the Top-15 Developing and Emerging Economies, 2008	152
5.8	Illustrative Examples of Innovations by Developing Economy Firms	153
5.9	Direct Instruments for Supporting Business R&D	154
5.10	General Science and Technology Instruments for Supporting Business R&D	154
5.11	Selected Research Universities from the World's Top 100 and 500, by Country	156
5.12	Instruments for Promoting Relevant R&D in Universities and Greater Commercialization of Knowledge and Interaction with Enterprises	158
6.1	Distribution of African University Graduates by Field of Study, 2005	173
6.2	Characteristics of Traditional and Lifelong Learning Models	175
7.1	Rankings of Economies for the Knowledge Assessment Methodology, Global Competitiveness, and Global Innovation Indexes, 2008–09	204
8.1	Diversity of Pragmatic Innovation Agendas	243
8.2	Possible Innovation Paths for Saudi Arabia	252
9.1	Leading Exporters of High-Value Commodities in Developing Countries	280
9.2	Moving Up the Value Chain in Tourism in Costa Rica, 1980s–Future	291
9.3	Competitive Industries and Innovation Systems in China, Costa Rica, Rwanda, and Vietnam	295

10.1	Training for Workers in SEZs in Selected Economies	308
10.2	Encouraging Innovation through SEZs	311
11.1	Progress toward Meeting Millennium Development Goals in Four Regions, 2006	337
11.2	Three Models for Enabling Businesses to Serve the Poor Economically	348

Foreword

Innovation, particularly technological innovation, is rightly seen as a key to economic and social development. For that reason, the World Bank Institute (WBI) is putting the question of innovation and its promotion at the very core of its work program.

As more and more countries begin to formulate policies that support innovation, they need to learn from the experiences and good policy practices of dynamic economies, especially those from the developing world. Although emulating the success stories and models of other countries is not easy, useful principles and illustrations drawn from the experiences of others can help inform effective approaches to innovation in the difficult institutional and business climates of low- and medium-income countries. This is precisely what this book, prepared by a WBI expert team, is aiming to do.

The book proposes a realistic approach to innovation. In the developing world, innovation is generally not something brand new but something new to the society in question, which, if broadly disseminated, brings significant economic, social, or environmental change. The book offers a comprehensive view of innovation policy, in which the government, acting as a gardener, supports the innovators by providing appropriate financial and other measures ("watering the plant"); by removing regulatory, institutional, or competitive obstacles to innovation ("removing the weeds and pests"); and by strengthening the knowledge base through investment in education and research ("fertilizing the soil").

The book suggests a gradual approach to implementing innovation policies, starting with localized successes in specific industries or geographic areas and, thus, preparing the ground for broader reforms. A key success factor is the integration of a vision for innovation into long-term development strategies. Such a vision allows a country to define priorities and implement them across ministries and throughout its territory with properly aligned policies and investments.

This book, which contains a host of examples and is written in a very accessible style, should be of great use for policy-making communities all over the world and for countries at widely different levels of development.

Sanjay Pradhan
Vice President
World Bank Institute
World Bank Group

Preface

This volume, prepared by the World Bank Institute (WBI), presents a conceptual framework for understanding and learning about the principles of innovation policies and programs in various policy contexts, with an emphasis on low- and medium-income countries. It is intended primarily for policy-making communities in charge of technology, industry, science, and education, as well as economics and finance—indeed, government as a whole, since innovation policy entails, by its very nature, a whole-of-government approach. The book contains a set of distinct and complementary chapters and provides both policy principles and a host of examples from countries at various levels of development.

The book was prepared by the WBI Skills and Innovation Policy Cluster under the leadership of Jean-Eric Aubert (consultant; former lead specialist). Contributors include staff members Derek Chen, Ronald Kim, Yevgeny Kuznetzov, Kurt Larsen, Florian Theus, Anuja Utz, and Justine White, and consultants Carl Dahlman, Patrick Dubarle, Thierry Gaudin, Thais Leray, and Désirée Van Welsum. Their specific contributions are indicated in the various chapters of the book.

This volume benefited from comments made by those who reviewed it, including Jean Guinet (Organisation for Economic Co-operation and Development), Ramesh Mashelkar (Council for Scientific and Industrial Research, India), Alfred Jay Watkins (World Bank), and Shahid Yusuf (World Bank).

Special thanks go to Derek Chen (WBI) and Janet Sasser (World Bank Office of the Publisher) for overseeing production and shepherding the book through the publication process.

Abbreviations

$	All dollar amounts are U.S. dollars unless otherwise indicated.
AGOA	African Growth and Opportunity Act
AIS	agricultural innovation system
ARVs	antiretroviral drugs
ATA	Aid to Artisans
ATI	Arco Technology Index
ATM	automatic teller machine
ATP	Advanced Technology Program
BC	before Christ
BOP	base of the pyramid
BRI	Bank Rakyat Indonesia
BRICs	Brazil, Russia, India, and China
BRS	Business Reporting System
CGIAR	Consultative Group on International Agricultural Research
CIMO	Integral Quality and Modernization Program (Mexico)
CIS	Community Innovation Survey
CNA	Confederazione Nazionale Artiglianato
CORFO	Chilean Economic Development Agency
CRI	Crops Research Institute
CSIR	Indian Council on Scientific and Industrial Research
CSR	corporate social responsibility
CUORE	Urban Operational Centers for Economic Renewal (Italy)
EAO	Economic Assessment Office
EIB	European Investment Bank
EIS	European Innovation Scoreboard
EPZ	export processing zone
ESC	Educational Service Contracting
EU	European Union

EU-27	Austria, Belgium, Bulgaria, Cyprus, Czech Republic, Denmark, Estonia, Finland, France, Germany, Greece, Hungary, Ireland, Italy, Latvia, Lithuania, Luxembourg, Malta, the Netherlands, Poland, Portugal, Romania, Slovakia, Slovenia, Spain, Sweden, and the United Kingdom
Eurostat	European Statistical Office
FDI	foreign direct investment
GAVI	Global Alliance for Vaccines and Immunizations
GCI	Global Competitiveness Index
GDP	gross domestic product
GII	Global Innovation Index
GNH	Gross National Happiness Index
GPS	global positioning system
HDI	Human Development Index
IASP	International Association of Science Parks
ICT	information and communication technology
IFC	International Finance Corporation
IK	indigenous knowledge
IPCC	Intergovernmental Panel on Climate Change
IPR	intellectual property rights
ISO	International Organization for Standardization
IT	information technology
ITIF	Information Technology and Innovation Foundation
KAM	Knowledge Assessment Methodology
KEI	Knowledge Economy Index
KIST	Kigali Institute of Science and Technology
LHC	Large Hadron Collider
LP/R	land pooling and readjustment programs
MDG	Millennium Development Goal
MEP	U.S. Manufacturing Extension Partnership
MFA	Multi-Fiber Agreement
MFN	most-favored nation
MIGA	Multilateral Investment Guarantee Agency
MNC	multinational company (or corporation)
MRTC	Malaria Research and Training Center
MSEs	micro- and small enterprises
NEU	Nueva Escuela Unitaria
NGO	nongovernmental organization
NIST	National Institute of Standards and Technology
NQF	National Qualification Framework
NTE	new technology enterprises
OECD	Organisation for Economic Co-operation and Development
POS	point-of-sale (terminal)

PPP	purchasing power parity
PRO	public research organization
R&D	research and development
RICYT	(Ibero-American Network on Science and Technology Indicators—Red Iberoamericana de Indicadores de Ciencia y Tecnología)
RTD	EU Research and Technological Development
S&T	science and technology
SERCOTEC	Technical Cooperation Service (Chile)
SEZ	special economic zone
SII	Summary Innovation Index
SMEs	small and medium enterprises
STCI	Science and Technology Capacity Index
TAI	Technology Achievement Index
TCM	traditional Chinese medicine
Tekes	Finnish Funding Agency for Technology and Innovation
TFP	total factor productivity
TK	traditional knowledge
TNC	transnational corporation
TPP	technological product and process
TVE	town and village enterprise
UAE	United Arab Emirates
UIS	UNESCO Institute for Statistics
UNCTAD	United Nations Conference on Trade and Development
UNDP	United Nations Development Programme
UNESCO	United Nations Educational, Scientific, and Cultural Organization
UNICEF	United Nations Children's Fund
UNICI	UNCTAD Innovation Capability Index
UNIDO	United Nations Industrial Development Organization
USAID	U.S. Agency for International Development
USPTO	U.S. Patent and Trademark Office
VC	venture capital
VET	vocational education and training
WBI	World Bank Institute
WEF	World Economic Forum
WFP	World Food Programme
WGC	World Gold Council

Overview
Innovation Policy: A Guide for Developing Countries

The presentation of innovation policy in this volume offers a detailed conceptual framework for understanding and learning about technology innovation policies and programs and their implementation in different countries. Inspired by the experience of both developed and developing countries, the book focuses on the latter's needs and issues.

The publication's main audience is the policy-making community. It includes not only those who are directly involved with technology, industry, science, and education but also those in charge of finance and economics, and indeed the top government leadership, which plays a crucial role in successful innovation policies.

This overview follows the organization of the volume, which is divided into parts and chapters. Before a summary of the individual chapters, however, the main messages that emerge from the volume as a whole are briefly presented.[1] The approach to innovation policy proposed in this volume revolves around the basic questions: Why? What? How?

Why? The Innovation Imperative

Technological innovation has always been at the heart of economic and social development. And as such, it is therefore essential to the further evolution of the developing world. Today, additional reasons make renewed attention to

---

This overview was prepared by Jean-Eric Aubert, with contributions from Carl Dahlman, Patrick Dubarle, Yevgeny Kuznetzov, Jean-François Rischard, and Justine White.

technology even more compelling. First, the world is in the midst of a serious economic crisis, and technology can be a means of relaunching or recreating economic activities worldwide. Second, major environmental challenges require wide-ranging changes in patterns of production and consumption. And third, the global technical system is undergoing a profound transformation based on information technologies and new technologies such as biotechnology and nanotechnology that are changing our world and our societies.

Innovation should be understood as the dissemination of something new in a given context, not as something new in absolute terms. While economically advanced countries naturally work at the technology frontier, developing countries have considerable opportunities for tapping into global knowledge and technology for dissemination in their domestic context. This ability will be decisive for initiating new activities, notably in service industries, for improving agriculture and industrial productivity, and for increasing overall welfare in areas like health and nutrition.

Innovation depends significantly on overall conditions in the economy, governance, education, and infrastructure. Such framework conditions are particularly problematic in developing countries, but experience shows not only that proactive innovation policies are possible and effective but also that they help create an environment for broader reforms.

What? The Government as a Gardener

Innovation can be approached from an organic and evolutionary perspective. An efficient innovation policy addresses the overall innovation climate, which goes far beyond traditional science and technology policy, and involves many government departments.

At the same time, government action can usefully focus on a few generic functions comparable to nurturing plants to help them grow. It can facilitate the articulation and implementation of innovative initiatives, since innovators need basic technical, financial, and other support (watering the plant). The government can reduce obstacles to innovation in competition and in regulatory and legal frameworks (removing the weeds and pests). Government-sponsored research and development (R&D) structures can respond to the needs and demands of surrounding communities (fertilizing the soil). And finally the educational system can help form a receptive and creative population (preparing the ground). For each of these functions, economically advanced as well as less advanced countries offer good practices that can be adapted to local contexts.

The firm backing of top leadership, such as the head of state or prime minister, is essential to the success of an innovation policy. It gives credibility to a national vision and facilitates the adoption of key measures for removing bureaucratic hurdles. It is also important to have efficient mechanisms that facilitate cross-departmental cooperation. By its very nature, innovation policy concerns parts of government that usually work independently. Agile and flexible agencies for implementing innovation policy measures may be necessary especially for supporting specific industries, technologies, or communities.

The institutional challenges to innovation policy should not be underestimated, as it intervenes in institutional settings that are already "crowded" with organizations that are supposed to fulfill—or claim to fulfill—its objectives and functions. Careful policy reviews and assessments, conducted with the help of the international community, can facilitate needed adaptations.

How? A Pragmatic Agenda

Since in most countries, particularly in the difficult institutional context of developing countries, implementing innovation policy is a challenging task, a long-term strategy should be inspired by a philosophy of "radical gradualism." That term refers to a sequence of finely tuned small, specific reforms and successful outcomes that paves the way for broader, institutional changes.

Depending on countries' technological competence and the quality of the business environment, governments will need to choose their goals. After focusing on prime movers and creating innovation endowments (well-defined technology centers, science parks, or export zones), they need to build critical masses of innovative and entrepreneurial initiatives by promoting industrial clusters, actively attracting foreign direct investment (FDI), and possibly even creating new cities. The multiplication of entry points in the economic system will facilitate broader reforms. In all cases, local communities and governments must be mobilized. This effort requires adequate incentives such as matching funds and administrative frameworks that include the delegation of power.

To materialize and advance this strategic process of change, policy initiatives targeted to specific industries, sites, or communities are best conceived through a collective vision and implemented in a holistic manner. They can thus fulfill the different "gardening" functions evoked above. Industries benefit from the necessary technological infrastructures, skill provision schemes, export networks, trade and intermediary professional structures, funding mechanisms, and the like. Technology sites, such as export zones or science parks, should combine the needed services and be well integrated in urban settings and well connected to the transportation infrastructure, including international airports. Local communities, even the poorest, have unique knowledge and entrepreneurial potential that can be exploited with appropriate support from surrounding actors such as research and education establishments, the business sector, and nongovernmental organizations. Acting in concert, with efficient local and global networks, is essential.

Innovation is fundamentally the task of the private sector and entrepreneurs. But history has shown that in moments of major transformations and crises, the role of governments has always been crucial. They alone can assume the launching of large-scale programs that help renew infrastructure while facilitating nationwide learning processes for innovative initiatives. Only they can legitimately impose and fund the adaptation of the educational, research, and other knowledge sources that are required to cope with deep and rapid technical change. This publication provides governments with ideas and tools to facilitate their tasks. A host of examples of policy actions from throughout the world are presented as a source of inspiration.

What Is Innovation?

In this volume, *innovation* means technologies or practices that are new to a given society. They are not necessarily new in absolute terms. These technologies or practices are being diffused in that economy or society. This point is important: what is not disseminated and used is not an innovation. Dissemination is very significant and requires particular attention in low- and medium-income countries.

Box O.1 provides examples of innovations in developing and emerging economies, ranging from the dissemination of new methods of eye care to the production of information technology (IT) components. Innovation, which is often about finding new solutions to existing problems, should ultimately benefit many people, including the poorest.

For understanding innovation, distinguishing high technology from low technology is not very useful, particularly in low- and medium-income countries.[2] High technology may not generate jobs and wealth, while low-technology developments and the exploitation of indigenous knowledge can lead to significant economic growth and improve welfare. The use of high technology in all sorts of products, processes, and services can be more important than producing it.

Innovation is distinct from research and in fact need not result from it. Innovations come from the entrepreneurs who make them happen and ultimately depend on a society's receptiveness. Innovation, therefore, is fundamentally a social process. The focus in this volume is on technological innovation, which is often accompanied by organizational and institutional innovation at both the micro- and the macrolevels.

The volume is a set of complementary chapters that form a structured whole. It offers a fairly exhaustive perspective on what innovation policy consists of and how it might serve concerned policy-making communities, from governments at the highest levels to managers of relevant organizations such as training institutions, R&D centers, or technological services.

Based on a better understanding of innovation, we now summarize the book contents, chapter by chapter. The volume is organized in three main parts that present the innovation policy concept, its functions, and the conditions of its implementation.

> **Box 0.1 A Few Examples of Innovations in Developing and Emerging Economies**
>
> - India's Aravind Eye Hospital deals with blindness in general and the elimination of needless blindness in particular in rural India. It reaches those most in need—the traditionally unreachable—through 20–25 weekly screening camps in villages. It also makes use of Internet kiosks in remote locations, where the information is sent electronically to a clinic for diagnosis. The Aravind eye-care system treats 1.4 million patients a year and, since its inception, has performed over 2 million operations and handled over 16 million outpatients.
>
> - To regain prominence as a leading center of learning, the Bibliotheca Alexandrina, in Alexandria, Arab Republic of Egypt, is playing a central role in the design, planning, and launch of a world digital library, in partnership with the U.S. Library of Congress and many other libraries around the world. This initiative, which includes digitizing its expertise, will make significant primary materials from cultures around the globe available on the Internet to people everywhere. These materials are to be accessible free of charge and in multilingual format.
>
> - The Malaria Research and Training Center in Bamako, Mali, is internationally recognized for its contributions to research on malaria and the improvement of public health standards. Its researchers participate in both international (National Institutes of Health, Institut Pasteur) and local networks. It works with traditional doctors to create a source of immediate care in the Bandiagara region and has helped reduce the mortality rates of young children significantly.
>
> - Intel's construction of a US$300 million semiconductor assembly plant in Costa Rica came as a surprise to many. Twelve years after the decision to invest was made, the initial investment had created many benefits, some of them unexpected. Intel's two plants employ 2,900, but the industry in Costa Rica now employs 12,000. The local support businesses for Intel alone reflects a base of 460 suppliers. The investment decision was the catalyst for a realignment of Costa Rica's competitive platform as an investment location, which led to newly secured FDI in other targeted sectors.
>
> - Tiny Estonia, a small Baltic state close to Finland, with a population of only 1.4 million, is leading an Internet revolution: its parliament has declared Internet access a basic human right. Estonia's well-educated, wired workforce and its liberal economic policies, low taxes, and low wages have helped make it an attractive business destination, especially for Sweden and Finland. It is also nurturing domestic innovation through key partnerships with Nordic neighbors. These include the development of devices such as doc@home, a hand-held electronic health kit that monitors blood pressure, stress, and weight and sends an alert to both patient and doctor in case of any sudden changes.
>
> *Source:* Justine White.

Policy Concept

Part one addresses the general approach to policy concepts, with a focus on two fundamental questions: (1) Why promote innovation? and (2) How can innovation be promoted? These chapters offer a historical perspective on innovation in economic and social development and show how government can promote innovative activity.

Chapter 1: Why Promote Innovation? The Key to Economic, Social, and Environmental Progress

Innovation has always played a decisive role in the economic and social development of countries: it is the main source of economic growth, it helps improve productivity, it is the foundation of competitiveness, and it improves welfare. Figure O.1 presents an example of the effect of innovation on the economies of two countries and shows that two-thirds of the differences in the growth performance of Ghana and the Republic of Korea over four decades are attributable to technology-related improvements.

In today's "poly-crisis" context, innovation is imperative. Innovation capabilities are seriously challenged both in the developed and in the developing worlds. Economically advanced countries need a more solid foundation than growth driven by financial speculation, as well as a truly innovative evolution of their economies and societies. Developing countries need ways to achieve broadly inclusive growth and innovation to benefit their many poor and not simply a narrow elite. More generally, adaptation to climate change, adjustment to limits of natural resources, and protection of biodiversity require fundamentally new patterns of production and consumption worldwide.

Finally, a more general reason to pay renewed attention to innovation is the current transformation of the world technological system in the wake of earlier transformations: the agricultural revolution in the Middle Ages and the Industrial Revolution in more recent centuries (see figure O.2). The four poles

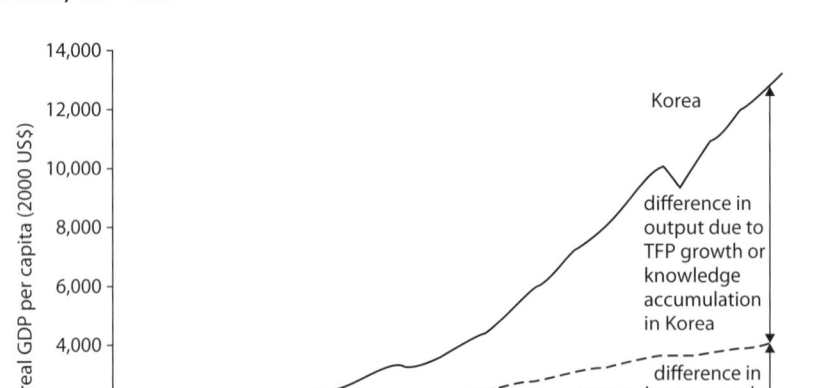

Figure O.1 How Innovation Contributes to Growth: A Comparison of Ghana and the Republic of Korea, 1960–2005

Source: World Bank 2007.
Note: TFP = total factor productivity.

Figure O.2 Major Technical Systems from the Middle Ages through the Present

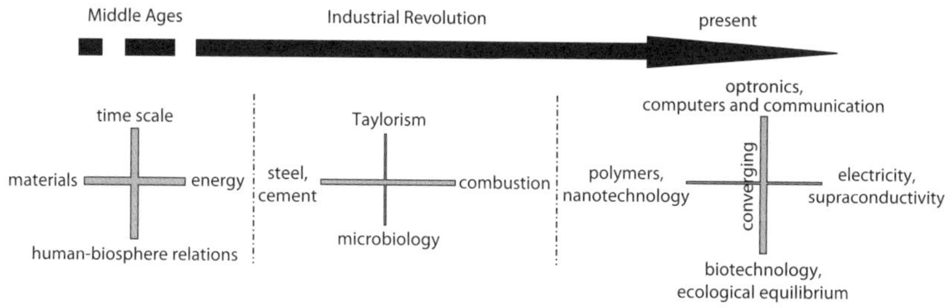

Source: Adapted from Gaudin 2009.

around which technological systems are structured—energy, matter, life, and time—are affected by these upheavals.

In the long term, all production systems are affected by such changes. They result in a cognitive revolution, which has today taken the form of a knowledge economy or knowledge society. The present situation is characterized by very rapid scientific and technical developments, and advances in science are making it possible to engineer new life forms and materials. The pervasive use of new technologies in all industries and activities requires new skills and new types of knowledge. Higher levels of education and greater flexibility in policies and institutions are necessary to take advantage of the innovation potential of such advances and to build the foundations of the so-called knowledge economy (World Bank 2007).

Chapter 2: How to Promote Innovation—Policy Principles

Governments have traditionally played an important role in promoting technology, sometimes by directly supporting the development of technologies (in space, defense, and the like) or more indirectly by creating a climate favorable to innovation through various incentives or laws. Every society has to find the ways and means to innovate that correspond to its needs and capabilities. Its innovation climate is largely determined by its overall macroeconomic, business, and governance conditions. Despite the nature of these conditions in low- and medium-income countries, well-designed and well-implemented innovation policies are very relevant. Moreover, they can be an efficient policy tool for triggering change and improvement in the country's overall framework conditions (this question is discussed in detail in chapter 8).

Figure O.3 depicts the diverse factors that influence developing countries' innovation capabilities. These countries can make considerable economic and social progress by tapping into globally available knowledge and technologies and adapting them to local contexts. Sources of foreign knowledge and technologies include trade activities such as imports of equipment and goods, multinational corporations, and skilled diasporas.

Figure 0.3 Determinants of Technology Upgrading in Developing Countries: Domestic Absorptive Capacity Both Conditions and Attracts External Flows

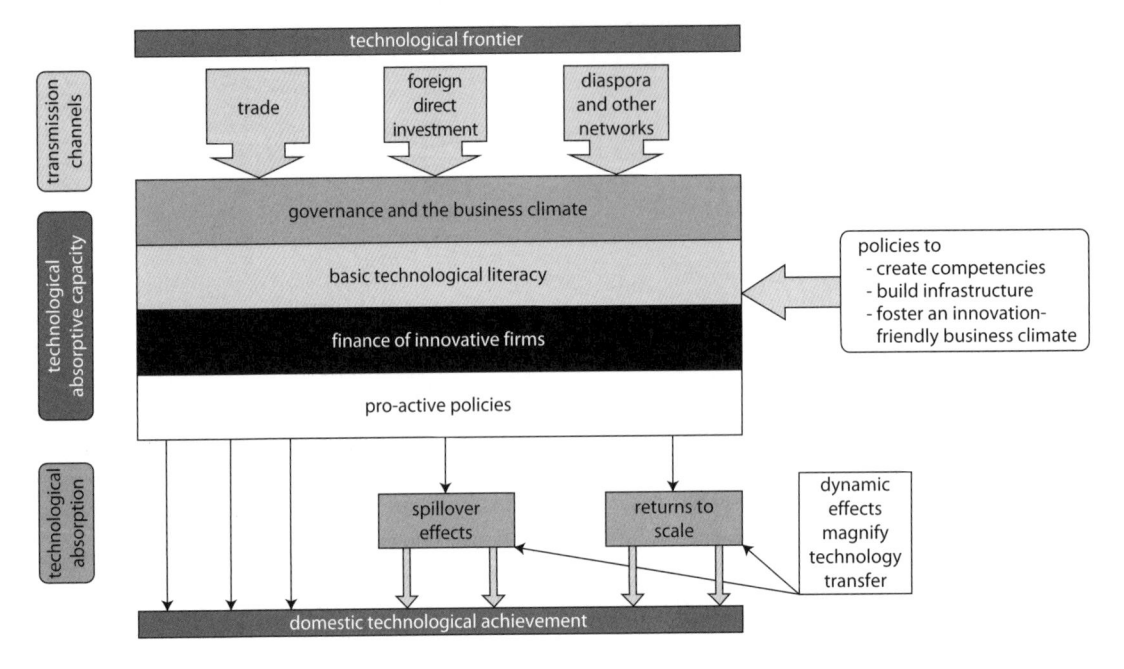

Source: World Bank 2008a.

Innovation processes germinate and develop within what are called "innovation systems." These are made up of private and public organizations and actors that connect in various ways and bring together the technical, commercial, and financial competencies and inputs required for innovation. It is on such systems that government innovation policies are focusing.

Avoiding misconceptions about the source and process of technological innovation, often wrongly perceived as deriving mechanistically from research and science, is important. Fundamentally, innovations are carried out by entrepreneurs who exploit existing knowledge and technology to propose new products or practices and disseminate them. The sources of their ideas are more likely to be users, suppliers, and customers than scientific research. Therefore, the role of governments is to facilitate this process by

- supporting innovators through appropriate incentives and mechanisms,
- removing obstacles to innovative initiatives,
- establishing responsive research structures, and
- forming a creative and receptive population through appropriate educational systems.

One may compare the tasks of governments to those of a gardener who should (a) water the plants, (b) remove the weeds and pests, (c) fertilize the soil, and (d) more broadly, prepare the ground so that plants can grow (see figure O.4). These four generic functions are detailed in part two.

Moreover, the government can intervene in areas of particular importance. In industrialized countries, for example, large-scale programs have targeted defense, space, and health, among others. This volume focuses on the promotion of competitive activities in agriculture, industry, or services; the development of innovative sites (industrial zones, technology parks, new cities); and the stimulation of innovation in, or for, poor communities. Issues and experiences relating to specific applications are discussed in part three.

It is clear from the above that innovation policy is broader than, and different from, science and technology policy, with which it tends to be merged. It also takes place as part of an overall trend toward knowledge-based economic strategies (see figure O.5). Innovation policy requires action in many different policy areas—education, trade, investment, finance, and decentralization, among others—and it is the right combination of interventions in these diverse domains that creates a fruitful innovation climate.

This approach to innovation policy reflects the evolving understanding of innovation policies in countries of the Organisation for Economic Co-operation and Development (OECD) over several decades (see box O.2). It explicitly recognizes the role of proactive and comprehensive government policies in establishing the overall framework and in fostering interaction among the actors, including different parts of government.

This fundamentally horizontal and interdepartmental innovation policy calls for a "whole-of-government" approach. It depends on the establishment of efficient government machinery able to ensure the needed coordination. Although its mechanisms must be adapted to existing institutional frameworks and to cultural backgrounds, models that are placing a powerful coordinating

Figure 0.4 Gardening Innovation

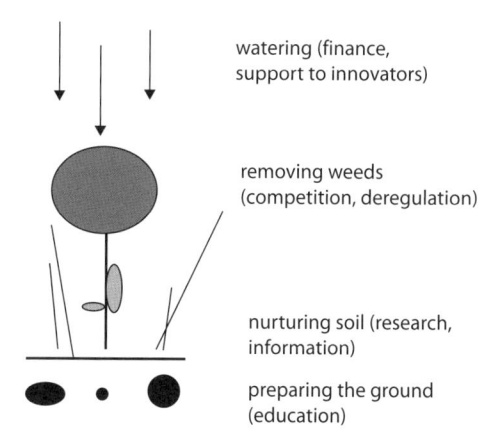

- watering (finance, support to innovators)
- removing weeds (competition, deregulation)
- nurturing soil (research, information)
- preparing the ground (education)

Source: Author.

Figure O.5 Innovation Policy in a Broad Perspective

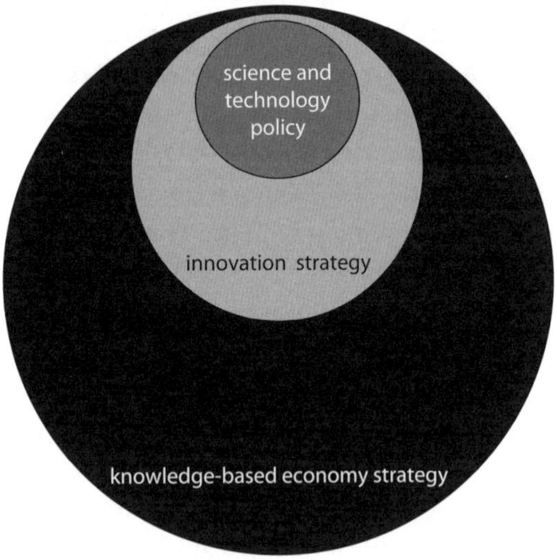

[Concentric circles diagram: innermost "science and technology policy", middle "innovation strategy", outermost "knowledge-based economy strategy"]

Source: Jean-François Rischard (personal communication).

Box O.2 Innovation Policies in OECD Countries—50 Years of Experience

Innovation policy has come into its own with some difficulty, as it was crushed between two ideologies with very active lobbies. The scientific ideology promoted the idea that technology derives naturally from science so that governments need do no more than build a good science base. The market ideology considered that innovation occurs naturally in a good business climate and that governments should concentrate on this aspect. They need only maintain an open, competitive environment and, in addition, fund public goods such as basic research, which the private sector is unable to finance.

Although these two views acted in coalition to promote their interests, governments felt the need to take specific measures to promote innovation. Their efforts took advantage of World War II initiatives and governments' strong involvement in the development of defense technologies.

Government efforts in the 1960s and 1970s were largely inspired by a linear model of innovation and the idea that science and research needed to be pushed toward technological and industrial applications; many policy initiatives therefore aimed at supporting enterprises in their R&D efforts or at improving university-industry collaboration. Concomitant large-scale space and defense programs facilitated the development of breakthrough technologies that were later used in civilian applications.

Recognition of the importance of interactions in innovation processes led to the concept of innovation systems, which was introduced in the literature in the late 1980s. This concept has been particularly fertile and has been variously understood. Most often, it defines the sets of interacting actors and institutions that provide the knowledge and financial resources required for the successful development of innovations.

Box 0.2 continued

Therefore, the first generation of innovation policy was replaced by a second generation in which innovation policy became more complex and aimed at facilitating interactions between the various actors and institutions involved in innovation processes: universities, research laboratories, banks for venture capital, and government agencies in charge of various sectors.

The boundaries of an innovation system legitimately include the "framework conditions" that encompass elements as apparently distant from the innovation process as the educational system or the macroeconomic environment. The OECD, for instance, explicitly includes framework conditions in its reviews of innovation systems. Thus, a third generation of innovation policy has appeared. It is inspired by a "whole-of-government" approach, in which all departments are potentially concerned.

Source: Author.

Figure 0.6 Model for a Strong Innovation Policy

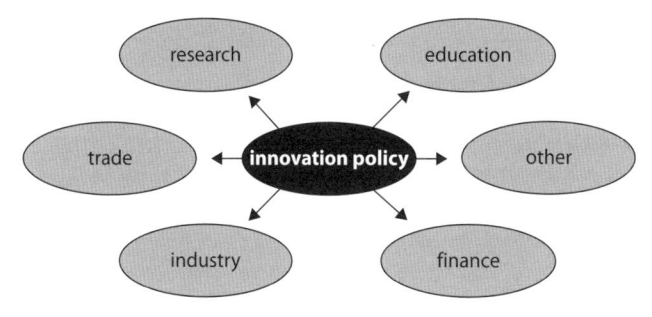

Source: Author.

body at the center of government allow innovation policy to have a pervasive influence (see figure O.6).

As *innovation* takes place primarily in local milieus with a concentration of knowledge, talents, and entrepreneurs; *innovation policy* is an important concern of sub-national governments that set up appropriate bodies (discussed in chapters 3 and 8).

Policy Functions

Part two addresses the four "gardening" policy functions described earlier. It discusses how government can provide basic support to innovative activity (chapter 3), reduce obstacles to innovation (chapter 4), sponsor appropriate R&D (chapter 5), and foster a receptive and creative population (chapter 6). This part also considers the important functions of policy evaluation and monitoring (chapter 7).

Chapter 3: Supporting Innovators

Supporting innovators effectively requires putting the necessary technical, commercial, and other services as close as possible to them. Such services should therefore be organized locally through the efficient mobilization of concerned authorities and with the active participation of concerned "clients." Services of strategic relevance for innovation policy include basic industrial services like promotion, marketing, and internationalization; technology extension services; metrology, standards, testing, and quality control; innovation in organization and management; and information and communication (see box O.3 for details of such services).

Supporting innovators also requires adequate financial support. Innovation expenses increase as projects develop and near commercialization. As such projects advance, government support should be increasingly based on the potential for commercialization and provided on a reimbursable basis.

Government measures in this policy area are many and varied. The difficulty in low- and medium-income countries is an overall lack of transparency

---

Box O.3 Business Services for Innovators

The following services potentially have strategic relevance for innovation policy:

Basic industrial services (promotion, marketing, and internationalization). Examples include assistance for direct investment abroad; assistance for inward investors; legal and financial assistance; financial services such as accounting and tax assistance; market information or other economic data; organization of and participation in trade fairs and other promotional events; partner search; and assistance for tenders of the European Union, World Bank, and other international organizations.

Technology extension services. Examples include assistance for patenting and licensing, for grant applications, for in-house R&D activities, and for subcontracting to research institutes; competitive intelligence (technological benchmarking, technology maps, information on emerging technologies); innovation diagnosis; review of current or proposed manufacturing methods and processes; participation in and organization of technology exhibitions; and technology brokerage.

Metrology, standards, testing, and quality control. Examples include calibration of equipment; quality certification; domestic standard; compliance with the International Organization for Standardization; technical assistance; demonstration centers and test factories; energy audits; materials engineering.

Innovation in organization and management. Examples include assistance for enterprise creation; interim management; logistical assistance; organizational consultancy, quality and training; productivity assistance; and incubation services.

Information and communication. Examples include advanced services for data and image transmission; assistance on communication strategies, telecom network connections and for the implementation of electronic data interchange systems; and database search.

Source: Patrick Dubarle.

and insufficient ability to evaluate projects. OECD countries often provide the business sector with fiscal incentives such as tax rebates to stimulate R&D and innovation-related efforts. Such incentives, which work best for medium- and large-scale industry, are generally not adapted to the situation of low- and medium-income countries, which lack sufficient accounting capabilities and have a large informal sector of small firms with no R&D expenses (for a more detailed discussion, see chapter 5).

A key issue is support for the incubation stages of innovation. While the financing of the initial stage, invention, is the responsibility of the public sector and the financing of the late stage is clearly the responsibility of the private sector, difficulties arise in the intermediary stages: prototype testing, product development, market research, and the like. For these middle stages, public-private networks or groups that can bring innovation projects to fruition by gradually mobilizing private money and management competencies, marketing opportunities, and other essential elements are critical.

Chapter 4: Improving the Regulatory Framework for Innovation
Removing obstacles to innovation means fighting anticompetitive and monopolistic practices, suppressing bureaucratic hurdles, and adapting the regulatory framework to support the search for and diffusion of novelty. It is a task that by nature should mobilize many areas of government—taxes, customs, procurement, and standards, for example—and requires vigilant action. This task is particularly necessary, but difficult, in developing country contexts.

The World Bank investment climate assessments and *Doing Business* surveys can help identify such obstacles. It is important to pay attention to those obstacles that are especially relevant to promoting innovation and entrepreneurship. Such obstacles can vary widely from transfer of pension rights for academics who become entrepreneurs to customs rules affecting technology imports or inappropriate safety regulations. Equally important is the establishment of durable institutional mechanisms that are able to improve the regulatory and legal framework in this regard. The maintenance of competitive pressure on firms (especially on state-owned firms in transition economies) and of all forms of incentives to innovate is also an essential element of innovation policy.

The design and implementation of effective procurement policies is a major instrument for promoting innovation. The experience of OECD countries offers a few valuable principles: define performance standards rather than set technical requirements; maintain fair competition in tendering procedures; and offer small and medium firms a share of contracts (perhaps 10 percent). Such principles could be usefully applied by low- and medium-income countries, particularly for infrastructure projects, which are generally financed largely by multilateral or bilateral partners.

In international commerce, fair-trade rules should be strictly applied. Developed economies should abolish the practice of taxing processed products (with added value) more than raw materials imported from developing countries, as it undermines the efforts of developing countries to climb up value chains.

Current international intellectual property rights regimes also need to be reconsidered. Regulations on access to technologies should be less stringent for developing countries, which cannot maintain costly protection systems or afford high licensing fees. Open-source regimes are also better adapted to the evolution and use of new technologies in software, genetic engineering, and related fields.

Chapter 5: Strengthening the Research and Development Base in Developing Countries

Developing countries should focus their research efforts on what has already been accomplished and take good advantage of it. OECD countries, particularly the largest, account for the bulk of R&D effort worldwide, although Brazil, China, India, and Russia are also becoming significant investors in R&D.

In developing countries, public and university laboratories are often ivory towers, cut off from local needs and poorly funded and staffed. Establishing a responsive research infrastructure depends principally on creating adequate competencies and laboratories with adequate funding mechanisms. These should ensure an appropriate proportion of stable financing with other funding from contracts with industry, communities, or the government. When research activities are partly dependent on external resources linked to explicit demands, the research structures are more attentive and more responsive to economic and social demands. Research structures should be linked to global centers of excellence and should work with local communities to satisfy basic economic or social needs.

Public research laboratories play a fundamental role in developing countries and should be equipped to respond efficiently to the need for technical research, technical assistance, certification, and quality control—functions that the business sector, which has low R&D capabilities in developing countries, is unable to perform. It is not advisable to privatize (former public) research structures to perform such tasks. For its part, the university sector should pursue high-quality research, and the results should be assessed through international peer reviews.

Incentives in OECD countries that facilitate collaboration by the university or public research structure with the business community, such as joint R&D projects partly funded by government agencies, could usefully be adapted to low- and medium-income countries if their transparency is ensured. Transferring intellectual property rights to universities or public laboratories that perform government-funded R&D (as in the United States under the Bayh-Dole Act) can be an effective incentive for engaging in innovation efforts, but such practices can also undermine long-term research efforts of collective interest and of a public good nature. The issue becomes more complex when multinational corporations are involved, as they often are in developing countries.

Promotion of R&D in the business sector is important for stimulating adaptive research as well as for helping firms face global competition successfully, a growing concern for a number of emerging and developing economies. Table O.1 summarizes the incentives and mechanisms at the disposal of governments, with their respective advantages and disadvantages.

Chapter 6: Fostering Innovation through Education and Training

No recipe can "make" innovators through education. Everything that facilitates the combination of the complementary competencies needed for innovation, such as engineering, design, and business, however, can help, especially in postsecondary education. Moreover, in addition to "hard" skills, people

Table O.1 Direct Support for Business Sector Research and Development

Instrument	Advantages	Disadvantages or shortcomings
Tax incentives for R&D	• Provides functional intervention, not picking winners • Offers less distortion, more automatic • Generally requires less bureaucracy to implement, although advisable to have monitoring and spot checks	• Has unclear fiscal costs in advance, may be high • Is difficult to ensure additionality • Is not very relevant for start-up firms that do not yet have taxable revenue streams • Is a blunt instrument, cannot target specific companies, although it can target specific sectors
Grants for R&D projects	• Allows specific targeting on case-by-case basis • Can control amount of subsidy granted • Can be given in tranches against defined goals • Can be structured as matching grants, which may help improve quality and efficiency	• Requires large bureaucracy to administer • May not select the best project • Is also difficult to ensure additionality
Accelerated depreciation for R&D equipment	• Reduces the capital costs of R&D projects	• Does not provide incentive for noncapital costs such as personnel and material inputs
Duty exemption on imported inputs into R&D	• Reduces cost of world-class inputs if country otherwise has high import duties	• Results in loss of tariff revenue • Is distortionary to the extent that it favors R&D over other activities
Venture capital to facilitate commercialization of research results	• Helps overcome financial market failure in making capital available to start-ups with no collateral or track record	• Requires detailed knowledge of sectors to evaluate technical and commercial prospects • Is often not successful because of limited deal flow and shortage of techno-entrepreneurs • Also requires developed stock market so that investors can sell off shares and reinvest in new projects

Source: Carl Dahlman.

need "soft" skills such as problem solving, communication, and teamwork and a good work ethic. These skills are important for innovation, as well as more generally in the economy, as innovators need to interact with both the business sector and the community.

The rapid expansion of knowledge-based industries has increased demand for more highly skilled labor. Because most new jobs will go to "knowledge workers," nurses in hospitals, farmers in automated stables, and workers in computerized factories will need to be able to manipulate symbols, read instruments, and interpret measures and data.

Today's workers and innovators therefore need a broad set of platform skills based on a good general education beyond primary schooling. This requirement implies interventions in primary and secondary education. Vocational training also plays a vital role in preparing workers for the labor market but has often received too little attention from policy makers. A country's youth must acquire—in addition to basic skills such as writing, counting, and the like—"functional literacy," a good "technological culture," and an ability to "think outside the box."

The timely acquisition of basic literacy conditions the effectiveness of subsequent lifelong learning, which individuals will need to function effectively in a knowledge economy. Lifelong learning requires a new pedagogical model, which may include customized learning, learning by doing, and teamwork.

On-the-job training assumes an important role in the lifelong learning system: it builds on the acquired soft and hard platform skills, adds specific skills necessary for the job, and helps upgrade skills continually. Especially for low-income countries, education policy should include skills development in the informal sector, which can represent 30 percent or more of nonfarm employment in a number of developing economies. An appropriate focus is improving traditional apprenticeship training, as it is responsible for more skills development than all other types of training combined in developing countries, particularly in the least developed.

The biggest challenge to educational reform is a deeply rooted model of schooling. That model, characterized by traditional teacher-dominated classrooms and strong emphasis on rote learning, determines practices both inside and outside the education community. A second challenge is to make educational strategies part of a broader innovation agenda, an effort resisted by vested interests such as existing institutions and teacher organizations. The challenges for most developing countries are more complex than for developed countries, as they must deal simultaneously with problems of provision and quality under serious financial and institutional constraints.

In investing in a well-educated workforce, low- and medium-income countries necessarily face the risk of a large-scale brain drain. Experience shows, however, that appropriate mechanisms can facilitate a "brain circulation" process by which talented migrants reconnect with their country of origin as efficient drivers of innovation in various forms: as creators of enterprises, openers of new markets, sources of venture capital, or facilitators of institutional reforms.

Chapter 7: Policy Evaluation—Assessing Innovation Systems and Programs

Like any government policy, innovation policy needs to be properly monitored and evaluated at two levels: the monitoring of innovation systems and the assessment of innovation programs and policies. To monitor countries' innovation capabilities at the macro level, a number of international bodies, including the World Economic Forum with its competitiveness indexes and the World Bank with its Knowledge Assessment Methodology, have developed benchmarking based on regularly updated databases. Benchmarking helps countries position themselves with respect to their competitors and observe their progress over time.

These macro-benchmarking approaches, however, have to be complemented by more detailed indicators that monitor and assess innovation systems, specifically, firms' resources and performance in research and innovation and their diffusion of specific technologies. These indicators should be systematically documented through the use of regular surveys, possibly limited to well-defined samples, but rigorously conducted.

Measuring the impact of policy programs as well as their relevance is indispensable. Industrialized countries have significant experience with measuring the impact of schemes such as tax incentives for business R&D or public R&D support on innovation efforts and performance. Quantitative methods, based on field experiments, are also being implemented specifically for use in the developing world. They help countries decide whether to scale up programs that prove effective.

Overall, the most appropriate methods for evaluating innovation policy are the peer review processes that were initially developed in economically advanced countries, notably by the OECD, and that are gradually and successfully being disseminated in low- and middle-income countries. Such national reviews can serve as a tool for shaping policy initiatives and triggering policy reforms (as discussed in chapter 8).

Policy Implementation

Putting in place an innovation policy is a daunting challenge, as economically advanced countries have learned in the past decades, especially because established agencies and departments supposed to carry out innovation policy functions have crowded the field. Implementing innovation policy is even more daunting in developing countries where the institutional context is more difficult, resources are necessarily limited, and managers able to carry out these programs and policy measures are lacking.

A long-term strategic approach, based on a clear long-term vision, for gradually implementing the necessary changes is therefore useful. This stepwise approach focuses on interventions in specific industries, sites, or communities. These chapters in part three first describe elements of the strategic framework (chapter 8) and then examine the promotion of competitive industries (chapter 9), the building of innovative sites (chapter 10), and the support of innovation in, and for, poor communities (chapter 11).

Chapter 8: Policy Implementation—The Art and Craft of Innovation Policy Making
The rationale for innovation policies is that they aim to boost technological change, which is considered the basic factor of economic growth, social development, and environmental adaptation. Countries differ considerably in their assets and capabilities, however, and developing countries are seriously affected by governance problems, lack of resources, insufficient infrastructure, and other constraints. It is therefore crucially important to provide orientations for making innovation policy work in different policy contexts, including the most difficult ones. This effort involves two complementary issues: the design of efficient and pragmatic policy agendas and the formation of institutional virtuous circles within an "evolutionary" perspective.

When designing pragmatic agendas for local contexts, policy makers should focus broadly on the sectors, sites, and groups of people with the greatest chances of successful development in view of their competencies, comparative advantages, and networking. Specific strategies will depend on the scientific and technological level of the country and the situation of its institutions and governance climate (see table O.2). In addition, it is important to distinguish between "prime movers' agendas," which entail starting from scratch with pioneer innovators, and "critical mass agendas," which largely entail attracting newcomers to a going concern. The objective in all cases is to favor a successful "self-discovery process" through appropriate combinations of public and private actors that take the best advantage of the situation, whatever its constraints and opportunities.

Clearly, government priorities and policy actions will differ considerably according to the country's technological competence and the nature of its business environment. For countries well equipped with R&D competencies and infrastructure and with a good business climate, it makes sense to pursue advanced research broadly along the frontier of technology, while facilitating—through encouragement of venture capital, technology brokering services, and training platforms—the development of innovation clusters in industries with international competitors.

For their part, low-income countries with limited knowledge endowments and a poor business and governance environment can focus their efforts on exploiting those endowments. They can tap into and adapt global knowledge and technology for their needs and support budding entrepreneurs through

Table 0.2 Country Contexts and Strategic Focal Points

Technology capabilities	Strong institutional framework	Tolerable and improving institutional environment	Weak institutions and investment climate
High (frontier technology creation)	Innovation leaders' agenda: development of proprietary technology through promotion of innovation clusters	Critical mass agenda: increase of value added of natural resources wealth and technology commercialization	Prime movers' agenda: leveraging pockets of dynamism
Medium (adaptation of technologies available worldwide)	Critical mass agenda: development of innovation clusters and high value-added supply chains	Critical mass agenda: development of innovation clusters and high value-added supply chains	Prime movers' agenda: leveraging pockets of dynamism
Low (adoption of technologies)	Creation of knowledge endowments through higher education and attraction of foreign technology and expertise	Exports as springboard agenda: development of nontraditional exports as entry points for institutional and technology assets	Creation of basic institutional infrastructure through a diversity of entry points

Source: Yevgeny Kuznetzov.

well-focused measures (technical assistance, mobilization of intermediaries, building of export networks, and the like). These activities do not preclude undertaking some frontier research, but because their R&D capability is limited, they must focus carefully on needs that cannot be addressed by existing knowledge. In other words, the issue for developing countries is to strike the right balance between using or attracting existing technology and knowledge, adapting them to local contexts, and pursuing focused research, including on frontier technology when appropriate. This balance will of necessity be country specific.

Various instruments are available for creating conducive institutional frameworks through virtuous circles, engaging actors in the self-discovery process, and building problem-oriented networks. Examples include well-designed matching funds, foresight exercises, federal contest funds, and other means that are not very costly but that do involve collective mobilization of motivations and the knowledge of targeted communities.

A well-defined strategy should then be articulated to move gradually from micro- to macroreforms (see figure O.7). Change often begins with effective microreforms, which then serve as models or sources of motivation for building a critical mass of initiatives through a combination of top-down and bottom-up actions. As the intermediate level between microreforms and structured national policy reform, the meso level is critical for scaling up these reforms because it creates the base for major reforms. Mass media should be actively mobilized throughout the process to generate public support.

Figure 0.7 Scaling Up Institutional Change, from Microreforms to National Reforms

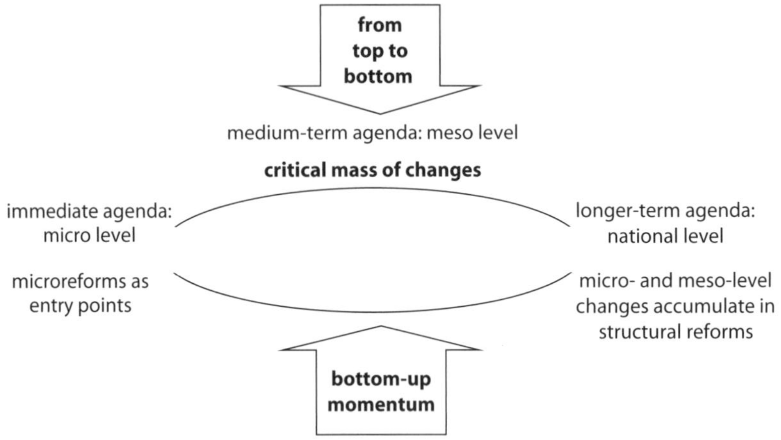

```
                    from
                   top to
                   bottom
                      ↓
          medium-term agenda: meso level
              critical mass of changes
immediate agenda:                    longer-term agenda:
   micro level                          national level

microreforms as                      micro- and meso-level
  entry points                       changes accumulate in
                                      structural reforms
                      ↑
                   bottom-up
                   momentum
```

Source: Yevgeny Kuznetzov.

Chapter 9: Promoting Competitive and Innovative Industries

Developing competitive industries is a key element in the approach to strategic policy making proposed above. As a source of wealth, competitive industries are a matter of national pride, and it is therefore important to understand how governments can efficiently intervene to promote them.

The goal is not to pick winners but to create a dynamic and receptive climate in which innovative initiatives in specific industries can be articulated and implemented. A competitive industry cannot simply be created: what is needed is not direct support so much as it is indirect interventions at determining points. All sectors are concerned: agriculture (of crucial importance in developing countries both as a source of exports and for subsistence), manufacturing (where low-income countries with low labor costs have a competitive advantage), and services (where a wide spectrum of opportunities involve tourism, information technology services, and creative industries, among others). As illustrated by the success stories featured in the chapter—coffee in Rwanda, textiles in China, tourism in Costa Rica, IT services in Vietnam—a holistic approach is necessary to ensure that all the activities in an efficient value chain are properly functioning and delivered.

The development of innovative and competitive industries implies, as a prerequisite, an adequate infrastructure as well as a friendly business environment. The government also works closely with the concerned trade and professional groups to ensure that key technological services are provided and to facilitate active cooperation with research, education, and other sources of knowledge to raise the technological level and the knowledge content of products and services.

The development of competitive sectors clearly requires the engagement of high-level leadership and the formulation of a mobilizing vision, collectively elaborated, to attract essential investment, remove obstacles, and launch pilot initiatives. Such actions should be directed toward creating a climate for broad reforms (as discussed in chapter 8).

Chapter 10: Building Innovative Sites
The successful development of specific sites—including techno-parks, industrial zones, or even new or renovated cities—depends primarily on the accumulation of a critical mass of talents and entrepreneurs, well connected to the global economy. The prerequisites are an efficient infrastructure, a lack of red tape, an attractive environment, and world-class knowledge institutions.

Technology and science parks are favored by policy makers, as they make innovation efforts highly visible. Experience shows, however, that few are successful. Success results from a series of conditions: a focused project, good positioning of specific technology and ambition, a clear and transparent agreement among partners (the business community, local and central authorities, and academic institutions), adequate integration in the urban structure (infrastructure, access), and good financing packages (including for start-ups).

For low- and medium-income countries, the creation of special economic zones, or industrial export zones, to which foreign subsidiaries are attracted with specific incentives, well-developed infrastructure, and a friendly business environment makes sense to the extent that it is part of broader national experimentation and learning processes, as China has shown in the past decades. Such zones require specifically designed mechanisms to facilitate transfer of technology and management competencies to local firms.

Innovative firms tend to develop today in what are called "industrial clusters," that is, concentrations of firms in loosely defined geographic areas, with complementary rather than competitive assets, which operate through networks. They tend to be spontaneous developments resulting from business initiatives. Governments, however, can play a decisive role as "brokers" by setting frameworks for dialogue and cooperation and developing incubating and training instruments.

Cities are becoming critical platforms for innovative activities and competitive centers in the global economy (World Bank 2008b). Essential to their success are a strong identity and a clear strategy for exploiting a comparative advantage or for creating it by attracting a critical mass of talent, while mitigating the factors that negatively affect their attraction as centers of innovation. A holistic view of the city is important.

Chapter 11: Stimulating Pro-Poor Innovations
Four billion people, a majority of the world's population, form the bottom of the economic pyramid. They have an annual income of less than US$3,000 in

local purchasing power. The promotion of pro-poor innovation, or inclusive innovation, is essential.

Innovation can be encouraged in poor communities in two ways: first, through the organization of formal links with the surrounding research, education, or business sectors; and second, through the exploitation of the specific knowledge and entrepreneurial drive present in such communities. Appropriate policy mechanisms are those discussed in the various chapters of part two, particularly in chapters 3 and 4.

Initiatives by businesses, academic institutions, or nongovernmental organizations (NGOs) to develop innovations in response to poor communities' needs exist throughout the world. Such initiatives require establishing close and durable connections with those communities and their innovative individuals and groups. The international community can be a great help through well-designed and -implemented support. A number of NGOs, for example, are assisting artisans in design, trade, and exports and are helping social entrepreneurs with funding and business management. When well designed, these programs have had very high social impact.

Poor communities have considerable resources in their traditional, indigenous knowledge. This potential remains unexploited except in a few areas such as pharmacology, generally to the benefit of multinationals. Systematic search, development, promotion, and protection (patent rights) of this potential are, however, not only possible but also fruitful as has been demonstrated in Africa, India, and elsewhere.

Poor communities in rural areas need help in ensuring their survival and preventing massive exodus to urban areas, even if urban concentration can pay off in the long term by raising gross domestic product per capita (see World Bank 2008b). Maintaining populations in rural areas calls for combining technological support; provision of equipment, seeds, and fertilizers for improving agricultural productivity; diffusion of health-care practices, schooling, and training efforts; and some infrastructure investments. Innovation policy thus becomes part of a broader, comprehensive plan.

Conclusion

Innovation is at the heart of economic development, social welfare, and protection of the environment. Today, the need for innovation is greater than ever, and the challenge to make these three objectives compatible is formidable.

Why Now?

Leveraging innovation is particularly important today, in what is the most severe global economic crisis since the Great Depression of the 1930s. By all indications, this crisis will last longer than most past crises because it is global in scope. No large region, therefore, will be able to lead a recovery by increasing its demand for imports. Moreover, the reaction to the excesses of the financial markets will be to price risk higher and to raise the price of capital. That increase will have a negative effect on all countries that rely on foreign capital, particularly capital-scarce developing economies. Taken together, these factors will lead to lower investment and consequently to lower growth. The higher cost of capital will also mean less investment in R&D because of its relatively long gestation period and risk. It will therefore be necessary to make more efficient and innovative use of existing resources and existing knowledge.

The Need for Green Technology
People around the world are increasingly aware of natural resource and energy constraints on growth and of the environment's limited capability to absorb pollution and CO_2 emissions. These conditions put a premium on innovations that can help conserve energy and resources and on the development of more resource- and energy-efficient technologies and non-carbon-based technologies. Yet, the demand for green technology comes precisely at a time when the capital needed for its development has decreased sharply. Clearly, more cross-national efforts to find innovative ways to deal with this and other issues of global public goods are urgently needed.

Innovation in a Time of Crisis
History has shown that times of crisis are also times of innovation, when institutional, mental, and other obstacles are more easily removed. The time is thus ripe for mobilizing creativity and entrepreneurship to meet the challenges ahead. Government and other leaders have a key role to play. Government can innovate in public goods and in finding ways to carry out its business more effectively. Most important, it should help provide the right environment for innovation. Although this volume has stressed the role of government and the need to adopt a whole-of-government approach to many aspects of innovation policy, most innovation occurs through firms, families, and individuals. Government needs to partner with the private sector and with individuals in support of innovation and to avoid interfering with the innovative efforts of firms or individuals.

While innovation remains fundamentally the work of private economic agents, governments facilitate the emergence and success of innovative initiatives by removing obstacles, by providing the necessary support to entrepreneurs, by investing in the needed technology and research infrastructure, and by carrying out appropriate reforms in education, the investment climate, and trade.

The Purpose of This Book
The volume describes the main elements of policy measures and offers an overarching strategic framework for implementing a pragmatic innovation

policy with a broad, long-term vision. This book argues that innovation policy should be at the core of government action and a focal point for mobilizing a country's agents of change. It is up to these public and private sector actors, working together, to determine what will best fit their specific context and leverage their country's innovation potential.

This book is meant to serve as a guide for policy makers, businesspeople, and the general public in developing countries and others interested in leveraging innovation to improve the performance and social welfare of their country, region, or organizational unit. It has presented a conceptual framework that includes a broad definition of innovation as the effective use of something that is new to a country, a region, a sector, or a firm. Innovation is the main source of increased performance—of getting more out of limited resources, of finding new ways to use existing resources and to mobilize people to produce better goods and services or to produce and deliver them more efficiently.

Finally, it is not possible to say, in a book of this size or even larger, which innovation policies might work best in widely different country and regional contexts. This volume simply aims to serve as a guide by providing a framework for thinking about and informing action in developing and emerging economies, to give helpful guidelines, and to provide concrete examples of what has been done various circumstances. Ultimately, it is up to the policy makers, entrepreneurs, and individuals in a given situation to determine what they can do to leverage the potential of innovation for addressing their needs. The authors hope that this book will help guide that process of trial and error, which is also an intrinsic part of innovation.

Notes

1. This book complements other practically oriented documentation prepared in other parts of the World Bank, notably, the Science, Technology, and Innovation Capacity Building Toolkit and the Technology Commercialization Handbook prepared by the World Bank Science and Technology Coordinator Unit to be available online.

2. The technology level of goods is determined, in international statistics, by the R&D intensity of industries that produce them. High-technology industries are defined as those that spend (approximately) more than 3 percent of their turnover on R&D (OECD standards; OECD 2010).

References

Gaudin, Thierry. 2009. "Foresight: A Global Agenda for the 21st Century." Unpublished background paper, World Bank Institute, World Bank, Washington, DC.

OECD (Organisation for Economic Co-operation and Development). 2010. *Main Science and Technology Indicators (MSTI): 2009* (2 edition). Paris: OECD.

World Bank. 2007. *Building Knowledge Economies: Advanced Strategies for Development*. Washington, DC : World Bank.

———. 2008a. *Global Economic Prospects 2008: Technology Diffusion in the Developing World*. Washington, DC: World Bank.

———. 2008b. *World Development Report 2009: Reshaping Economic Geography*. Washington, DC: World Bank.

Introduction

One can readily name many recent innovations from developed countries, from the Toyota Prius (a sophisticated energy-saving hybrid car), to the Iphone, the Global Positioning System (GPS), or Wal-Mart, among many other products or services. But what about innovation in emerging and developing economies, which is the subject of this book?

Innovation, Did You Say?

Innovation is about finding new solutions to existing problems, as well as offering opportunities of new activities. It should ultimately benefit many people, including the poorest. Here are some examples that illustrate the range and success of such innovations:

- India's Aravind Eye Hospitals deal with blindness in general and the elimination of needless blindness in particular in rural India. They reach those most in need—the traditionally unreachable—through 20–25 weekly screening camps in villages. They also make use of Internet kiosks in remote locations in Madurai to screen people's eyes under the supervision of a paramedic. The information is then sent by the Internet to a clinic for diagnosis. The Aravind eye-care system treats 1.4 million patients a year, and since its inception, it has performed over 2 million operations and handled over 16 million outpatients.

- To regain prominence as a leading center of learning, the Bibliotheca Alexandrina, in Alexandria, Egypt, is playing a central role, including through its digitizing expertise, in the design, planning, and launch of a world digital library, in partnership with the U.S. Library of Congress and

---

This introduction was prepared by Jean-Eric Aubert.

many other libraries around the world. This initiative will make significant primary materials from cultures around the globe available on the Internet to people everywhere. It will include manuscripts, maps, rare books, musical scores, recordings, films, prints, photographs, architectural drawings, and other significant cultural materials. These materials are to be accessible free of charge and in multilingual format.

- The Malaria Research and Training Center in Bamako, Mali, created in 1992, is internationally recognized for its contributions to research on malaria and the improvement of public health standards. Its researchers participate in both international (National Institutes of Health, Institut Pasteur) and local networks. It works with traditional doctors to create a source of immediate care in the Bandiagara region and has helped reduce the mortality rates of young children significantly.

- Intel's construction of a US$300 million semiconductor assembly plant in Costa Rica came as a surprise to many, especially in view of the country's small size and the fierce competition for attracting such an investment. Twelve years after the decision to invest was made, the initial investment had created many benefits, some of them unexpected. Intel's two plants employ 2,900, but the industry in Costa Rica now employs 12,000. The local support businesses for Intel alone reflect a base of 460 suppliers and US$50 million–$150 million in local purchases each year. The investment decision was the catalyst for a realignment of Costa Rica's competitive platform as an investment location, which led to newly secured foreign direct investment in other targeted sectors.

- Tiny Estonia, a small Baltic state close to Finland, with a population of only 1.4 million, is leading an Internet revolution: its parliament has declared Internet access a basic human right. Estonia's well-educated, wired workforce and its liberal economic policies, low taxes, and low wages have helped make it an attractive destination, especially for Sweden and Finland. It is also nurturing domestic innovation through key partnerships with its Nordic neighbors. These include the development of devices such as doc@home, a hand-held electronic health kit that monitors blood pressure, stress, and weight and sends an alert to both patient and doctor in case of any sudden changes.

These examples show that developing and emerging countries have considerable creative potential. Innovations are fundamentally brought about by private entrepreneurs with a clear vision, strong networking, and the ability to mobilize all sorts of resources, including at a global level. As these examples make clear, however, government has a key role to play in creating a conducive environment by articulating national objectives, establishing an attractive business climate, funding appropriate research, providing infrastructure, putting in place well-designed regulations, and ensuring the provision of a well-educated workforce.

These examples also remind us that innovation should be understood as something new to a given context that improves economic performance, social well-being, or the environmental setting. It can be new to the firm (or the organization), new to the economy, or new to the world. From this perspective, the issue for developing countries is to strike the right balance between using or attracting existing technology and knowledge, adapting them to local contexts, and pursuing focused research, including on frontier technology when appropriate.

What Is This Book About?

This book is meant to serve as a guide for policy makers, businesspeople, and people at large in low- and medium-income countries, along with others interested in leveraging innovation to improve the performance and social welfare of their country, region, or organizational unit. Its aim is to help successful innovative firms multiply, to increase the number of sectors that perform well, and to facilitate the process of priority setting in policy making. This volume has three main parts:

- The first presents the rationales and the main principles of innovation policy: why governments should promote innovation, how they should approach it, and what types of institutions and instruments are effective.
- The second part details the basic functions that governments should fulfill to create a climate favorable to innovation: support to innovators, removal of obstacles, strengthening of research and development structures, and adaptation of education and training. In addition, it gives elements for evaluating innovation systems and policies.
- The third part discusses policy implementation issues. It proposes a strategic framework with pragmatic agendas and stepwise approaches adapted to the context of low- and medium-income countries, and it details focused applications of innovation policy: how to promote competitive industries, how to build fertile sites, and how to help poor communities.

This volume provides general principles of action, illustrated by many and diverse examples from various policy contexts. It is ultimately the role of policy makers in a given concrete situation, however, to determine how they can leverage the potential of innovation to address their needs.

Part I
Policy Concept

1

Why Promote Innovation? The Key to Economic, Social, and Environmental Progress

The main reason for governments to pay attention to innovation, particularly in the developing world, is that innovation is the key driver of economic development and the principal tool for coping with major global challenges, notably those induced by climate change. Moreover, the fundamental technical change that our economies and societies are undergoing requires major adaptations at the same time that it is opening broad opportunities.

Innovation is a new and better product or service or a new and more efficient, or less costly, way of producing, delivering, or using that product or service. Innovation is important because it provides a means for getting more output or welfare from limited resources. Innovation may be new to the world as a whole, new to a country, new to a sector, or new to an individual. These distinctions are important, particularly from the perspective of developing countries, because of the tremendous amount of knowledge that they are not using. If countries or firms devise better policies for acquiring and exploiting that knowledge effectively, they can greatly improve their growth and welfare.

This chapter begins by providing a long-term view of the role of innovation in mankind's economic development and an overview of the major technological transitions the world has experienced over the last thousand years. It then summarizes how economists have attempted to quantify the role of innovation in growth and gives some empirical estimates of its importance. It examines

---

This chapter was prepared by Carl Dahlman, with a contribution from Thierry Gaudin on the technical systems section.

the potential for developing countries to catch up with the technological frontier, identifies a small group of countries that have been quite successful in that effort, and highlights the innovative elements of their strategies. The following section takes a more detailed look at the diffusion of some basic welfare-enhancing technologies and major innovations. The chapter concludes with some implications for developing countries.

Innovation and Societies: The Long-Term View

In this section, the role of innovation in the development of economic and welfare is examined. The section particularly focuses on the technical transitions that have taken place over time, including two centuries of technical progress.

Innovation in Economic and Welfare Development

Innovation has been critical for the rise in population and in per capita income and welfare. In world history, the first major technological innovation was probably the development of agriculture as far back as 9000 BC, followed by the development of pottery in about 6000 BC. Other important innovations were the development of the plow and irrigation between 5000 BC and 4000 BC, which facilitated growth of the world population. The development of metallurgy and writing dates from around 3000 BC. The development of mathematics dates from about 2000 BC (figure 1.1). World population

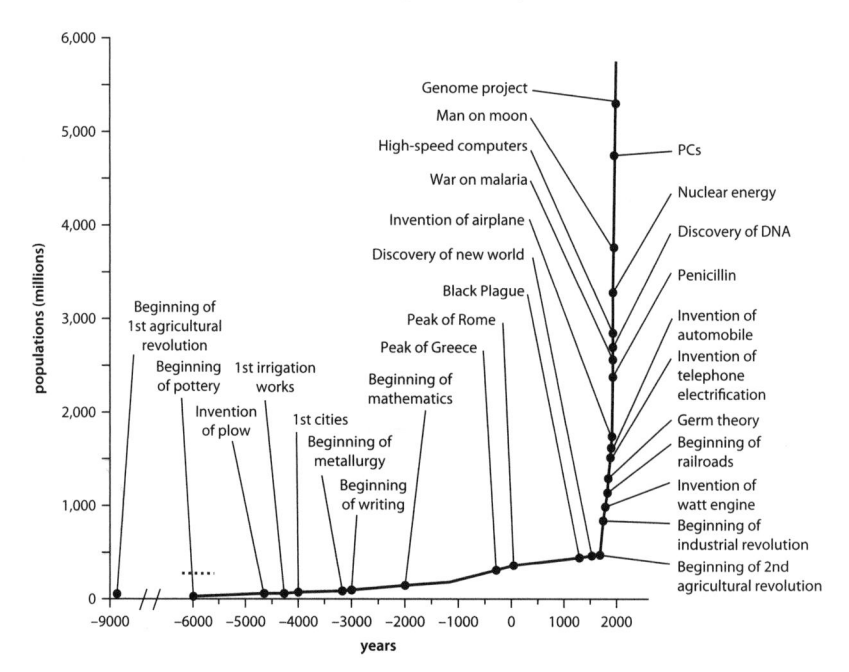

Figure 1.1 World Population Growth and Major Technological Events, 9000 BC to Present

Source: Commission on Growth and Development 2008, 108, based on Fogel 1999.
Note: DNA = deoxyribonucleic acid; PCs = personal computers.

continued to increase very slowly time. Life expectancy during the Greek and Roman empires averaged about 20 years, not much more than in the preceding few millennia.

The relationship among innovation, population, life expectancy, and growth can be seen more clearly over the past two millennia (figures 1.2 and 1.3).[1] For the first 1,400 years of the past two millennia, the world's population grew very slowly. Although some privileged elites had much higher income during this period, average per capita incomes hovered around $400 (in 1990 international dollars).[2] This figure is sobering in that it is roughly the same as that of today's poorest countries.

Then, between 1400 and 1500, something remarkable began to happen. Global population and per capita income began to increase simultaneously (figure 1.3). This growth resulted from the convergence of many factors: better hygiene, more efficient ways to harness wind and water power to augment human and animal energy, and advances in agricultural techniques

Figure 1.2 Average Life Expectancy over the Past Two Millennia

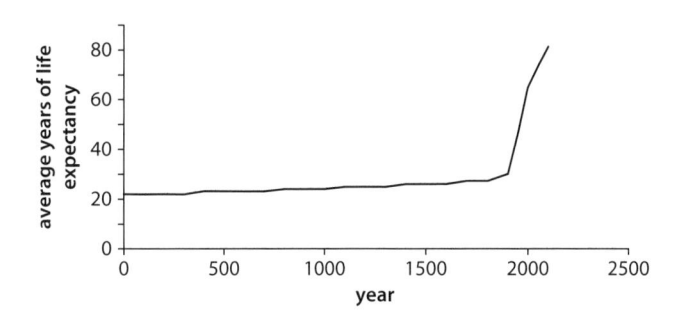

Source: Compiled from Ross 1997 and Lee 2003.

Figure 1.3 Growth in Population and GDP per Capita in the Past 2,000 Years

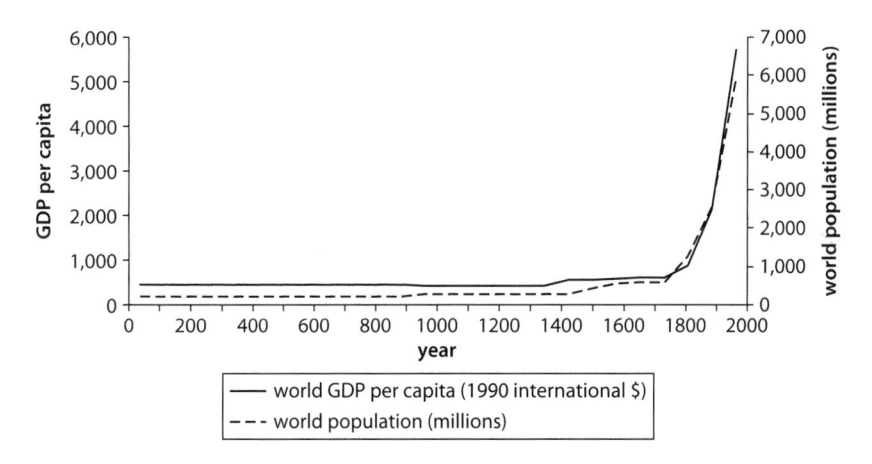

Source: Maddison 2006.

like irrigation, improved seeds, and multiple cropping. In addition, advances in shipbuilding and navigation technology, including the astrolabe and the compass, led to increased trade, thereby expanding markets and specialization.

What is even more remarkable, when viewed from a long-term perspective, is how suddenly, even apparently exponentially, both population and per capita incomes began to rise from the 1800s onward (figure 1.3). This tremendous growth was in large part led by the development of the steam engine, which first enabled humankind to harness fossil fuel energy for productive tasks. This augmentation of power brought about the Industrial Revolution, with a corresponding proliferation of productive activity and expansion of products and services brought to market.

Transitions in the Technical System

Systemic transitions have taken place throughout history. Before the Industrial Revolution, which started in the 18th century, another technological economic and social revolution occurred in the 12th and 13th centuries in Europe. Still another occurred during the 7th and 6th centuries BC in the Middle East, India, and China along the so-called Silk Road.

In such systemic changes, daily life is profoundly transformed, and the ruling class replaced. The changes work their way through society over more than a century, the pace limited only by the human factor: change cannot proceed faster than the speed of human adaptation to the new technologies. Succeeding generations define the rhythm of adaptation. In the early Middle Ages as well as during the Industrial Revolution and the present one, which may be called the "cognitive revolution," the change in technologies can be described in terms of four poles, usually presented on a symbolic cross (see figure 1.4).

In the Middle Ages, the basic innovation in materials was the use of iron in agriculture, not only for ploughs. All sorts of tools were developed, which defined the technical environment of the peasant until the industrialization of agriculture in the 20th century. In terms of energy, water mills became nonspecialized sources, used not only for baking but also for carpentry, textiles,

Figure 1.4 Structure of the Global Technical System

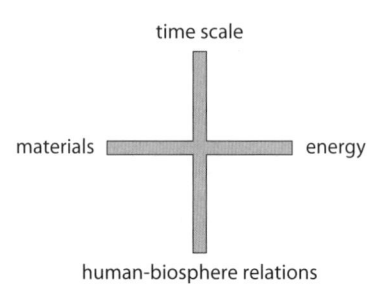

Source: Author.

and beverages. The social time scale was defined by the sound of the belfry's tolling bells, which gave its rhythm to the life of the countryside. Finally, relations between humans and the biosphere became more systematic, with seed selection and cattle breeding. In all these fields, research was mostly driven by the monasteries, where tests were carried out and experimental results analyzed, stored, and diffused to other monasteries in manuscripts (Gutenberg printing came only in 1450). At that time, universities were just emerging (Bologna, Oxford, Paris). A few were involved in technology, such as Oxford's work on measuring time.

In the Industrial Revolution, these four poles were again activated for innovation (see figure 1.5). But the vertical axis moved up an order of magnitude in finesse and complexity. The chronometer measures a tenth of a second rather than the hours of the belfry's bells as in the Middle Ages, and Pasteur's microscope looks at cells and microbes. In fact, the Industrial Revolution includes a series of inner technological revolutions (which may be called the second-order revolutions), which go along with infrastructure and institutional changes. Table 1.1 summarizes such changes.

This industrial age, however, is also the result of a disruption that is now coming to an end. During the 17th and early 18th centuries, the overexploitation of European forests led to their exhaustion, and the economy had to turn to nonrenewable sources of energy: coal and, in 20th century, oil. The transition was easy, but it disregarded the equilibrium between humans and the biosphere that had for millennia been the sacred rule of survival.

With the cognitive revolution, the order of magnitude of the time scale shifts from one-tenth of a second to one-billionth of a second (a 100 million times thinner) in a first stage and probably even a million times thinner again (the femtosecond 10^{-15}) with optical commutation. Materials are now elaborated at molecular level for polymers and even at atomic level (one billionth of a meter) with the development of nanotechnologies. Biotechnology, by manipulating genetic codes, also reaches that level of detail. Perhaps to stimulate the understanding (and financing) of these fields of research, politicians have promoted the term *converging technologies* as a nano-bio-info-cogno complex (see figure 1.6).

Figure 1.5 The Industrial Revolution in Europe, 1750–1970

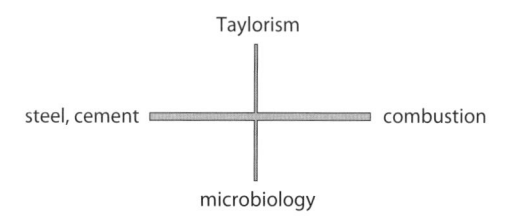

```
                    Taylorism
                        |
                        |
    steel, cement ══════╬══════ combustion
                        |
                        |
                   microbiology
```

Source: Author.

Table 1.1 Industries and Infrastructure of Each Technological Revolution, 1770–1970

Technological revolution	New technologies or redefined industries	New or redefined infrastructures
First industrial revolution: From 1771 in Britain	Mechanized cotton industries, wrought iron, machinery	Canals and waterways, turnpike roads, water power (improved water wheels)
Age of steam and railways: From 1829 in Britain, spreading to continental Europe and the United States	Steam engines and machinery made from iron for many industries, including textiles, railways, steamships; iron and coal mining playing a central role in growth	Railroads, national telegraph mainly along railway lines, universal postal service, worldwide sailing ships, great ports and depots, city gas
Age of steel, electricity, and heavy engineering: From 1875 in the United States and Germany, overtaking Britain	Cheap steel, full development of steam engines for steel ships, heavy chemistry and civil engineering, copper and cables, canned and bottled food, paper packaging	Worldwide shipping in rapid steel steamships, worldwide railways, steel bridges and tunnels, worldwide telegraph, national telephone, electrical networks for lighting and industrial use
Age of oil, the automobile, and mass production: From 1908 in the United States, spreading to Europe	Mass-produced automobiles; cheap oil fuels, petrochemicals; internal combustion engine for automobiles, transportation, tractors, airplanes, war tanks; electricity and electrical home appliances; refrigerated and frozen foods	Networks of highways, ports, airports, oil pipelines; universal electricity for industry and home; worldwide analog communications (telephone, telex, and cablegram wire and wireless)

Source: Adapted from Perez 2003, 14.

Figure 1.6 The Cognitive Revolution, 1980–2180

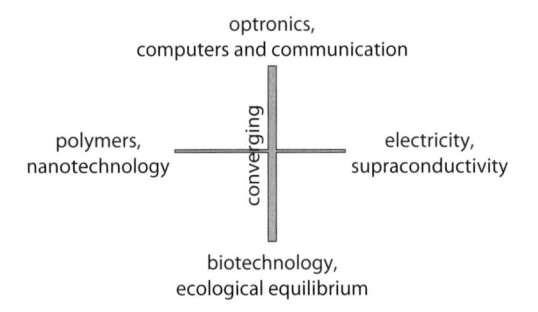

Source: Author.

What is new is the order of magnitude and the speed of change. Change has occurred so quickly that the average citizen does not realize that his cellular phone computes in nanoseconds, as does her laptop, and that a car's global positioning system (GPS) can transform the signals from the satellite to a position on Earth with a degree of precision of less than one meter in that time scale.

Also new is the ecological challenge. Global constraints on growth are becoming more apparent. The rapid rise of commodity prices in 2008 in general and of oil in particular drew attention to the pressures of excessive demand on limited resources. Demands on environmental resources, such as water, and levels of air pollution have also been very high. A particularly serious problem is the impact of increased greenhouse gases on global warming. The Intergovernmental Panel on Climate Change (IPCC) appointed by the

United Nations concluded in its final report in 2007 that global warming was unequivocal, that it was likely caused by human activity, that it would have serious negative impacts on a wide range of areas, and that adaptation and mitigation strategies were critical to managing these risks (IPCC 2007) (box 1.1).

Clearly, if humans do not succeed in rebuilding a sustainable equilibrium with nature, what is called civilization will inevitably collapse.[3] The question concerns not only nonrenewable mineral resources but also biodiversity now under strong pressure from human activities. The point is clear: nature can survive without humans, but humans cannot survive without nature.

Box 1.1 Innovation Is Essential to Tackling Climate Change

Climate change presents the world with a completely new set of challenges. It will require fundamental changes in the way we live. Diffusing known technologies worldwide and creating new and more effective ones will be essential to mitigating and adapting to climate change. At its current rate, climate change will transform the world into a vastly different place by the end of the century. Temperatures could rise to more than five degrees Celsius warmer than in preindustrial times. More frequent and intense storms, floods, and droughts will inflict heavy damage on human health and habitat and on biodiversity. Island nations and inhabited coastlines could be submerged, and up to 50 percent of species could become extinct. In the best-case scenario, temperatures are unlikely to be stabilized at less than two degrees Celsius above preindustrial temperatures. Most of the costs of climate change—some 75–80 percent—will be borne by developing countries. These countries are particularly reliant on natural resources, their populations live in exposed areas in precarious conditions, and they are ill-equipped to adapt financially or institutionally. Agricultural productivity will likely decline, particularly in the tropics, and 1–3 million more people are likely to die from malnutrition each year.

Dealing with climate change requires immediate action. Greenhouse gas concentrations in the atmosphere will increase average temperatures for centuries. As they cannot be reduced, future mitigation cannot make up for a lack of effort today. Moreover, the effects of climate change are already being felt. Even under an optimistic scenario of a change of two degrees Celsius, the impacts could be catastrophic for the most vulnerable populations.

Tackling climate change in a cost-effective and timely way will call for technology and innovation. A world with an increased temperature of two degrees Celsius will require greenhouse gas emissions to be 50 percent below 1990 levels by 2050. Because the world population is growing and poor countries are becoming richer, the world will need to change in a fundamental way how it produces energy, how that energy is used in transportation, buildings, and industry, and how forests, land use, and agriculture are managed. Existing technologies can buy time if they can be scaled up. Greater energy efficiency, management of energy demand, and diffusion of low-carbon electricity sources such as wind, hydro, and nuclear could produce half the required emission cuts. To satisfy future global energy demand, however,

continued

> **Box 1.1 continued**
>
> will require improving the performance of low-carbon technologies and developing technological breakthroughs. Promising technologies include carbon capture and storage, second-generation biofuels, and solar photovoltaics.
>
> Adapting to climate change will require increasing agricultural productivity and more "crop per drop." Research will be needed to develop resilient crops adapted to new environments and to manage water systems more effectively. Adaptation will also require a deeper scientific understanding of how climate change affects local environments and application of this knowledge to the design of new types of coastal protection systems, urban environments, and disaster-response communication systems.
>
> The challenges of climate change will require technology diffusion and innovation efforts in all countries. High-income countries will need to push the technology frontier, and developing countries will need to build their capacity to absorb, adapt, and diffuse existing technologies, as well as to create technologies appropriate for their local environments.
>
> *Source:* World Bank 2009.

Technology and Economic Growth

Regrettably, the benefits of historical advances have not spread equally. Major civilizations, such as China, and India, which were at the forefront of technical progress in earlier centuries, missed the Industrial Revolution (see box 1.2 for the case of India).

Divergences among Countries

Since the 1700s, per capita incomes have been diverging across countries and regions (figure 1.7). The benefits of increased per capita income were first concentrated in England, during the Industrial Revolution, and then spread to Western Europe and soon thereafter to the United States. By the end of the 1800s, the United States began to overtake Europe in many areas of industrial production.

Figure 1.7 raises the question, What accounts for the dazzling performance of the United States? Starting with the railroad, U.S. growth was largely supported by a vast internal market that allowed broader exploitation of transportation and communications advances. Embracing these technologies brought significant cost reductions through extensive economies of scale and scope. The United States was also rich in natural resources, including navigable rivers, arable land, timber, and minerals. Yet, more important than these contributing factors, the foundation of U.S. economic growth was a fabric of institutions and an economic and institutional regime that supported entrepreneurship, experimentation, and risk taking. Indeed, the United States may

> **Box 1.2 India, an Early Innovator**
>
> In India, ever since the Indus Valley civilization of about 5,000 years ago, innovation has been part of Indian culture and the basis of its civilization. India's prominent innovations have included remarkable town planning, the use of standardized burnt bricks for dwellings, an interlinked drainage system, wheel-turned ceramics, and solid-wheeled carts. The dockyard at Lothal is regarded as the largest maritime structure ever built by a Bronze Age community. The discovery of zero and the decimal-place value system by Indians dates back to the Vedic. Later pioneering work in algebra, trigonometry, and geometry deserves a mention. Innovations in medicine aim not only at the cure of diseases but also, and more importantly, at the preservation of health. The system of Ayurveda as well as advanced innovations in surgery including laparotomy, lithotomy, and plastic surgery are noteworthy. The iron pillar at Delhi, which has remained rust-free until today testifies to India's achievements in metallurgy some 1,500 years ago. Early Indian civilization was characterized by scientific thought, capabilities, and techniques at levels far more advanced than others. When the scientific and industrial revolutions took place in the West, however, India, with its highly feudalistic structure, was undergoing a period of stagnation. The lack of development during this period was a result of a hierarchical approach, irrational subjective thinking, and the buildup of superstitions and superficial ritualism. The earlier great traditions were allowed to decay. It was when its society was in this state that India came under colonial domination.
>
> *Source:* Ramesh Mashelkar, personal communication 2009.

Figure 1.7 Per Capita GDP for Selected Countries or Regions, 1480–1998
1990 international dollars

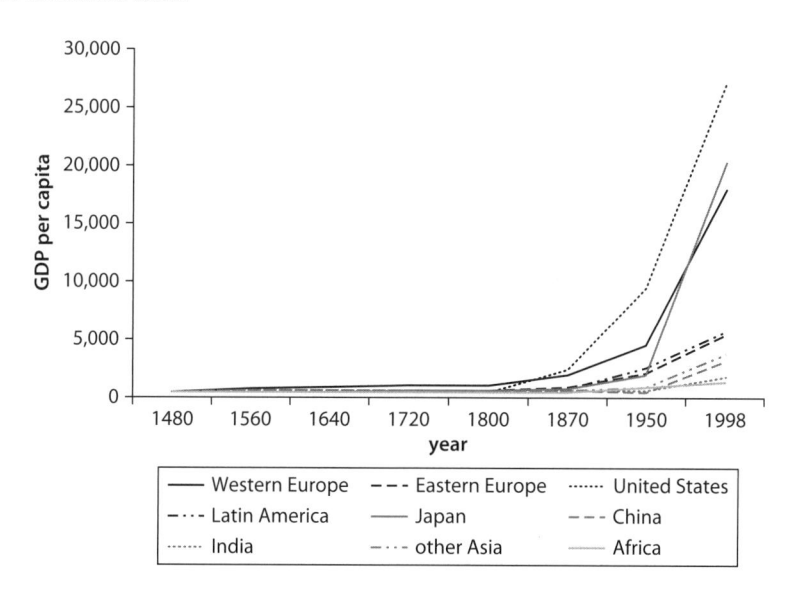

Source: Maddison 2006.

be said to have invented the process of invention itself, when Thomas Alva Edison created the first industrial research and development (R&D) laboratory. After Edison, many large U.S. companies created industrial R&D labs. By 1900, there were more industrial research laboratories in the United States than in Europe.

Calling attention to R&D as the basis of U.S. economic growth may lead some to think that the developing world needs to create more research capability to address unequal economic growth. While greater capacity for research can help, the innovation needs of developing countries are both simpler and more complex: simpler, because to a large extent they can increase productivity by making effective use of existing knowledge;[4] more complex, because the key requirements of technology-driven development are not simply new knowledge. Economic development requires education, combinations of technical skills, and a whole series of institutions, networks, and capabilities that enable the effective use of existing knowledge, all of which must be part of, or even precede, any serious effort to create new knowledge.

Economic Analysis of Innovation

The economics profession has been somewhat slow to acknowledge the importance of innovation for economic growth. Even Adam Smith, writing in the middle of the Industrial Revolution, was not fully aware of the fundamental nature of the changes in the economic paradigm around him.

In earlier economic models, output (Q) was expressed as a function of capital (K) and labor (L), and technology was assumed away (see equation 1.1):

$$Q = f(K, L). \tag{1.1}$$

Economist Robert Solow in 1957 became famous for noting that increases in capital and labor did not fully account for economic growth. There was another factor (A), which represented technical change and enhanced the productivity of capital and labor. Thus, technology was inserted as separate factor (A), which augmented the productivity of capital and labor, as in equation (1.2):

$$Q = A f(K, L). \tag{1.2}$$

Technology, however, was assumed to be exogenous. It took nearly three decades before Paul Romer in 1986 modeled technology not as exogenous manna from heaven but as the result of explicit effort. Thus, the new growth theory modeled technology (T) as the result of explicit inputs, namely, research and development ($R\&D$) and human capital (HC), as in equation (1.3):

$$A = f(R\&D, HC). \tag{1.3}$$

A good deal of empirical work has been done on the relationship between growth and the reduced form of the basic growth equation, where *R&D* and *HC* are substituted for technology, as in equation (1.4):

$$Q = f(K, L, R\&D, HC). \qquad (1.4)$$

This approach has tended to work better for analysis of developed countries than for developing countries for two main reasons: first, developing countries do not do much R&D; and, second, the principal ways in which developing countries produce products or processes that are new to them is by importing knowledge that already exists in developed countries. Thus, these growth equations need to incorporate the imports of capital goods and components, as well as imports more generally, foreign direct investment (FDI), and other channels for accessing existing global knowledge.

To some extent, knowledge is what lies behind total factor productivity (TFP), which is the residual for the growth in output that is not explained by the growth in inputs.[5] Many elements other than the underlying technology, however, affect the efficiency with which factors are used. These include the quality of factors themselves (such as the age of the capital equipment) as well as utilization rates and other ingredients that affect the efficiency with which they are used. A large literature has focused on the determinants of TFP, modeled to include inputs into the creation of knowledge (such as R&D and education), as well as access to foreign knowledge, human capital, physical infrastructure, the financial system, trade, the institutional regime (such as property rights, the rule of law, competitive pressure), and geography (climate, disease, distance from markets).

A good review of the literature on the determinants of TFP can be found in Isaksson (2007). It concludes that capital accumulation is a very important determinant of growth, not only because of capital deepening but also because more recent equipment tends to embody more productive new technology. Human capital in the form of education and health is also important—with health more important for countries at lower levels of development. Openness to foreign knowledge is more important than R&D for developing countries for the reasons noted above. R&D is more important for developed countries at the frontier, although developing countries may also need to undertake some R&D to absorb foreign knowledge. Finally, the review finds that competition, the rule of law, and the enforcement of contracts are all positively related to greater TFP growth.

A major debate in the economic literature has centered on whether capital accumulation or technical change are more important for growth. The findings depend very much on the methodologies used and the level of development of the countries studied. A very careful study of 112 countries over the period 1970–2000 used different growth models and grouped countries by the World Bank's four income classifications (Hulten and Isaksson 2007). It also distinguished the old tigers (Hong Kong, China; the Republic of Korea; Singapore; and Taiwan, China) from the new tigers (China, Indonesia, Malaysia, and Thailand). Hulten and Isaksson make the important point that the typical growth-accounting model in equation (1.2) is useful for analyzing the contribution to growth of capital deepening as compared to technology. That analysis, however, needs to be supplemented by a parallel analysis of growth levels. Their results show that using the conventional analysis of the type in equation (1.2), capital deepening, explains more than half the growth rate of output per worker in the majority of countries. Only in both types of the rapidly growing tiger economies is the contribution of TFP growth greater (see table 1.2).

Differences in levels, however, are explained mostly by differences in TFP. As table 1.3 shows, the level of TFP in low-income countries is only 20 percent of that in high-income countries, while that of lower-middle-income countries is 43 percent, and that of upper-middle-income countries is 63 percent. Moreover, the breakdown of the level of output into the capital deepening and the TFP components shows that the share of TFP growth (column 5 in table 1.3)

Table 1.2 Conventional Breakdown of Sources of Growth, 1970–2000

Indicator	Average annual growth of GDP per worker	Average annual growth of capital-labor ratio	Average annual growth of total factor productivity
Low income	0.17	0.25	−0.07
Lower-middle income	1.01	0.61	0.40
Upper-middle income	0.99	0.59	0.40
New tigers	3.79	1.70	2.09
Old tigers	4.89	2.37	2.52
High income	1.95	1.00	0.95

Source: Hulten and Isaksson 2007, 29.

Table 1.3 Level of Productivity in Countries of Various Incomes, 1970–2000

Indicator	Level of GDP/worker relative to high-income countries (percent)	Level of TFP relative to high-income countries (percent)	Log of GDP per worker	Log of capital-labor ratio	Log of TFP
Low income	6.05	19.84	7.76	2.61	5.55
Lower-middle income	22.46	43.41	9.08	3.14	5.93
Upper-middle income	44.47	63.30	9.76	3.45	6.31
New tigers	8.50	23.57	8.09	2.78	5.31
Old tigers	49.53	67.24	9.83	3.48	6.35
High income	100.00	100.00	10.57	3.81	6.77

Source: Hulten and Isaksson 2007, 30.
Note: TFP = total factor productivity.

is always greater than that of capital deepening (column 4) for all groups of countries.[6]

The bottom line of this analysis is that innovation, as roughly proxied by TFP (or what cannot be explained simply by factor inputs), is the major contributor to the differences in development levels across countries. Moreover, while capital deepening is more important in explaining the growth of countries at lower levels of income, TFP growth accounts for more than half the growth in those economies (the tiger economies) that have grown the fastest (table 1.2). And the level of TFP accounts for the bulk of difference in GDP per worker for all countries (table 1.3).

Innovation and Emerging Economies

Given the large stock of knowledge in the world and its rapid expansion, developing countries would seem to have tremendous potential for moving up rapidly to the world technological frontier. It would seem to be even easier now that transportation and communications costs have been falling continuously and that the world is more globally integrated through trade and other forms of exchange. The share of imports and exports in global GDP, for example, increased from 40 percent in 1990 to 61 percent in 2006 (World Bank 2008).

A Macro View: Fast-Growing Economies

In the past 55 years, only six economies—Hong Kong, China; Japan; Korea; Malta; Singapore; and Taiwan, China—have made the transition from developing to developed economies. That they have done so shows that it is possible. That they are so few indicates that it is not as easy in practice as in theory. A broader sample of rapid catch up can be obtained by examining developing countries that have had at least 25 years of consecutive growth above 7 percent since 1950. The Growth Commission Report of 2008 found only 13 economies that had achieved such high rates of growth. These were the six mentioned above along with Botswana, Brazil, China, Indonesia, Malaysia, Oman, and Thailand. With the exception of Botswana and China, that rate of growth has not been sustained, preventing others from making the transition to the per capita income levels of developed countries.

According to the Growth Commission Report, five main elements accounted for the rapid growth of the economies listed above:

- They fully exploited the world economy.
- They maintained macroeconomic stability.
- They mustered high rates of savings and investment.
- They let markets allocate resources.
- They had committed, credible, and capable governments.

It is noteworthy that the first factor is exploitation of the world economy. The second, third, and fourth are largely the basics of development theory (although there is some difficulty squaring some of the interventionist policies of many of the governments with the principle of letting markets allocate resources). The fifth highlights the role of government, including its role in overcoming market failures, again with some tensions over reliance on the market.

It is worthwhile looking more closely at the first factor because it is essentially about the importance of innovation and technology in the development of those economies:

> One, they imported ideas, technology, and know-how from the rest of the world. Two, they exploited global demand, which provided a deep, elastic market for their goods. The inflow of knowledge dramatically improved the economies' productive potential; the global market provided the demand necessary to fulfill it. *To put it very simply, they imported what the rest of the world knew, and exported what it wanted.* (Italics added) (Commission on Growth and Development 2008)

The first component is tapping global knowledge, which is innovation in the broad sense used in this volume. The second has to do with exploiting economies of scale beyond the confines of limited domestic markets and building on comparative advantage. Both are very much among the policy options and strategies that countries can pursue in their development strategies.

A Micro View: Firm-Based Innovation Surveys

The preceding section has pointed to the critical role of innovation in explaining economic growth. It has also shown how tapping into global knowledge is particularly important for explaining rapid innovation in the fastest-growing economies. This section takes a more detailed perspective. It reports on the results of firm-level surveys on innovation, which are consistent with the analysis at the macroeconomic level. It then examines the dissemination of three specific social technologies across countries. Their dissemination is still very limited in many areas, and this discussion gives a rough notion of how much less-developed countries could improve their welfare if they adopted these technologies more quickly. Finally, this section summarizes and comments on some more general trends in the dissemination of technologies.

Detailed surveys of innovation offer some highly relevant insights into the frequency and type of innovation, its determinants, and its impact on the productivity or growth of firms. These surveys have been developed and implemented most systematically in the European Community as part of the Community Innovation Surveys, which are currently in their sixth round. Surveys generally following the same methodology have also been carried out in Latin America, in Argentina, Brazil, Chile, and Uruguay (Crespi and Peirano 2007).[7]

Innovation survey studies also generally find a positive relation between product innovation and increased labor productivity, although it depends on the type of innovation being measured. The most comprehensive study to date, based on firm-level innovation data for 20 countries in the Organisation for Economic Co-operation and Development (OECD),[8] found that *product* innovation has a positive effect on labor productivity but that *process* innovation has a negative or insignificant effect, at least in the short run. While the latter finding was initially surprising, it appears that it takes time for the firm to adjust to and learn the new production technology before it starts to see the full benefits.

Microdata on firm productivity point to potential gains from the dissemination and effective use of existing knowledge. Table 1.4 presents the very high dispersion of value added per worker across nine representative industrial sectors in Brazil (Rodriguez, Dahlman, and Salmi 2008). Particularly striking is the size of the difference between the most and the least efficient firms—peaking at 300,000 times in the machinery and equipment sector. The average for all nine sectors is an amazing 57,000 times. Adjusting the minimum by "eyeballing" the distributions of dispersion and taking as the maximum the value when the distribution begin to have some density gives a conservative measure less influenced by outliers. That adjusted maximum averaged 53 percent of the distance to the recorded maximum. Even with these conservative adjustments, it appears that if average productivity could be raised to the adjusted maximum level, average productivity would increase by a factor of 10.[9]

With a similar methodology, the average level of productivity was estimated to rise by a factor of five in India. It is surprising that the productivity dispersions are, on average, twice as large in Brazil as in India, considering that dispersions in the latter already exceed those in most of the countries to which it has been compared (Dutz 2007).

Table 1.4 Productivity Dispersion in Brazil's Industrial Sectors, Mid-2000s
value added per worker

Sector	Maximum/minimum	Adjusted maximum as % of maximum	Adjusted maximum/mean
Food and beverage	12,900.07	57.22	9.42
Textile	1,169.01	67.31	5.99
Apparel	79,103.56	31.60	9.14
Leather and footwear	65,897.30	73.33	4.81
Chemicals	9,879.34	61.91	7.83
Machinery and equipment	315,929.99	37.98	33.83
Electronics	6,658.67	52.03	10.00
Auto parts	689.60	64.88	4.17
Furniture	26,916.31	35.06	7.88
Average	**57,682.65**	**53.48**	**10.34**

Source: Rodriguez, Dahlman, and Salmi 2008; calculations based on the World Bank Investment Climate Survey of Brazil.
Note: The top and bottom 1 percent of the sample were discarded to winnow out false readings from data errors.

This analysis suggests just how much national output could be raised—at least in principle—if all Brazilian or Indian firms were to adopt existing technology that other firms are already using. Obviously, moving to these higher-productivity technologies is not costless. The firms that use them now are likely to be much larger; they are using other modern equipment; they generally employ more up-to-date management practices; they use better inputs; and they have better educated and more highly skilled workers. Yet the larger point is that at least some firms are using these production technologies, while those that are not are operating far below their more efficient counterparts. Much more must and can be done to disseminate and effectively employ existing knowledge across the board.

Dissemination of Technology

The dissemination and use of existing technologies are key for economic and social development. Two issues are discussed below: first, how simple technologies can improve welfare and, second, how the speed of diffusion has increased over time.

Simple Technologies That Can Significantly Increase Welfare

The three technologies that can significantly improve welfare are vaccines, access to clean water, and access to sanitation:

- *Basic vaccines.* Basic vaccines can make a big difference to children's health: DPT (diphtheria, whooping cough, tetanus) and measles. In developed countries, they are given to virtually all children as part of basic preventive pediatric medicine. In developing countries, the immunization rates, compared to those of high-income countries, vary widely from 66 percent in Sub-Saharan Africa and South Asia to 94 to 101 percent for other developing regions such as Latin America (table 1.5).

- *Access to clean water.* Access to clean water can be achieved through many relatively simple technologies. Although the technology is well known and there has been some improvement over the past decade, 20 percent of the total population of low- and middle-income countries still lacks access to clean water. The figure varies by region, as well as between urban and rural dwellers. In rural areas, 30 percent of the population on average does not have access to clean water, compared to only 7 percent among urban dwellers (table 1.6).

- *Access to sanitation.* Sanitation also enhances basic welfare and prevents the spread of many diseases; it can be made available through many simple technologies. The average rate of access in low- and middle-income countries is barely 50 percent (up from just 33 percent a decade earlier) and

Table 1.5 Percentage of Children Worldwide Who Received Basic Vaccines and Ratio to High-Income Countries, 1993 and 2003

Location	DPT[a] 1993	DPT[a] 2003	Measles 1993	Measles 2003	Ratio to high-income countries[b] DPT	Ratio to high-income countries[b] Measles
Region						
East Asia and Pacific	83	83	79	83	0.87	0.90
Europe and Central Asia	80	89	84	91	0.94	0.99
Latin America and the Caribbean	78	90	82	93	0.95	1.01
Middle East and North Africa	85	91	84	92	0.96	1.00
South Asia	59	63	59	61	0.66	0.66
Sub-Saharan Africa	49	59	51	61	0.62	0.66
High-income countries	88	95	83	92	1.00	1.00
World	**71**	**76**	**71**	**75**	**0.80**	**0.82**

Source: World Bank data.
Note: Percentages refer to children ages 12 to 23 months.
a. Immunization to protect against diphtheria, pertussis (whooping cough), and tetanus.
b. Data from 2003 only.

Table 1.6 Percentage of Rural and Urban Population with Access to Clean Water, 1990 and 2004

Location	Total 1990	Total 2004	Rural 1990	Rural 2004	Urban 1990	Urban 2004
Region						
East Asia and Pacific	71.8	78.5	61.4	69.8	97.3	91.9
Europe and Central Asia	91.7	91.7	83.4	79.8	97.0	98.7
Latin America and the Caribbean	82.8	91 0	50.0	73.0	92.6	96.0
Middle East and North Africa	87.5	89.5	78.9	80.8	96.1	96.3
South Asia	70.6	64.4	64.9	81.3	88.6	93.6
Sub-Saharan Africa	48.9	56.2	36.1	42.4	81.9	80 1
World	**76.4**	**82.7**	**63.2**	**72.2**	**95.2**	**94.5**
Countries						
High income	99.8	99.5	99.1	98.5	99.8	99.8
Low and middle income	72.1	79.9	60.6	70.5	93.3	92.8
Low income	64.3	75.0	56.7	69.4	87.0	88.1

Source: World Bank data.

averages just 34 percent for the rural population compared to 74 percent for the urban population (table 1.7). Again, there is wide diversity across regions and within countries.

Clearly, the potential for increasing growth and welfare is tremendous even if countries simply bring existing technologies to those who do not yet have them.

Table 1.7 Percentage of Rural and Urban Population with Access to Sanitation, 1990 and 2004

Location	Total 1990	Total 2004	Rural 1990	Rural 2004	Urban 1990	Urban 2004
Rural						
East Asia and Pacific	29.7	50.6	15.3	36.1	65.5	72.4
Europe and Central Asia	86.1	85.0	72.0	70.3	93.7	93.0
Latin America and the Caribbean	67.4	77.1	35.4	48.7	80.7	85.7
Middle East and North Africa	69.9	76.2	52.0	57.9	87.1	92.3
South Asia	17.4	37.2	6.3	26.6	50.3	62.7
Sub-Saharan Africa	31.5	37.2	23.8	28.2	52.4	53.3
World	**44.4**	**57.0**	**22.8**	**37.7**	**77.2**	**79.4**
Countries						
High income	100.0	100.0	100.0	100.0	100.0	100.0
Low and middle income	36.2	51.4	17.9	34.4	70.3	74.4
Low income	21.3	38.3	11.6	28.5	49.6	60.5

Source: World Bank data.

Time Lags in the Dissemination of Major Technologies

A detailed study of the dissemination of major global technologies showed two key trends relevant to the present discussion.[10] The first is that the speed at which major innovations disseminate across countries has increased over time. Thus, while key innovations developed between 1750 and 1900 took, on average, slightly more than 100 years to disseminate to 80 percent of the countries surveyed, those developed between 1900 and 1950 took an average of 61 years; those developed between 1950 and 1975, an average of 24 years; and those developed between 1975 and 2000, an average of 16 years (see table 1.8).

The second is that while the dissemination of technology to the capital and major cities of developing countries has increased, dissemination within countries remains very slow. This pace holds true even for some relatively old technology, such as electricity and paved roads, and in the case of the three social welfare–enhancing technologies cited above. Thus, affordability and skills constrain the adoption of the old technologies that could make a big improvement in people's lives.

New Global Perspectives

Innovation agendas in the developed and in the developing world will differ significantly. The drivers for innovation in the developed world have been centered on getting more (performance and productivity) from less (physical, financial, human capital) for more (profit, value to the share holder). In contrast, the drivers in the developing world are to get more (performance, productivity) from less (cost) for more and more (people). In other words,

Table 1.8 Rate of Dissemination of Major Technologies, 1748–2000
years

Technology	1748–1900	1900–50	1950–75	1975–2000	Number
Transportation					21
Shipping (steam)	83				57
Shipping (steamMotor)	180				93
Rail (pass.)	126				99
Rail (freight)	124				153
Vehicle (private)	96				123
Vehicle (commercial)	63				109
Aviation (passenger)		60			103
Aviation (freight)		60			
Communications					
Telegram	91				77
Telephone	99				156
Radio		69			154
Television		59			156
Cable TV		50			98
Personal computer			24		134
Internet use			23		151
Mobile phone				16	150
Manufacturing					
Spindle (ring)	111				50
Steel (OHF)	125				50
Electrification	78				155
Steel (EAF)		92			91
Synthetic textiles		36			75
Medical (OECD only)					
Cataract surgery	251				19
X-Ray*		93			27
Dialysis*		33			29
Mamography			33		18
Liver transplant			28		29
Heart transplant			28		27
CatScan			18		29
Litho triptor				15	26
Average (excluding medical)	106.9	60.9	23.5	16.0	
Average (including medical)	118.9	61.3	25.7	15.5	

Source: Cited in World Bank 2008.
Note: The table indicates the number of years elapsed from discovery or invention until the technology had reached 80 percent of reporting countries.

innovation in the developing world has to focus on "inclusive growth"—hence the importance of "inclusive innovation" policies (see chapter 11).

Emerging and new drivers for innovation need to be stressed. Developing countries such as China, Brazil, and India as well as other emerging economies will continue to grow. They will focus on domestic competition–led growth and increasingly reduce their dependence on developed markets. As a result, during this decade 2–3 billion people will become part of the aspiring middle class, putting enormous pressure on resources—fossil fuels, commodities, and water. The combination of contradictory forces—that is, pressure to conserve resources, addition of 2–3 billion consumers, and further globalization—will require new, hitherto unheard of models of innovation.

The key innovation priority for developing countries is to acquire and use knowledge that already exists, which is less costly and less risky than creating new knowledge. While some of this knowledge is protected by intellectual property rights and therefore would have to be purchased, an enormous amount is in the public domain. Therefore, policies that facilitate access to global knowledge are critical. How well developing countries use this form of innovation will depend not only on their policies but also on the support of the country's institutions and the effectiveness of those institutions and the people in them. Because productivity is dispersed within sectors, raising average productivity to local best practice (or, even better, to *global* best practice by acquiring more knowledge from abroad) can generate high returns.

More effort must go into applying innovative approaches to global public goods, such as health, the environment, and global warming, and for developing new and better ways of dealing with global health problems like malaria, HIV/AIDS, and pandemics. And as tables 1.5, 1.6, and 1.7 indicate, the scope for expanding preventive health techniques in developing countries is immense.

Finally, innovative solutions must be part of any strategy for addressing the pressure on physical resources, including the need for clean air and water, and for mitigating global climate change; current development models are not sustainable and better ones must be found. Although elements of sustainable development exist, lack of information, money, and, in some cases, even concern for global problems is preventing wide adoption. Unfortunately, political systems, with their short-term orientation, have a strong tendency to push difficult problems to the future. Political leaders who wish to be reelected often avoid hard choices for fear of alienating voters and simply pass the problems on to the next generation of decision makers. There may be tipping points, however, beyond which it is more costly and more difficult to act. The present generation, therefore, should shoulder more responsibility for dealing with these problems by bringing to bear the full array of innovative solutions at its disposal.

Notes

1. At the broadest level, average per capita income is a good summary measure of the effective application of knowledge to production of goods and services, although in comparisons across countries it is necessary to be mindful of cases where rents from the sale of natural resources such as oil bias per capita income upward.

2. See Maddison (2006) for a millennial historical overview.

3. See Jared Diamond's *Collapse,* on the end of past civilizations.

4. As pointed out by Gershenkron (1962), the advantage for late industrializers is that they can draw on the technology and experience of developed countries. However it is not easy to replicate what other countries have done, as evidenced by the very small number of countries that have made the transition from low to high incomes.

5. This is what Abramowitz (1956) famously called the" residual of our ignorance."

6. Hulten and Isaksson use different estimates for the distribution of capital and labor shares. Those reported in Tables 1.2 and 1.3 are based on the usual ratio of two-thirds capital, one-third labor. Table 1.3 is based on a Cobb-Douglas function with constant returns to scale and assumes the same shares across all countries.

7. They follow the OECD's *Oslo Manual* of 1992, with revisions in 1997 and 2005 which continually broadened the definition of innovation.

8. See OECD 2008. "Innovation in Firms: Findings from a Comparative Analysis of Innovation Survey Microdata," chapter 5.

9. While it is a thought-provoking exercise to analyze dispersion in productivity within sectors, it must be noted that in some cases the variance in productivity levels may be caused by other factors, such as economies of scale and greater capital intensity.

10. Comin and Hobijn (2004) traced the diffusion of 100 key technologies between 1750 and 2003 in 157 countries.

References and Other Resources

Abramovitz, Moses. 1956. *Resource and Output Trends in the United States Since 1870.* New York: National Bureau of Economic Research.

Comin, Diego, and Bart Hobijn. 2004. "Cross-Country Technology Adoption: Making the Theory Face the Facts." *Journal of Monetary Economics* 51 (1): 39–83.

Commission on Growth and Development. 2008. *The Growth Report: Strategies for Sustained Growth and Inclusive Development.* Washington, DC: World Bank.

Crespi, Gustavo, and Fernando Peirano. 2007. "Measuring Innovation in Latin America: What We Did, Where We Are and What We Want to Do." Paper prepared for the Conference on Micro Evidence on Innovation in Developing Countries, UNU-MERIT, Maastricht, The Netherlands, May 31–June 1.

Diamond, Jared. 2005. *Collapse: How Societies Choose to Fail or Succeed.* New York: Viking.

Dutz, Mark, ed. 2007. *Unleashing India's Innovation: Toward Sustainable and Inclusive Growth.* Washington, DC: World Bank.

Fogel, Robert. 1999. "Catching Up with the Economy." *American Economic Review* 89 (1) (March): 1–21.

Gershenkron, Alexander. 1962. *Economic Backwardness in Historical Perspective.* Cambridge, Massachusetts: Belknap Press.

Hulten, Charles, and Anders Isaksson. 2007. "Why Development Levels Differ: The Sources of Differential Economic Growth in a Panel of High and Low Income Countries." Geneva: United Nations Industrial Development Organization.

Intergovernmental Panel on Climate Change. 2007. *Climate Change 2007.* Synthesis Report. http://www.ipcc.ch/pdf/assessmentreport/ar4/syr/ar4_syr_spm.pdf.

Isaksson, Anders. 2007. "Determinants of Total Factor Productivity: A Literature Review." Geneva: United Nations Industrial Development Organization.

Lee, Ronald. 2003. "Age Structure and Dependency." In *Encyclopedia of Population*, ed. Paul George Demeny and Geoffrey McNicoll, 542–45. New York: Macmillan Reference USA.

Maddison, Angus. 2006. *The World Economy: A Millennial Perspective.* Paris: Organisation for Economic Co-operation and Development.

OECD (Organisation for Economic Co-operation and Development). 2005. *Oslo Manual: Guidelines for Collecting and Interpreting Innovation Data.* Paris: OECD.

———. 2008. *Science Technology and Industry Outlook 2008.* Paris: OECD.

Perez, Carlotta. 2003. *Technological Revolutions and Financial Capital: The Dynamics of Bubbles and Golden Ages.* Cheltenham: Edward Elgar.

Rodriguez, Alberto, Carl J. Dahlman, and Jamil Salmi. 2008. *Knowledge and Innovation for Competitiveness in Brazil.* Washington, DC: World Bank.

Romer, Paul M. 1986. "Increasing Returns and Long-Run Growth." *Journal of Political Economy* 94 (5): 1002–37.

Rosenberg, Nathan. 1984. *Inside the Black Box: Technology and Economics.* Cambridge, UK: Cambridge University Press.

Ross, J. A. 1997. *International Encyclopedia of Population.* New York: McGraw Hill.

World Bank. 2008. *Global Economic Prospects 2008.* Washington, DC: World Bank.

———. 2009. *World Development Report 2010.* Washington, DC: World Bank.

How to Promote Innovation: Policy Principles

After the presentation in chapter 1 of the rationale for promoting innovation, the next step is to consider the ways and means. The developed economies have now had some 40 or 50 years of experience with innovation policy, as distinct from, say, science and technology policy (a brief history of innovation policy in the countries of the Organisation for Economic Co-operation and Development [OECD] can be found later in the chapter). Based on that experience but also on the specific features of developing economies, it is possible to define a few basic principles that should inform the efforts of policy makers as they seek to promote innovation in their own countries. The following principles are explored in this chapter:

- Take a broad view of innovation and its forms and sources.
- Go beyond traditional science and technology policy and adopt a "whole-of-government" approach.
- Create receptive and mobilizing climates.
- Put in place efficient institutions and instruments.
- Adapt to the societal context.

The end of the chapter discusses ways to adapt these principles to the specific characteristics of developing countries.

This chapter was prepared by Jean-Eric Aubert, with contributions from Thierry Gaudin and Carl Dahlman. Jean-François Rischard also provided useful inputs.

Take a Broad View of Innovation

Chapter 1 makes clear the importance of taking a broad view of innovation as something that is new relative to a given context. Innovation may be new to the country in which it appears, to the region or the sector in which it takes place, or to the firm that develops or adopts it. What matters is the diffusion of this relative novelty as a source of wealth, jobs, and welfare, a particularly relevant factor for the developing world. Innovation policy should, in priority, aim to capture global knowledge and technology and to adapt and disseminate them in local contexts.

Another important point is that all potential types and sources of innovation should be considered and addressed by innovation policy, not only science- and research-driven innovation. It is true that key changes in societies and economies have been and continue to be brought about by technological advances deriving from science and research efforts. Other types of innovation, however, including those in the economically advanced economies, derive from sources other than research and development (R&D) yet have a considerable impact as the origin of new industries, jobs, and income. For instance, although the cultural and creative industries, such as those related to the media, certainly make use of technologies, sometimes sophisticated ones such as electronics, their novelty lies in offering a new service, better packaging, and the like.

Similarly, many innovations in logistics, service delivery, and supply chains make use of technologies like information technology (IT) but are fundamentally managerial in nature. These, too, have considerable importance for economic growth and the improvement of living conditions (Wal-Mart is a typical example). To that list may be added innovations that are entirely social in nature—with no technological foundation; yet these too can have an enormous impact. An emblematic example is microcredit, introduced by Muhammad Yunus initially in Bangladesh, which has since spread throughout the world.

A third important point is to understand that the development of any new industry requires a complex set of activities and competencies that go far beyond technology or R&D. The example of the wine industry in South Africa—which, of course, requires technological competency in the production process but also management competency and investment in complementary activities for tourism and export—illustrates the point (see figures 2.1 and 2.2).

Adopt a "Whole-of-Government" Approach

Innovation policy has come into its own with some difficulty, as it was crushed between two ideologies with very active lobbies. The *scientific ideology* promoted the idea that technology derives naturally from science, so that governments need do no more than build a good science base. The *market ideology* considers that innovation occurs naturally in a good business climate and that

Figure 2.1 Process Components of the Wine Industry in South Africa

unique cepages and blends
maceration
"soft" equipment
barrel ageing
yeasts
viniculture
quality testing
temperature controls
hygiene

Source: Mytelka 2004.

Figure 2.2 Organizational and Marketing Elements of the Wine Industry in South Africa

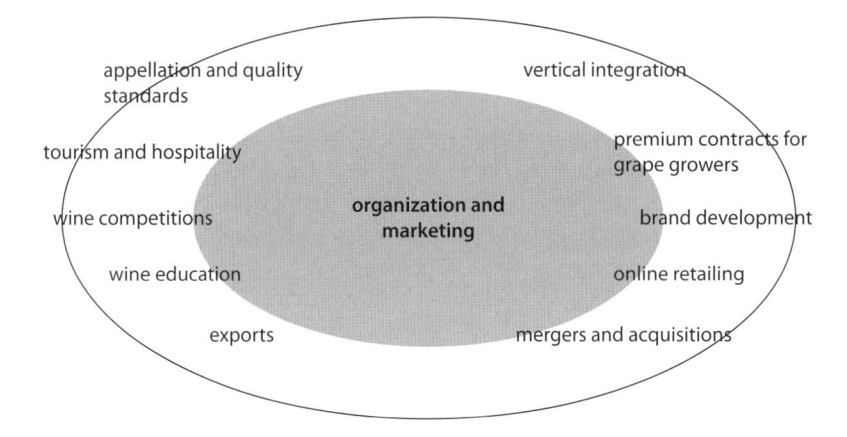

appellation and quality standards
vertical integration
tourism and hospitality
premium contracts for grape growers
wine competitions
organization and marketing
brand development
wine education
online retailing
exports
mergers and acquisitions

Source: Mytelka 2004.

governments should therefore concentrate on creating an environment conducive to business. Governments, according to this latter view, need only to maintain an open, competitive environment and, in addition, fund public goods such as basic research that the private sector is unable to finance. Although these two views have acted in concert to promote their interests, governments have nevertheless felt the need to take specific measures to promote innovation. Their efforts took advantage of World War II initiatives and governments' strong involvement in the development of defense technologies.

As noted in box 2.1, government efforts in the 1960s and 1970s were largely inspired by a linear model of innovation and the idea that science and research efforts needed to be pushed toward technological and industrial applications; many policy initiatives therefore aimed at supporting enterprises in their R&D

Box 2.1 A Brief History of Innovation Policy in OECD Countries

In the first part of the 20th century, innovation policy as such did not truly exist. It was gradually developed as a way to promote the industrial competitiveness and social welfare of countries, as a complement to actions taken by governments to develop defense technologies, as initiated in World War II. Innovation policy has emerged gradually as a policy distinct from both science and industry policies. The evolution of government efforts to encourage innovation over the second part of the 20th century can be summarized as follows:

1950s: This decade saw the building of modern science systems in the industrialized world. In some countries, piecemeal measures were occasionally adopted to reduce identified weaknesses in the innovation process, including the creation of the National Research and Development Corporation in the United Kingdom (1949), the aim of which was to facilitate the promotion and diffusion of inventions from public laboratories and universities. Among others, France established sector technical centers to help industries with technical research, assistance, and information, and Germany set up the Fraunhofer system of applied R&D.

1960s: Two trends were noticeable. First was launching of large-scale programs in areas such as space, nuclear technology, and oceanography, in countries such as France, the United Kingdom, and the United States. This reflected the need for strong government engagement in strategic fields perceived as important to the national interest, beyond the defense sector, strictly defined. Second was the emergence of the concept of innovation policy as distinct from science policy. The seminal report in this area is the Charpie Report published in the United States at the request of the Department of Commerce in 1967 (U.S. Department of Commerce 1967). It stated clearly the need to act on diverse factors affecting the innovation climate—university-industry relations, venture capital, procurement policies, tax incentives, and competition laws—with particular attention to small enterprises and individual inventors, presented as the main source of innovation. This report had a relatively limited impact on concrete policy actions, but it was important from a conceptual viewpoint and was used in several countries to give innovation policy a specific identity and separate it from science policy.

1970s: This decade saw a proliferation of government measures to promote innovation in the form of civilian technology programs, R&D incentive schemes for in-house efforts in the business sector, or university-industry collaboration. This was particularly evident in Europe and Japan, which were concerned with an increasing technology gap with the United States. The oil crisis of 1973–75, and the subsequent economic slowdown, also led to renewed interest in innovation policies. It nonetheless remained difficult to capture the field of innovation and the nature of innovation policies, as demonstrated by a major survey undertaken by the OECD in 1973–76. It revealed the extent to which measures reported by governments were influenced by the general institutional setting in which governments operate. For instance, the United States, with its very decentralized system, reported the innovation policy measures of many different agencies, whose measures mattered from their specific viewpoints. For instance, the Small Business Administration was strongly involved in support for small firms, perceived as a key source of innovation (as in the Charpie Report); the National Science Foundation for its part supported basic research; and the various sector agencies (defense, commerce, interior, and so on) all had technology-related programs. At the other end of the spectrum, there were the

Box 2.1 continued

countries (notably the large European countries) that reported a limited set of measures, which specifically complemented their science policy, their industry policy, or their education policy. The nature of these measures helped show the role innovation policy can play and the way it operates among other established policies.

1980s: Two major trends emerged. First was the development of regional technology and innovation policies, owing to an increasing perception that innovation flourishes in sites with a concentration of talent, knowledge, and resources. It was therefore considered important to build critical mass, and major programs were set up to build science parks or "technopolises" (in Japan, for example). The need was also felt to act as closely as possible to entrepreneurs and potential innovators in order to help them efficiently. Hence, territorially decentralized innovation policy initiatives proliferated, often encouraged by central governments through various schemes (such as decentralized antennas of central innovation agencies, or matching funds provided to local governments). The second major feature of this decade was the emergence of the notion of national innovation systems, which emphasized the interactions among key actors and communities (research, business, education) as a source of the innovative dynamism of countries and the need for governments to strengthen such systems through appropriate policy actions.

1990s: Inspired by this concept, as well as the acceleration of the globalization process, the spread of information and telecommunication technologies, and the emergence of new technologies such as biotechnologies, governments systematically engaged in building innovation policies that encompassed established policy fields. In the traditional science policy field, efforts were made to connect basic research more closely to applications. In education and, notably, university policy, attention was given to developing interest and competence in innovation among youth. In industry policy, horizontal actions to boost innovation efforts were perceived as an efficient way to replace traditional policies to "pick winners," which were criticized for their inefficiency and ideological inadequacy. The Nordic countries have probably been the most active and the most coherent in adopting this approach, because it is easier to implement in societies with a strong communitarian and consensual mode of governance and economies of relatively small size. In the mid-1990s, Finland, for example, created two key institutions for promoting innovation: Tekes, the technology agency in charge of supporting innovation directly with a very significant budget; and the Science and Technology Policy Council, chaired by the prime minister, with the active participation of all ministers (including finance), which seeks to improve the innovation climate in all relevant policy fields and is directly inspired by the concept of the innovation system.

2000s: The notion of innovation policy has become very fashionable, and all countries have adopted it, as evidenced by the development and proliferation of OECD innovation system and policy reviews. Initially pioneered in the mid-1980s, such reviews now respond to strong demand and replace the earlier science, and then science and technology policy, reviews implemented from the early 1960s. The demand for these reviews comes not only from the "old" OECD members but also, massively, from the transition economies that have recently joined the OECD, as well as dynamic economies from different parts of the world, such as Chile or China.

Source: Author.

efforts or at improving collaboration between universities and industry. Concomitant large-scale space and defense programs facilitated the development of breakthrough technologies that were later used in civilian applications.

Recognition of the importance of interactions in innovation processes led to the concept of *innovation systems*, which was introduced in the literature in the late 1980s. This concept has been particularly fertile. Although variously understood, most often it defines the sets of interacting actors and institutions that provide the resources (knowledge, finance, and the like) required for the successful development of innovations.

The first generation of innovation policy was therefore replaced by a second generation in which innovation policy became more complex and aimed at facilitating interactions between the various actors and institutions involved in innovation processes: universities, research laboratories, banks (for venture capital), and government agencies in charge of various sectors (industry, health, and agriculture, for example).

It is inherently difficult, however, to define precisely the boundaries of an innovation system, and figure 2.3 suggests why this is so. Some legitimately extend the frontier of the system to what are known as the "framework conditions" that encompass elements as apparently distant from the innovation process as the educational system or the macroeconomic environment. The OECD, for instance, explicitly includes framework conditions in its

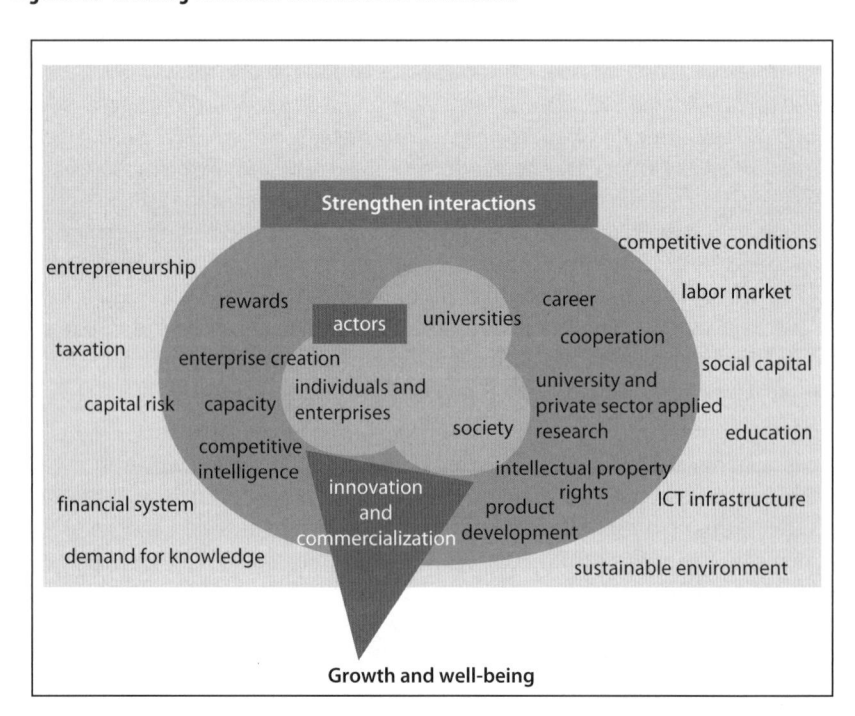

Figure 2.3 Creating Favorable Conditions for Innovation

Source: Andersson, Djeflat, and de Silva 2006.
Note: ICT = information and communication technology.

reviews of innovation systems. Thus, a third generation of innovation policy has appeared, inspired by a "whole-of-government" approach, in which all departments are potentially concerned.

Specific features of innovation systems in developing countries are depicted in figure 2.4. An innovation may come from abroad or from other users in the same country, or it may be created by public or private R&D labs or firms in the same country (first column). The innovation may be transferred in various ways, ranging from investment or formal purchases of technology, capital goods, components, or products to movement of people and informal sharing of information by people or through information-enabled networks (second column). It may be transferred to users: firms, government, public institutions, social organizations, or individuals (third column). Dissemination occurs through market mechanisms such as the growth of more efficient firms, as well as through informal networks and special institutions or programs such as technological information centers and productivity and extension agencies (fourth column).

The broader economic and institutional regime is a key determinant of the innovation climate. These influences include a country's macroeconomic

Figure 2.4 Schematic of the Innovation System in a Developing Country

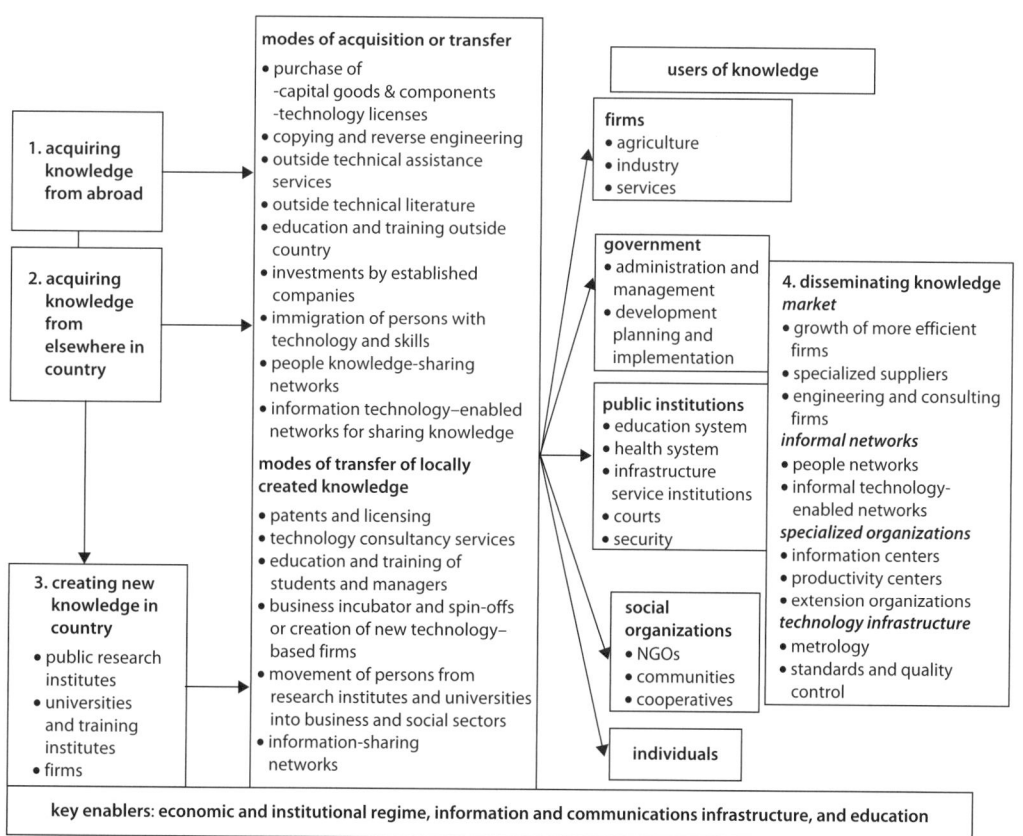

	modes of acquisition or transfer	users of knowledge
1. acquiring knowledge from abroad	• purchase of -capital goods & components -technology licenses • copying and reverse engineering • outside technical assistance services • outside technical literature • education and training outside country	**firms** • agriculture • industry • services
2. acquiring knowledge from elsewhere in country	• investments by established companies • immigration of persons with technology and skills • people knowledge-sharing networks • information technology–enabled networks for sharing knowledge	**government** • administration and management • development planning and implementation
	modes of transfer of locally created knowledge	**public institutions** • education system • health system • infrastructure service institutions • courts • security
3. creating new knowledge in country • public research institutes • universities and training institutes • firms	• patents and licensing • technology consultancy services • education and training of students and managers • business incubator and spin-offs or creation of new technology-based firms • movement of persons from research institutes and universities into business and social sectors • information-sharing networks	**social organizations** • NGOs • communities • cooperatives **individuals**

4. disseminating knowledge
market
• growth of more efficient firms
• specialized suppliers
• engineering and consulting firms
informal networks
• people networks
• informal technology-enabled networks
specialized organizations
• information centers
• productivity centers
• extension organizations
technology infrastructure
• metrology
• standards and quality control

key enablers: economic and institutional regime, information and communications infrastructure, and education

Source: Author (Carl Dahlman).
Note: NGOs = nongovernmental organizations.

conditions (inflation, interest rates, exchange rates), the business environment (rule of law), the quality and effectiveness of government (including whether regulation is appropriate or excessive), and competition policy. The quality and efficiency of the physical infrastructure and the information and communications infrastructure, as well as the education and skills of the population and the workforce, also affect innovation.

Create a Receptive and Mobilizing Environment

There is a need to adopt an organic and evolutionary approach to innovation, rather than a mechanistic one. From that perspective, a government should see itself as creating an overall climate that helps innovative initiatives flourish and grow and focus on fulfilling a few key generic innovation policy functions:

- Supporting innovators by appropriate incentives and mechanisms
- Removing obstacles to innovative initiatives
- Establishing responsive research structures
- Fostering a creative and receptive population through appropriate educational systems.

One may compare the tasks of governments to those of a gardener who should water the plants (i.e., provide finance and support to innovators) remove the weeds (i.e., through competition and deregulation), fertilize the soil (research and dissemination of information), and, more broadly, prepare the ground in which the plants can grow (promote education) (see figure 2.5). In addition, governments can efficiently mobilize support for well-defined technologies through large-scale programs or for well-defined sites that concentrate talents and entrepreneurship.

Figure 2.5 Government Roles in Encouraging Innovation

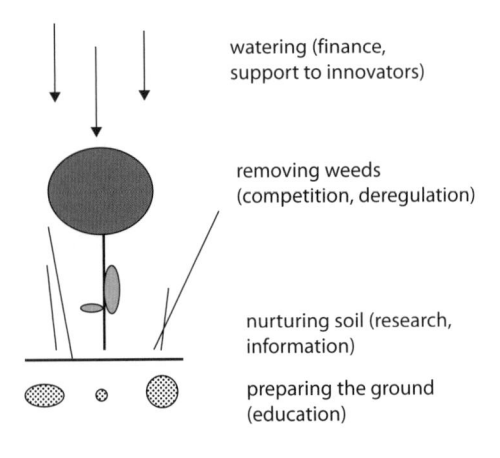

watering (finance, support to innovators)

removing weeds (competition, deregulation)

nurturing soil (research, information)

preparing the ground (education)

Source: Author.

Some aspects of these different policy elements deserve detailed comments. For a number of experienced policy makers, the need is not so much to stimulate innovation processes as it is to create receptive environments that will elicit the creativity of the other actors. For instance, the Six Countries' Program conceived of innovation policy in three parts: a technical culture policy, a policy to remove obstacles to innovation, and large-scale, meaningful programs (Gaudin 1995).[1] Those three elements constitute the core of innovation policy. Fertile sites and innovation groups also play an important role in this respect.

Promoting Technical Culture
Research policy and the transmission of research through teaching are major elements of innovation policy. However, research produces only research results. Although research results accumulate and increase the knowledge available, knowledge is not innovation. Innovation occurs when someone (the innovator) assimilates and uses the knowledge to do something new. Therefore, the cultural aspects of dissemination of knowledge and know-how are important.

Moreover, since most innovations come from a combination of ideas, links between technical disciplines are important. These disciplines operate in silos, however, each with its own specialized language.[2] Therefore, cross-fertilization is vital to innovation, and there is a need for a technical culture policy, at least to initiate a dialogue between isolated technical dialects.

How can this dialogue be established? The simplest way is for go-betweens with a general understanding of science and technology to approach firms and labs and bring together specialists who can learn from each other. A more complete model, found in the 165 prefecture labs set up after World War II in Japan, made critical resources available to small firms: documentation, testing, and measurements facilities; quality control; and information on standards and prototype elaboration.

Cultural activities also help. As applied to technology, these activities include exchanging information on technological advances in other countries and organizing technological fairs and exhibitions at which companies present their novelties to customers and colleagues. The Internet, of course, makes such communication easier, although direct human contact remains necessary.

Removing Obstacles to Innovation
A second part of innovation policy concerns the removal of obstacles, the most difficult part of innovation policy and the one most governments try to avoid. A description of the obstacles innovators face helps clarify the reasons for government inertia.

Immaturity of the Innovation. The first major obstacle is that the final user may not be ready for the innovation. For instance, in 1902, a British chemist produced DDT and commercialized it in his drugstore as a new insecticide.

He was well known and received a prize in London for his invention. To persuade his customers that his product served as an insecticide, he mixed it with naphthalene. After a time, he considered that DDT was too costly to produce and that naphthalene could be sold as easily without it. He abandoned his project, and DDT was reinvented in Switzerland 36 years later by Ciba-Geigy, a large chemical firm, and marketed for agriculture.

Apart from the issue of the maturity of the invention and its adaptation to the market, which defines the gap between an invention and an innovation, innovators usually face two other major obstacles: bureaucracy and vested interests.

Bureaucracy. As Parkinson observed, bureaucracy is universal and concerns all big structures, whether public or private. The basic source of bureaucracy is failure. When an incident occurs, the bureaucracy generates a new process to cope with the problem. It never removes the old processes, however, which linger on "just in case." From the viewpoint of the individual entrepreneur, bureaucracy simply creates an enormous waste of time and energy that may threaten his or her innovation. From the viewpoint of innovation policy, overcoming bureaucratic complications and reluctance is necessary, and rewarding bureaucrats who exercise good judgment in facilitating innovations would be important.

Vested Interests. The third obstacle, the coalition of vested interests, is even greater. Owing to the natural evolution of an economy, the most successful companies grow. Usually, a stable situation is reached when a small number of dominant firms (an oligopoly) control the market and organize a lobby. In countries where antitrust legislation is absent or inactive, this concentration of economic power is reinforced and leads to a monopoly. For the simple technologies of daily life, which are not industrial but operated by craftsmen, the defense of vested interests takes the shape of guilds. Control of the marketplace by a community of storekeepers (bazaars or shopping centers) may also organize resistance to novelty. All these forces defend their positions and often view innovations as threats or disturbances. Support from the public bureaucracy, often part of the same ruling class, strengthens the resistance of these vested interests.

From the viewpoint of innovation policy, it is necessary to implement efficient antitrust and small business legislation. In addition, laws conceived for the earlier industrial technology system may need to be reviewed. For instance, intellectual property laws may not adequately apply to new technologies such as software, drugs, and copyrights on music, literature, and films. Instead of stimulating creativity, as they are supposed to, they reinforce monopolistic positions, slow the diffusion of culture, and hinder the curing of illnesses in the poorest countries. Overcoming the obstacles to innovation, then, may be the most effective part of innovation policy, but it requires enormous courage.

Developing Meaningful Programs
Large-scale programs have always played a fundamental role in the promotion of innovation. The first and foremost example is the military, where efforts are not motivated by profit or money but by other concerns. Similarly, innovators are motivated more by a search for meaning or the realization of a dream than by profit. Many innovators earn money to innovate but do not innovate to earn money.

In any event, thanks to generous resources and other supporting actions (notably, regulatory adjustments), large-scale programs help mobilize the creativity of scientists, the energy of entrepreneurs, and the finance of venture capitalists. Moreover, such programs facilitate a considerable learning process beyond the new technologies that they help develop and disseminate.

Moreover, one should not underestimate the influence of public procurement on the constitution of a creative community. In the United States, for instance, most public procurement has been driven by military goals, which escape the usual accounting constraints. Without such constraints, investigation and testing have been much easier; civilian products came later. Many other big programs lead to such spillovers. For instance, the recently built LHC (Large Hadron Collider), the world's biggest particle accelerator, gave the contracting firms experience and know-how in the area of supraconductivity. Supraconductivity is also developed for medical instrumentation and may serve to transport electricity in the future.

Building Fertile Sites
Innovation tends to develop in microclimates with an accumulation of talent, entrepreneurs, and knowledge. Like certain biological processes, this concentration favors a natural dynamism. The phenomenon is illustrated by famous locations such as Silicon Valley, Italy's industrial districts in traditional sectors, or Bangalore in IT services, as well as many lesser-known sites throughout the world. Cities and regions with strong knowledge assets or large creative classes become innovative sites naturally and have a definite advantage in global competition. Well aware of the importance of these concentration effects, governments attempt to recreate such sites artificially by establishing technopoles, science parks, special economic zones, and the like.

The global resources for innovative projects in the developing world are of absolutely crucial importance. The acceleration of the globalization process in the past decade or two has brought opportunities linked to telecommunications, trade, and foreign direct investment and has considerably changed the conditions of innovation. The innovation process should be conceived and managed in a "glocalization" perspective, with the local and global dimensions intermingled. It has been noted that most of the technologies in use in the developing world come from the economically advanced countries. But other crucial resources—scientific, managerial, and the like—also come from abroad, along with financial support. Acknowledging the global inputs in innovation in developing countries does not mean that indigenous innovators play a limited role—quite the contrary.

Innovative Groups

The groups of people that bring innovation to fruition are another important element of innovative climates. The innovator is generally not a single person. One of the basic works on this subject is a study of 200 big companies by Roberts (1991) of the Massachusetts Institute of Technology in the late 1960s. He finds five different roles being played in every innovation. In most cases, these roles are not officially defined by the management but appear as the innovation advances as a spontaneous, self-organized process:

- *The inventor.* Inventors produce the idea, which often is an association of ideas. Their motivation recalls the motivation of artists or even of prophets. They want their idea to come to life and change things. They think their invention will improve people's lives, and they are often motivated by generosity.
- *The entrepreneur.* Entrepreneurs face challenges. They are energetic and take charge of the project. They are persuasive: they make their case to the management, the bankers, the retailers, and the customers and deal with the unavoidable difficulties. Their role is to make the innovation succeed.
- *The facilitator.* Facilitators are more humble actors, often accountants, but their role is essential: they anticipate problems, overcome obstacles, and smooth the path of progress. They know about the practical aspects of the effort involved and prepare the logistics for making the project flow.
- *The godfather.* Godfathers are well-known individuals with influence, inside and outside the organization. Their role is to protect the innovation during the early stages of its maturation. This role is important because at this stage, when the innovation is still largely an idea and has not been completely endorsed, it is fragile.
- *The information gatekeeper.* Information gatekeepers play a very important, though often neglected, role, one that may even be forgotten by both the actors and the observers: they circulate information. They are not exactly researchers but rather those who keep abreast of developments in science and technology and alert actors about potential opportunities.

Developing countries may have a slightly different set of key actors: both inventors and entrepreneurs are, of course, needed. Godfathers are also needed, but they are likely to be influential figures in the government—even a head of state who is a champion of innovation—and powerful enough to remove regulatory, informal, and other obstacles to the innovative undertaking. Finally, there are foreigners, or diaspora members, who generally bring in key technological elements as well as financial resources and sometimes managerial competencies.

Put Efficient Institutions and Instruments in Place

Implementing innovation policies requires efficient institutions and instruments. Three points deserve particular attention.

Flexible Agencies with Local Offices

Supporting innovation requires flexible agencies that are able to act nimbly. They need to deal with many different types of technical, financial, or commercial means to mobilize, or provide, the support that potential innovators require. Such agencies should be closely attuned to local needs and communities with appropriate, decentralized offices. Examples abound of such agencies in the developed countries (Finland's Tekes or the French OSEO, for instance).

Central Coordinating Bodies

Moreover, it is important that government have a strong and legitimate body that can mobilize the relevant departments. A key example is the Finnish model of the Science and Technology Policy Council, chaired by the prime minister. All key ministers participate, along with top representatives of the business, labor, and civic communities. This institutional arrangement is in accord with the whole-of-government approach but differs from the model formerly adopted by many countries, and still in use in some, in which innovation policy is seen as bridging science and industry and sometimes, in a more sophisticated form, as bridging science, education, and industry. In any event, viewing innovation as the tail of an R&D sequence undermines its role by making it simply a policy segment that aims to add value to the results of science and technology.

At the same time, innovation policy cannot claim to address directly the fundamental elements of the overall economic and social system such as education, macroeconomic conditions, and financial structures, which certainly influence innovative capabilities; if so, it would appear as the overall government policy for economic management. However, a broad approach is important in the design of the institutions for innovation policy, rather than one focused simply on support of innovation processes (see figures 2.6 and 2.7).

Financial Incentives and Stimulating Instruments

Innovation policy necessarily acts within an established institutional setting, a setting already crowded with other organizations that consider themselves as

Figure 2.6 Traditional Layout of Innovation Policy

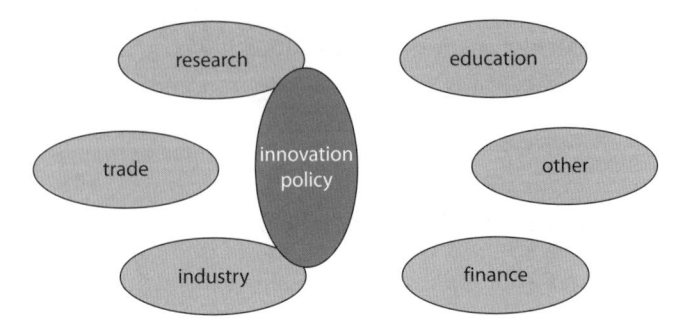

Source: Author.

Figure 2.7 Comprehensive Layout of Innovation Policy

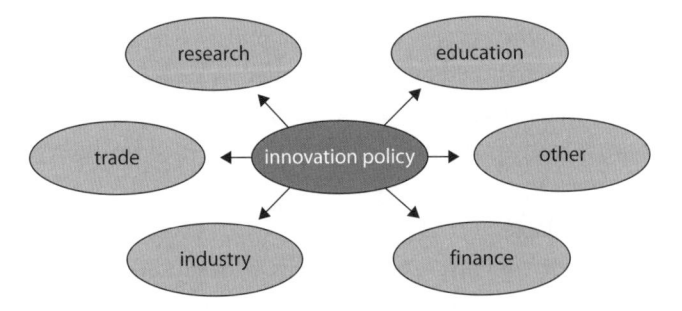

Source: Author.

legitimate actors in policy fields directly related to innovation. They often do not understand the need for additional or different measures.

Therefore, key innovation policy instruments often act as "rudders," that is, they operate with limited financial resources but reorient broader masses of resources through clever incentives or influence the behavior of established institutions. Typical financial instruments for such purposes are matching funds, provided according to well-established criteria to key actors that mobilize their own resources. Such mechanisms may give businesses matching funds on the condition that they invest an equivalent amount in an R&D project developed in collaboration with university or public laboratories. Similar matching-fund tactics can also be used to mobilize local government resources, for instance, to establish local offices to support innovators or set up technology centers.

A second instrument is aimed at stimulating changes in the perspective and behavior of established communities and institutions and requires few resources. One example is awareness raising through fairs or media campaigns; another is legal intervention and control, such as audits of institutions or overcoming obstacles to innovation, such as monopolistic, corporatist, or rent-seeking behaviors.

Adapt to Societal Specificities

The need to adapt innovation policy to a country's specific comparative advantages—natural endowments, trade positioning, cost of labor, or indigenous knowledge, for example—is obvious, as is the need to respond to most pressing problems. Less obvious, however, is the approach to sociocultural specificities. Success stories in economic development show the crucial importance of the acculturation of "imported" technical or organizational devices. Such acculturation requires a true and deep process of appropriation by the concerned people and also needs to respect the inherent and irreducible cultural differences among civilizations and countries.

There are striking differences, for instance, between Eastern and Western civilizations (World Bank 2007). These can be imputed in part to different cognitive processes, with implications for relationships to the world, as well as for social organization. Two different postures can be identified: in the East, people tend to think in terms of putting objects in the whole context in which they are placed; in the West, they tend to separate them from the context and focus on them individually (Nisbett 2003). These different ways of thinking imply differences in various domains of human activity, including medicine, law, science, human rights, and international relations. In science and technology, the Western approach to reality favors a scientific search for causality in understanding phenomena, while the Eastern mind favors holistic combinations of existing elements. With regard to the legal and institutional environment, Western societies are concerned with the establishment and observance of the rule of law as the basic means of protecting the individual, while Eastern societies tend to emphasize informal relationships regulating collective groupings, such as the Chinese *guanxi* (that is, a cluster of interactions among people). Such differences lead to two clearly different economic systems, with contrasting features (see table 2.1).

These sociocultural disposals made the Asian countries very receptive to imported technology and methods, while finding in themselves unique development models onto which they found indigenous innovation capabilities and economic growth.

Table 2.1 East-West Contrasts in Socioeconomic Systems

Indicator	West	East
Innovation	Importance of science based innovations	Technology and production-driven innovations
Educational values	Individualism, exploration of the unknown	Collective values, imitation
Institutions	Public/private systems regulated by the rule of law, formal contracts	Connection-based systems with informal relations

Source: Author.

For each nation, very specific behavioral features regulate the economy and society that have all sorts of implications for other aspects of life: business management, education and training, and government-citizen relationships, among others. In the developing world, understanding these unique features determines the success of technology transfer and the adoption of modern management methods (D'Iribarne and Henry 2003). Cultural characteristics present both strengths and weaknesses, and the policy implications are clear: build on natural strengths while being conscious of the weaknesses.

Policy Conclusions for Developing Countries

Based on the argument of this chapter, innovation policies in developing countries should take into account of their specific features. Several points need to be emphasized: technology strategy, institutional issues, the legal framework, countries' specific needs and assets, agents of change, reforms, and cultural and behavioral characteristics.

- *Technology strategy—tapping into global knowledge and technology for dissemination in the local economy.* Low- and middle-income countries should emphasize adapting global knowledge to local needs, while the R&D structure should focus on adaptive research in close contact with local needs and users. Those countries should also give priority to establishing a dense network of offices and mechanisms for facilitating the diffusion and adoption of new technologies and practices among peasant and other communities.

- *Institutions—minimal equipment.* Developing countries in general, and low-income countries in particular, tend to have a mediocre innovation climate, including poor governance, limited infrastructure, inadequate education, and lack of managers. Middle-income countries may at least have specific areas (cities, for examples) whose institutions function at the level of those in high-income countries. In these difficult contexts, however, there is a need for at least minimal policies and mechanisms for supporting innovation, starting with an autonomous agency able to act flexibly on all types of issues, including (a) the direct support of innovative projects through provision of technical, financial, and other needs; (b) the removal of regulatory or informal obstacles to innovative efforts, such as customs procedures or rules on university-industry cooperation; and (c) the stimulation of change through demonstration projects, such as programs to familiarize schoolchildren with science and technology.

- *Legal framework—minimal rules of the game.* Most countries, including the poorest, need to reorient their established structures for research, education, and the like. It is important to adopt the types of policy rules outlined above, such as matching funds or minimal contract funding. When new institutions are needed, clear rules of the game inspired by those principles should be imposed. Similarly, it is essential to have a solid infrastructure for norms, standards, and quality control to ensure proper commercialization of products for either the internal or the external market.

- *Policy focus—specific needs and assets.* Sectors such as agriculture and tourism are typically among those that should receive close attention and adequate support in all areas, including technology, trade, management, and logistics. Poor communities also deserve particular care and can benefit greatly from well-tuned, but not necessarily costly or extensive, support.

- *Agents of change—using global connections for leveraging change in the domestic context.* Dependence on foreign technology, the importance of foreign actors for accessing global markets, the potential role of diasporas, and the relative weight of foreign aid in the government budget are all factors that can influence change and help reverse the institutional and behavioral inertia that affects domestic activity.

- *Reform approach—acting on specific sites and stimulating broader reforms via success stories.* Since it is inherently difficult to engage reforms nationwide, government policies should concentrate on specific sites or sectors, given that there are always assets to exploit. A well-articulated government action—with an appropriate package of measures—will help ensure success and build trust and confidence in society. When a critical mass of such projects becomes visibly successful, a positive association process leads to broader reforms. It then becomes possible to reshape institutions gradually in line with global standards.

- *Cultural and behavioral characteristics—respecting cultural and behavioral specificities.* Like the economically advanced countries, the developing world has its specific characteristics. The idea that "one size fits all" is now widely rejected, but beyond that there is a need to understand specific motivations and behavior as people innovate, create new things, adapt their institutions, and manage their businesses. These cultural specificities differ not only from one country to another but also within a single country among its provinces, cities, and villages.

Notes

1. The Six Countries Program was created by leading innovation policy makers of European countries in the mid-1970s. The group used to hold two meetings a year, and it is still active; the three types of policy described below result from a meeting held in 1980.

2. For example, the basic vocabulary for daily life in a foreign country requires 600 words, a successful novelist would use approximately 6,000 words, and a good dictionary for a language would include around 60,000 words. The number of terms used in contemporary technology is around 6 million, a hundred times the language of the dictionary.

References

Andersson, Thomas, Abdelkader Djeflat, and Sara Johansson de Silva. 2006. "The innovation system and related policy issues in Morocco." Research report, International Organisation for Knowledge Economy and Enterprise Development (IKED), Malmö, Sweden.

Aubert, Jean-Eric. 2005. "Promoting Innovation in Developing Countries: A Conceptual Framework." Policy Research Working Paper 3534, World Bank, Washington, DC.

D'Iribarne, Philippe, and Alain Henry. 2003. *Le Tiers Monde qui réussit.* Paris: Le Seuil.

Gaudin, Thierry. 1995. Qu'est ce que la politique d'innovation? Paris: Editions de l'Aube.

Mytelka, Lynn. 2004. "Building Innovation Systems in Natural Resource-Based Industries." PowerPoint presentation at the United Nations University Conference, "Global Innovation Policy Dialogue," Beijing, October 16–20.

Nisbett, Richard. 2003. *The Geography of Thought: How Asians and Westerners Think Differently and Why.* New York: Free Press.

OECD (Organisation for Economic Co-operation and Development). "OECD Reviews of Innovation Policy." http:// www.oecd.org/sti/innovation/reviews.

Roberts, Edward B. 1991. *Entrepreneurs in High Technology: Lessons from MIT and Beyond.* New York: Oxford University Press.

"Six Countries Program on Innovation Policies." http://6cp.net.

U.S. Department of Commerce. 1967. *Technology Innovation: Its Environment and Management* (Charpie Report). Washington, DC: U.S. Department of Commerce.

World Bank. 2007. *Building Knowledge Economies: Advanced Strategies for Development.* WBI Development Studies Series. Washington, DC: World Bank.

Part II
Policy Functions

3

Supporting Innovators

Innovators are, first and foremost, entrepreneurs, and they need marketing intelligence and basic support, as the product or process they seek to introduce is new. Yet linking knowledge and creativity to the market requires skills they often do not possess. Their potential customers are increasingly demanding and globalized, and quality control is crucial, especially in less developed countries where competition among producers is often insufficient. These entrepreneurs will require good industrial organization if they are to succeed in solving the development issues raised by the generation of new products. More generally, they need accurate information and the capacity to communicate about innovation.

Innovations, in the form of new products and processes, are introduced in the marketplace because of these dynamic entrepreneurs, the development of successful projects, and efforts to respond to demand from consumers and from manufacturing and service industries. They require market-relevant investment in research and development (R&D), creativity, and often cooperation with institutions of higher education and firms. Because the outcomes of innovation are highly uncertain, however, firms and entrepreneurs are often reluctant to invest sufficiently in R&D, and risk-averse behavior often stifles creativity. In addition, the lack of an enabling environment and the difficulty of appropriating the economic benefits of investment in innovation hamper the development of collaboration between firms and between firms and research institutions.

Governments are therefore often called on to bridge the gap and address these issues. Public policies that support innovation have most often been based

---

This chapter was prepared by Patrick Dubarle.

on the assumption that market failures lead to significant underinvestment in research and innovation across the economy. Other sources of suboptimal outcomes include potential innovators' inability to act in their own best interests, institutional rigidities that prevent institutions from contributing to innovation, and network and coordination problems. Yet innovation is increasingly considered a product of systems that involve not only firms but also institutions and intermediaries. Innovation policy and support for innovators therefore need to respond to these systemic failures.

At a more practical level, the generation and diffusion of new technology and knowledge, government efforts to transfer technology, and the educational system's ability to produce science and engineering graduates influence a country's innovation capability. Moreover, the absorptive capacity of firms is crucial for translating innovative ideas into productivity gains. The proximity of firms to each other helps bind these various dimensions into an innovation system. As a result, support for innovators is often the result of initiatives by local or regional governments, which have more knowledge and better information about local firms with high potential and can better assess the risks linked with local or regional innovation than the national government. Innovation is also increasingly considered a crucial driver of regional development.

Central and subcentral governments have a range of business assistance programs to support innovators. The following five sections discuss particular examples of these programs with illustrative practices from both industrialized and developing and emerging economies:

- Sector-oriented entities and technology transfer centers, serving mainly new and small enterprises
- National or regional small business policies that seek to meet the needs of firms at various stages of the innovation process (design, development, diffusion)
- Access to equity and (venture) capital to help develop new products and processes and to mitigate the risks of commercialization
- Government support of clusters and networks as they become major actors in innovation
- New policy approaches to intermediaries and bridging institutions.

Provision of Business Services

The public sector makes certain business services available to companies, generally in return for payment. To a certain extent, these services can be considered (partially) public goods, because they add to the country's or region's endowment, induce learning, and generate positive externalities. They are particularly important in developing countries because the market is often unable to offer the necessary service infrastructure owing to low demand, lack of supply, and information asymmetries. These business services aim to increase the competitiveness and market opportunities of user firms—notably of innovative firms—by transferring new knowledge to firms and triggering their learning processes to improve their organization of production and their relation to the market. Such services are expected to contribute to the speed and quality of economic development.

Often, private enterprise does not supply these services for a number of reasons: the necessary expertise may not be available in the social environment in which firms operate; the investment needed to produce the required service is high, and return on this investment may be slow to materialize; the private sector may be ill placed to provide these services, because they may rely on what is essentially a public good, such as knowledge; and, finally, such market failures are particularly widespread in low- and medium-income countries.

Business Services That Support Innovation

The following services have strategic relevance for innovation policy.

- *Basic industrial services (promotion, marketing, and internationalization):* assistance for tenders of the European Union (EU), World Bank, and other development organizations; assistance for direct investment abroad; assistance for inward investors; legal and financial assistance; financial services, including accounting and tax assistance; market information or other economic data; organization of and participation in trade fairs and other promotional events; partner search.

- *Technology extension services:* assistance for patenting and licensing, grant applications, in-house R&D activities, and subcontracting to research institutes; competitive intelligence, including technological benchmarking, technology maps, and information on emerging technologies; innovation diagnosis; review of current or proposed manufacturing methods and processes; participation in and organization of technology exhibitions; technology brokerage.

- *Metrology, standards, testing and quality control:* calibration of equipment; quality certification; domestic standard; ISO (International Organization for Standardization) compliance; technical assistance; demonstration centers and test factories; energy audits; materials engineering.

- *Innovation in organization and management:* assistance for enterprise creation; interim management; logistical assistance; organizational consultancy, quality and training; productivity assistance; incubation services.

- *Information and communication:* advanced services for data and image transmission; assistance on communication strategies, telecom network connections, and the implementation of electronic data interchange systems; database search.

Government bodies or independent public agencies can provide these services, along with public-private partnerships that can, at least in theory, combine the advantages of the legitimacy and neutrality of public bodies and the business efficiency and management styles of the private sector. Consortia and specialized agencies sponsored by industrial associations, as well as private companies acting according to government guidelines or within government projects or subsidization schemes, can also provide such services.

The performance of the different providers will depend on several factors, some of which relate to their institutional nature. Legitimacy and ability to build consensus may favor government agencies or associations. When services imply disclosure of sensitive information or honest brokerage between potentially conflicting interests, public actors may again be perceived as offering better guarantees of neutrality and confidentiality. Although the public nature of the provider may often be an essential feature, a variety of providers can be instrumental in effective business service policies. It is generally acknowledged that the private sector delivers too few services, while political actors deliver too many.

As different types of service are often needed, firms find one-stop shops practical. Usually, the more packages of services are targeted to specific types of firms, the more likely such services are to be useful. Excellent services or service providers can help concentrate resources, drive efficiencies (especially in government and quasi-government agencies), and clarify the market for small and medium enterprise (SME) clients. A number of criteria can be used to do so (see box 3.1). Networking is the most efficient way for business services to maximize their contribution, and experience with cooperation substantially widens the scope of possible links. It is possible, but difficult, to transfer best practices. They are more easily reproduced when there is no need for substantial interaction with the local environment and when some strategic functions, especially those that are crucial for guaranteeing quality standards, are controlled by a central body.

Industrialized countries offer many examples of multipurpose services that are provided publicly (by government) or collectively (by industry associations). In the United States and the United Kingdom, which have a tradition of networking (for example, by industry and research associations) and constantly diminishing subsidies, those who provide business services tend to become more market oriented and inclined to take less risk when deciding to support projects. In Germany, many technology centers are part of national organizations, and science and technology policy is closely linked to institutions such as the Fraunhofer Society. As for France, the regional network of technical centers there is funded through the payment of a specific tax, public aid, and services revenue. In Italy, such services include market information, testing, and export support, which are often provided on a regional basis. In Spain, publicly provided multipurpose

> **Box 3.1 Priorities for Business Services Support Schemes**
>
> Policies in support of business services require a *significant degree of consensus*. They are by no means an "obvious" policy option, which can be decided simply on technical grounds, as they must take account of the socioeconomic structure of the context and be an explicit part of an accepted economic strategy.
>
> Such policies also require a *significant degree of participation by clients* in the supplier-company relationship. By definition, consumers take an active role to some degree in the production of the service. To the extent that they seek to reduce costs, firms will look for routine or "compulsory" services, such as accounting. More proactive, but still cost-oriented strategies, will require specialized services that provide technical, financial, or training skills. Such services can be defined as "strategic."
>
> The functions of business service providers involve certain skills that need to be available to the provider's organization, even if they are not part of it: *awareness building*, which requires sophisticated "industrial marketing" skills; *problem framing*, which implies the ability to provide tailored diagnoses, based on comprehensive knowledge of company behavior and organization; *problem solving*, which implies the ability to carry out specific improvement projects, based on technical expertise; *search for resources* (financial and nonfinancial), which requires expertise and connections with public and private institutional sources; and finally *alliance building*, which implies the ability to identify and create innovative links between companies and between companies and other actors, as well as the credibility necessary to guarantee the value and the trustworthiness of the partners.
>
> *Source:* Bellini 1998.

services take the form of technology and business development centers. In these countries, as well as in Australia, Ireland, Japan, and Portugal, studies indicate participation rates of some 20–30 percent of the universe of SMEs, but there are very significant differences among sectors and locations (very different rates are reported in the United Kingdom and the United States, for example).[1]

In emerging and less developed economies, private supply of business services is generally in very early stages. In China, the public sector takes the lead in providing collective and support services to provide for the innovation needs of firms and other actors in local and regional innovation systems. Figure 3.1 shows that Shanghai's R&D public service platform seeks to address a wide range of services similar in principle to those found in developed countries. These services cover the innovation development process, from the sharing of scientific information to the technology testing and transfer services that support entrepreneurship and management. In Mexico, although the public research centers of Mexico's National Council for Science and Technology agency, CONACYT, remain institutionally under presidential authority, its degree of autonomy with respect to the orientation and organization of its activities has recently risen, enabling it to increase the share of self-financing in its total

Figure 3.1 Shanghai R&D Public Service Platform

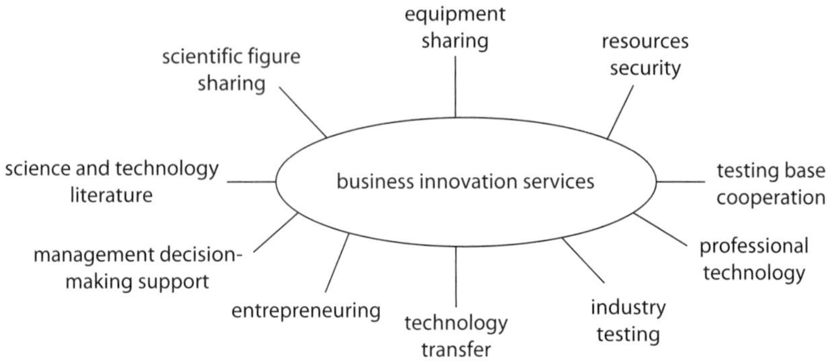

- equipment sharing
- scientific figure sharing
- resources security
- science and technology literature
- business innovation services
- testing base cooperation
- management decision-making support
- professional technology
- entrepreneuring
- technology transfer
- industry testing

Source: Shanghai Municipality Science and Technology Commission 2006.

budget. This growth has led to a more market-oriented approach and greater cooperation with the private sector and other institutions to which CONACYT provides technological services.

Specialized Service Infrastructure

Specialized service infrastructure comprises a number of components, including basic investment promotion services, technology extension services, standards and metrology, productivity centers, and information and communication services.

Basic Investment Promotion Services. Attracting foreign direct investment (FDI) or enhancing domestic investment requires a wide range of efforts, such as the identification of suitable inward investment prospects and the active servicing of the strategic needs of foreign-invested firms once they are established. Skills development, recruitment services, and identification and upgrading of local suppliers are crucial not only to attract investors but also to create synergy with the local environment.

For these reasons, investment promotion agencies need to be in a position to ensure the cooperation of the different entities in charge of strategic resources such as infrastructure, training and skills resources, and SME promotional bodies. National agencies should supervise regional entities not only to ensure an appropriate degree of efficiency through regular audits and continuous monitoring but also to avoid duplication of effort, incentive "wars," and costly interagency competition.

Studies of successful agencies in developed and developing countries show that investment promotion programs entail a vast number of activities, such as establishing the policy context and the priorities and form of interventions, building up a promotional campaign for potential investors, meeting the needs of interested investors, and implementing a strategy based on past promotional activity.

Industrialized countries practice a highly sophisticated form of investment promotion designed to achieve strategic industrial or regional development objectives. For example, Scottish Enterprise, the chief investment promotion agency in Scotland, coordinates initiatives that encourage entrepreneurship through efforts to attract and retain inward investment. A Scottish Enterprise audit in 2003 estimated that the agency added some £1.6 billion to Scottish gross domestic product (GDP) over three years, as a result of its activities in 2001–02. In Italy, the Piedmont Agency for Investment, Export and Tourism is organized as a one-stop shop for companies investing in the Piedmont. It is also in charge of regional investment contracts, a financial instrument unique to Italy, which foster the internationalization of the region through increased investment. Within this framework, research entities, science parks, and innovative companies can apply for specific grants.[2]

Many emerging countries have established similar investment promotion agencies that often have a good performance record. In Thailand, for example, the Board of Investigators, the agency responsible for attracting foreign inward investment, has designed a strategy that builds on the country's ability to provide cost-effective local inputs and on the competitiveness of domestic parts manufacturers. The availability of a large pool of labor that can be trained, natural resources, and government protection for fledging industries have also been instrumental in contributing to the increase in FDI. At the same time, differences in incentives for central and peripheral regions have helped reduce the pressure on the capital and on congested areas. Given this favorable context, FDI increased from less than 0.6 percent of GDP in the 1980s to an average of 1.5–2 percent of GDP in the 1990s and early 2000s.

Technology Extension Services. The aim of technology extension is to create small but profitable improvements by extending established technology to smaller firms. While the designs of technology extension organizations differ, all have relations with small firms and with sources of technology. Technology extension programs either provide resources that enable firms to identify needs and find appropriate technological solutions or identify and provide solutions through targeted assistance.

Particularly well known is the U.S. Manufacturing Extension Partnership (MEP), a network launched in 1988 that covers the entire country, with some 400 offices offering public and private industrial assistance. Technical assistance is often provided by the engineering applications programs of local universities, where engineering staff work with clients at the clients' site. Some of these university programs are industry specific. Others are "teaching factories" to which clients travel to receive assistance. While MEP was intended to bring leading-edge technology to clients, in practice it focuses on giving help for more traditional technologies and management.

The success and longevity of MEP rely on a combination of public and private funding. On average, partnership financing is ensured by state (35 percent), federal (35 percent), and private funds (30 percent). Firms receiving assistance pay at most 40 percent of the cost. MEP assists about 25,000 firms a year and generates US$280 million in revenue (Shapira 2007). A five-year pilot study and an unpublished update show that clients assisted by MEP have up to 5.2 percent higher productivity growth than comparable firms not served by MEP (NIST 2007).

In Japan, about 170 technology upgrading centers (*kosehtsushi*) provide support for small firms. Unlike extension services in the United States, they deliver only technological services. Other services (management or financial) are offered by other agencies. *Kosehtsushi* centers conduct (very applied) research; have labs for training, testing, and examining products for compliance with industry standards; and promote technology diffusion. Most services are free of charge for SMEs. Each year, 900,000 tests are carried out, and around 3,900 technological advisers are mobilized to meet the 500,000 problem-solving requests addressed by client firms. Prefectures and local governments provide most of the funding; the private sector contribution is limited (6 percent of the total). The strength of the *kosehtsushi* is due to the stable relationship established by the centers' personnel and staff with clients and their knowledge of SME needs. Users seem to rank *kosehtsushi* centers' services higher than those provided by universities on their ability to perform promised services and to communicate about them. Success at diffusing technical knowledge to clients, however, is considered equal (Izushi 2005).

Emerging economies also recognize the need for efficient extension services. For example, Chile's Technical Cooperation Service, SERCOTEC, a branch of the Chilean Economic Development Agency (CORFO), is charged with the promotion of micro and small enterprises. Central to its strategy to assist SMEs are its Web site and online advice provided at no charge to 30,000 registered firms and its support to CORFO's mainstream activities. SERCOTEC has partnered with many other institutions to give expert advice and diffuse information to clients. The system is low cost, is easy to implement, and requires low maintenance. CORFO also operates the Technical Assistance Funds, which aim to integrate modern business management techniques and new commercialization technology and strategies.

Standards and Metrology. The globalization of value chains—with a multitude of firms acting as interconnected suppliers, intermediaries, and marketers—has occurred in parallel with the drive toward the standardization of practices and procedures. Firms' interactions along the value chain require meeting agreed standard business practices in contracting, accounting, project management, and the communication of product design and engineering information.

Standards would be meaningless in the absence of the ability to measure precisely the various attributes—chemical, electrical, physical, and so forth—of outcomes at each stage of the value chain, using common modes of measurement, with the assurance that the measured magnitudes are correct within agreed tolerances for error. Metrology is thus the foundation of standardization processes, maintained through a carefully linked hierarchy of metrology agencies: some are autonomous and responsible only for metrology, while others are embedded in organizations with related responsibilities (UNIDO 2002).

In the United States, the National Institute of Standards and Technology (NIST) is an example of good practice because of its wide variety of services, its focus on research, and its systematic self-assessment processes. Through its regional network, NIST provides access to technical and standards databases and sets excellence guidelines for U.S. subcontractors and manufacturers. It offers calibration services, special tests, and a measurement assistance program to monitor parameters and ensure appropriate quality control. NIST also funds industrial and academic research and offers grants to encourage work in precision measurement, fire research, and materials science. In addition, NIST has a traceability policy, provides answers to clients' requests, and sells standard reference material (NIST 2007).

In less developed countries, metrology, standardization, and industrial quality systems are integrated only to some degree, and their services are often limited. Standards are modeled primarily after ISO standards, but quality certification is slow and insufficient. These countries need to increase their capacities in metrology, testing, and quality assurance to underpin their ability to innovate and export. At the same time, they face new challenges, such as an accelerated market cycle, new regulatory demands for a sustainable society, and the shift toward global markets.

In these countries, weaknesses in the standard-setting and accreditation processes are major problems. In South Africa, for example, shortages of human resources in the field of standardization have hampered the country's participation in international standard setting. A fund for bridging the standardization gap has been set up to mobilize efforts, especially in the information and communication technology (ICT) sector. In Brazil, technical regulations are decentralized through different line ministries and regulators. Inmetro, the national standards agency, maintains an updated technical regulations database on its Web site and makes available regulations and government resolutions on products subject to compulsory certification.[3] The country's agencies that certify quality management systems operate independently, and controversies sometimes arise about the "subjective" character of their assessments.

Productivity Centers. These centers are broadly focused and geared more to industrial than to strictly technological development. They work with firms to promote efficiency and productivity in manufacturing and change their focus to fit the changing nature of the problems to be studied. They are generally initially funded by the central government to promote awareness of the need to enhance productivity. Most campaigns focus on the positive relations between employment and productivity growth to combat fears that increased productivity will displace workers.

The Japan Productivity Center, founded to bring together labor, management, and academia, merged with the Socio-Economic Congress of Japan in 1994. The principles of the new organization, known as the Japan Productivity Center for Socio-Economic Development, are that productivity gains increase employment, that labor and management must work together, and that productivity gains should be shared by labor, management, and the public (UNIDO 2002).

Productivity centers can provide private firms with vital information and services. For example, the Hong Kong Productivity Council provides information on international standards and quality and provides training, consultancy, and demonstration services to small firms at subsidized rates. Serving over 4,000 firms a year, the council acts as a technology import, diffusion, and development agent for the economy's main industrial sectors. It identifies relevant new technologies in the international market, builds its expertise in those technologies, and then introduces them to local firms.

In Mexico, state centers stress the management and organization of small firms. For example, the Instituto Poblano para la Productividad Competitiva, located in Puebla, aims at accelerating the growth of firms by helping small enterprises become medium sized and medium-sized enterprises become large. To this end, the institute establishes a mentor relationship with firms that pay a fee to join. At present, 3,150 SMEs are registered. The concept is based on the idea that talented entrepreneurs often fail to act in ways that maximize their talent. The program tries to help firms amass an appropriate combination and organization of skills. In 2007, it trained 3,000 microenterprise leaders, chief executive officers, and business people, with a view to creating 1,230 new positions and conserving 1,600 others. It has identified 150 champion SMEs. Over two years, 100 microenterprises became SMEs; over three years, 40 small firms moved to the medium-sized group, and 10 medium-sized firms became large ones.

Information and Communication Services. Providing information services requires technically competent specialists. These services are the least dependent on prior targeting of specific groups of firms. Serving as an "intelligent" gateway to globally available, searchable knowledge bases, they offer a truly generic service of potential use to all. As such, they are the service organizations that come closest to providing a public good that has universal value. Many information centers also routinely produce materials to disseminate the results of their continuing research.

There are advantages to centralizing these activities in organizations with special capabilities for carrying them out. Offshoots of the National Association of Chambers of Commerce and Industry, France's regional agencies for scientific and technological information, for example, advise on SMEs' development projects in their technological and competitive environment. They help firms exploit information (technological intelligence, regulatory regimes, standards, markets), advise them on intellectual property and innovation, and warn them about counterfeiting risks. They also sponsor innovation workshops.

Most development agencies worldwide have established information services on their Web sites. In Singapore, for example, the "technopreneurs" (technology entrepreneurs) service portal is a platform for information exchange between technopreneurs and investors. Technopreneurs can obtain information regarding, and even create links with, business angels, venture capitalists, investment bankers, business consultants, and other relevant agents. Aspiring technopreneurs can also put their business plans on the Web site where investors can easily access this information. The portal even provides a complete guide to the various support services available to high-technology start-ups. Because it was sufficiently publicized, the portal has contributed significantly to overcoming the information deficiencies that tend to deter new ventures (UNIDO 2002).

Entrepreneurship and New Innovating Firms

In theory, all firms are concerned with innovation, but in practice policies tend to focus on particular categories of firms. Assistance to large firms can stimulate their commitment to precompetitive research and facilitate their involvement in large-scale R&D projects, but direct support to big business operations can distort market competition. The situation is different for small and new firms, which are at a disadvantage because of their size and problems of access to input markets. While governments tended in the past to underestimate the role of SMEs in innovation, they have rebalanced their priorities in the past decades, significantly increased support for small firms, and added preferential benefits for SMEs to their programs. This shift of emphasis has two sources.

First, innovation increasingly takes place in small new companies. Recent research by the Organisation for Economic Co-operation and Development

(OECD) on three global industries (ICT, automotive, and pharmaceuticals) clearly shows that in major global industries, the role of SMEs has not diminished (OECD 2006d). In fact, they are often the source of new ideas that are integrated into other products or brought to the market in their own right by large firms. Second, there is significant untapped potential for developing new products and processes in small businesses. Although SMEs play an important part in national economies, notably in employment, they have limited access to technological expertise, have difficulty mobilizing large-scale resources, and are generally slow to adopt new technology. These limitations have a negative effect on their potential for growth and, in many cases, their survival. Furthermore, small enterprise managers are often not aware of new technology, do not recognize the potential for improvements, or lack the financial, organizational, and managerial capabilities to incorporate new technology or to obtain external advice from consultants. For consultants and technology providers, the costs of reaching small firms with relevant information are relatively high, as are the costs of tailoring equipment to their needs. As a result, technology markets suffer from problems of information asymmetry, transaction costs, and lack of scale economies. These factors warrant policy intervention, both to improve the infrastructure for technological services and to encourage their use. They also imply adapting assistance to the different phases of the life cycle of new products and processes from design to maturation and internationalization and providing a local framework for incubating new firms.

Policy Initiatives in Support of Small, Innovative Firms
The establishment of new businesses is increasingly seen as a primary source of the revitalization and expansion of the local and regional economic fabric. Beyond the start-up phase, support to innovators takes into account subsequent stages of the firm's life cycle, including the globalization stage. In most industrialized countries, governments increasingly aim to provide comprehensive support from incorporation to internationalization. In the United Kingdom, for instance, the main goal of innovation policy is to help more businesses start up and survive. Through coaching and mentoring, free advice, and guidance, the goal is to increase the number and quality of new businesses by enabling people with an interest in starting a business to take the step and helping those from underrepresented groups and disadvantaged communities overcome the barriers they face. It is also to ensure that U.K. businesses, especially high-productivity innovative businesses, are able to identify and successfully exploit opportunities in overseas markets. The policy targets SMEs that are new to exporting, are innovative, and are between one and five years old. Cofunding for certain export projects may be provided in addition to information and advice.

Recent policy initiatives in industrialized countries offer some examples of best practice. The United Kingdom's Small Business Research Initiative, for instance, aims to raise productivity and business innovation by providing R&D contracts to technologically based small business.[4] The government is also working to embed innovation in public sector procurement policy.[5] The Netherlands has devoted attention to bridging the gap between SMEs' use of knowledge and innovation by granting special vouchers to small firms (see box 3.2). In France, a new scheme reduces taxes and social charges for small innovative firms less than eight years old that devote more than 15 percent of their total expenditures to R&D, providing that they are truly new ventures, and not the result of restructuring or the extension of preexisting activities, and have an ownership structure that reflects their independence from larger firms. The Republic of Korea has recently expanded technical and financial assistance for SMEs and start-ups by introducing new policies for accepting technology as collateral (knowledge asset) for bank loans, providing SMEs with subsidies for employing R&D personnel, and making available technical information and services to SMEs.

Emerging economies share these concerns, although their budgetary efforts on behalf of SMEs vary considerably. Since 1999, China has provided grants to small firms through a fund for small technology-based firms. Beneficiaries are requested to match the grant amount. This is not China's only program for SMEs, but it is the only one with an innovation focus. In Brazil, the federal government created a number of new programs targeting the SME sector in the late 1990s to help small business with technology transfer and innovation through loans and training and reinforced these initiatives in the framework of the 2004 law on innovation. Malaysia also adopted an integrated approach to increasing local SMEs' capabilities for technology acquisition and global competitiveness. Its SME Development

---

Box 3.2 Knowledge Vouchers

The Netherlands has observed that the general quality of business knowledge is good but that companies, especially SMEs, do not fully exploit it. The government therefore established knowledge vouchers (also called innovation vouchers, research vouchers, or simply vouchers) as a special incentive for linking SMEs to knowledge providers. The knowledge voucher is a coupon that entitles SMEs to a number of free consultancy or research visits to large, knowledge-intensive organizations (companies, research institutes, educational institutions). The vouchers have been a success, and many firms have used them.

Source: OECD 2007a.

Plan (2001–05) emphasized the strengthening of advisory services, the creation of new ones, and the fine-tuning of existing broad-based programs.

Incubating Firms

The business incubator is the instrument most widely used to support these various policy initiatives. To nurture the development of firms, business incubators offer, on a temporary basis and at relatively low cost, the use of shared premises, capital equipment, and business and technological services.[6] Incubators have diverse sponsors and stakeholders, including government agencies, universities, chambers of commerce, and nonprofit organizations. Private for-profit agents also sponsor business incubators, generally as part of a business estate venture. The convergence of innovation and enterprise policy and business estate initiatives is an area over which local authorities have significant control. Incubators increasingly tend to specialize (see box 3.3) so that they can provide tailored responses

---

Box 3.3 Types of Incubators

- *General/mixed-use incubators:* The main goal of these incubators is to promote regional industrial and economic growth through general business development. While they include knowledge-intensive firms, they also include low-technology firms in services and light manufacturing. A main focus of support is local and regional access to technical, managerial, marketing, and financial resources.

- *Economic development incubators:* These are business incubators with specific economic objectives such as job creation and industrial restructuring. Often the result of local government initiatives, their main goal is to help create new firms and nurture existing firms that create jobs. In some countries, they may target specific groups such as youth, the long-term unemployed, women, and minorities. In the United States, examples include "empowerment and microenterprise" incubators.

- *Technology incubators:* The primary goal of technology incubators is to promote the development of technology-based firms. Usually located at or near universities and science and technology parks, they are characterized by institutionalized links to knowledge sources such as universities, technology-transfer agencies, research centers, national laboratories, and skilled R&D personnel. They may also target specific industrial clusters and technologies, such as biotechnology, software, or information and communications technologies. A main aim is to promote technology transfer and diffusion while encouraging entrepreneurship among researchers and academics. In some countries, technology incubators not only focus on new firms but also help existing technology-based small companies to thrive.

Source: OECD 2006c.

to a wide variety of small innovative firms ("gazelles"), small firms in specific sectors or clusters, microenterprises in need of mentoring, and small firms with a narrow customer base, among others.

According to the EU, supporting incubators is a cost-effective way for national and subnational authorities to facilitate the development of entrepreneurship. The impact of business incubation has been highly favorable, as 90 percent of firms in incubators are still active after three years. Furthermore, the 900 business incubators operating in Europe have helped create 29,000 firms annually, a rate higher than that for nonincubated enterprises.

Support to incubators is often justified on the grounds of systemic market failures (because of weak links in the innovation system), which can impede commercialization and diffusion of technologies by new firms. In addition, entrepreneurs face significant obstacles for starting businesses: high fixed and entry costs, lack of access to equity capital, insufficient technical and market information, and weak management skills. Incubator services can address most of these issues and thus help reduce uncertainty and increase chances for survival. When located in science parks, incubators can provide a significant stimulus to local development and help stabilize job creation. They are also a means of enhancing returns to public R&D spending by promoting commercialization and diffusion. To be efficient, incubators nevertheless need to respect a number of principles: flexibility, quality of management and services, local support, and sound financing (see box 3.4).

In the United Kingdom and the Netherlands, business incubators were developed in the late 1970s. They took the form of "managed workplaces," whereby small firms were located in unused buildings and offered common services as a means of regenerating declining regions. In France, local governments and community actors have sponsored business incubators to stimulate local job creation. Over the period 2000–2003, the French government, in partnership with the EU and regional and local authorities, provided €25 million to 31 incubators. While these have performed relatively well, they have so far failed to attract significant private investment. In Italy, business incubators are a recent development and generally target the creation of manufacturing and innovative firms in the southern part of the country and in depressed industrial regions of the north.

In a number of emerging countries, in the wake of the creation of science parks and the renewal of science and technology policies, incubators have gained in popularity. In China, the inclusion of innovation centers and incubators in the Torch programs (see box 3.5) has led to a considerable increase in their number. They have been particularly effective for linking actors—entrepreneurs, researchers, financers—and for supporting firm spin-offs. The creation of 40 university science parks has also encouraged the establishment of incubators close to universities.[7] In the Persian

> **Box 3.4 Good Practices for Business Incubators**
>
> - Maintain the building and the surrounding environment.
> - Deliver high-quality, reliable central services, such as telephone answering, mailing, conference, and meeting facilities.
> - Provide technical support, either physical or online assistance.
> - Keep the workspace flexible, so that businesses may expand if they wish to do so and so that businesses of different sizes can be accommodated.
> - Ensure security for the business.
> - Establish flexible terms of occupancy, with easy conditions for entry and exit.
> - Develop meeting opportunities to encourage businesses, especially young ones, to learn from one another. Experience shows that social interaction can lead to greater trading opportunities. Workspace managers can facilitate this interaction.
> - Work toward achieving high occupancy rates, following the lead of commercial workspaces. Those funded from public sources may place more weight on moving tenants out after perhaps two years to make space available for new businesses seeking their first location. There is a clear trade-off between commercial returns and social objectives, which policy makers should recognize.
> - Make careful tenant selections to avoid clashes or to focus on particular "types" of tenants, such as those in technology sectors.
> - Consider excluding charges for support services in the rent. Some tenants value this support highly, whereas others prefer less support and lower rent. Normally, this issue is resolved by not including support services in the rent.
>
> Given the above considerations, it is not surprising that the most successful workspaces are public-private partnerships.
>
> *Source:* OECD 2006c.

Gulf, the Bahrain Business Incubator, which provides capacity building and training to young entrepreneurs, has been very successful. Its services focus on counseling, assessment of project viability, and arrangement of links with banks. This model is currently being replicated in other locations in Bahrain as well as in Kuwait, Lebanon, Saudi Arabia, and Syria. Since its inception in 2006, Mexico's program has created 254 business incubators, most of which specialize in intermediate technologies, and has led to the creation of 10,042 firms. An infrastructure for "business accelerators" has also emerged in Mexico since 2004.

In certain countries, incubators are seen as a crucial instrument for internationalization and innovation. Singapore describes itself as a global "entropolis," a hub where entrepreneurs and enterprises converge, create innovations, forge partnerships, and create value in manufacturing and service industries. Singapore's policy is based on the understanding that relations are the essence of business. That country implements the policy mainly through the establishment of foreign incubators.

> **Box 3.5 Singapore: Incubators Underpinning a Relationship Hub**
>
> In Singapore, the number of business incubators and accelerators increased from 37 in 2001 to 101 in 2005. Foreign business incubators and accelerators increased from 3 to 35 during the period and now nurture and support foreign enterprises from Europe, the United States, and the Asia-Pacific and, more recently, from emerging growth areas such as Dubai and Saudi Arabia. The current 101 incubators have more than 1,100 enterprises. One is the China Torch Center, established in 2003 by China's Ministry for Science and Technology to facilitate the internationalization of Chinese enterprises. Another is the Japan External Trade Organization, which set up business support centers in 2001 to help start-up SMEs from Japan establish and grow their operations in Singapore.
>
> Foreign incubators aid Singapore in establishing itself in regional growth patterns, as the country seeks to influence the behavior of Singaporean and foreign companies and make Singapore into a natural nexus of business ideas and deal making. The countries setting up incubators in Singapore are also showing recognition that internationalization requires more than assisting domestic companies in exporting from their domestic base.
>
> Another example of Singapore's focus on internationalization is the Vietnam Singapore Investment Park, located in Vietnam and managed by Singapore. Singapore encourages domestic firms and others to locate operations at the business park in Vietnam to leverage Singapore's and Vietnam's complementary strengths: Singapore's in R&D, advanced manufacturing, and logistics and Vietnam's in low-cost manufacturing and market potential.
>
> *Source:* Singapore Economic Development Board 2006.

Finance for New and Innovative Firms

Governments increasingly recognize that business innovation is more than just research and development. They know that providing incentives, promoting a good environment through diversified business services, and nurturing innovators are necessary but not sufficient. Beyond access to R&D and physical facilities such as incubation, commercialization of technology requires access to adequate capital for dealing with the uncertainties of the innovation process and providing a robust financial base. Early development of new products and processes generates little and often no profit. Bridging the financing gap is therefore crucial for new firms or for autonomous development of innovation projects.

From R&D to Venture Capital

Finance for innovation usually comes from internal sources (cash flow), but when substantial investment is required, external investment may be needed. Owing to the highly uncertain nature of such projects, outside investors may not have confidence in entrepreneurs' ability to manage risky ventures, or they may find it difficult to identify good projects. Long-term perspectives may give way to "short termism." This myopia in the innovation market warrants government intervention and the use of public money to provide grants and incentives to innovating firms and entrepreneurs.

A key constraint to the successful commercialization of research outcomes is the lack of early-stage investment capital, because private venture capitalists are reluctant to invest in the uncertain stages of new product development. Indeed, the preseed and initial seed financing stages present great policy challenges in all countries. Difficulties such as the lack of institutionalized markets may also impede the placements of initial public offerings on the stock market. After the initial public offering, new technology-based firms may encounter further obstacles in raising second and subsequent tranches of finance. When such firms wish to grow significantly, they sometimes have to mortgage the company to exploit the opportunities afforded by rapid early growth in demand (see figure 3.2).

To reduce these constraints and induce venture capitalists to finance projects that transform the research outputs from universities or public labs into commercial success, industrialized countries have improved stock market regulations and intensified support for venture capital by allocating larger budgets to venture capital, especially for SMEs and technology-based start-ups, by providing tax incentives to nonresident investors, and by forming partnerships with private venture capitalists.

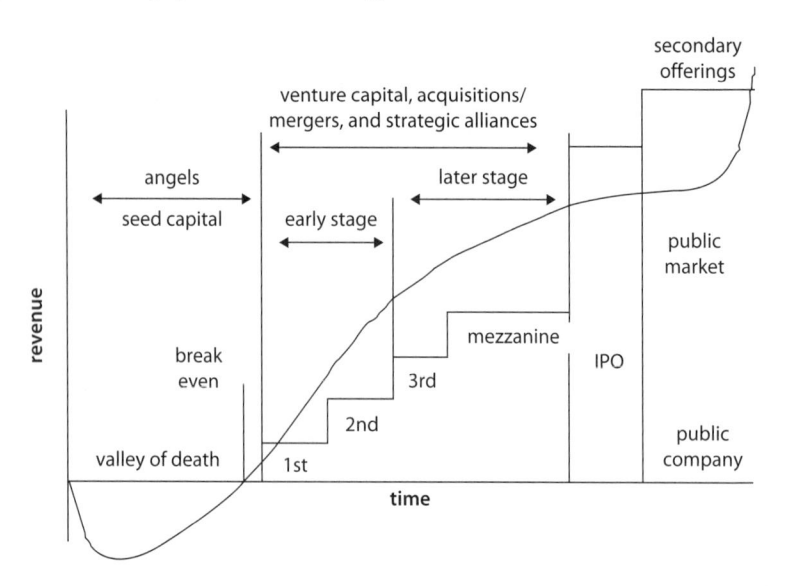

Figure 3.2 Financing Cycle for New Technology-Based Firms

Source: Cardullo 1999.
Note: IPO = initial public offering.

Examples of Policy Measures

Most countries have various schemes for new firms and SMEs, including general guarantee schemes, matching of investments made by small business investment companies with public loans, seed capital schemes, or schemes that enhance equity.[8] Public venture capital can reveal lucrative investment opportunities to potential suppliers, and many industrialized countries have had positive experience with public venture capital programs.

Public Venture Capital Funds. Several European and Asian countries, for example, have established public venture capital funds. In Germany, the Kreditanstalt für Wiederaufbau and Tbg have such programs. Combined with the Länder (state) programs, this arrangement ensures a relatively even regional spread of funds. By contrast, funds in the United Kingdom are heavily concentrated around London.

Experience with venture capital programs provides helpful lessons (OECD 2006c):

- Public venture capital programs work only when interaction with the private venture capital market is strong.
- Venture capital is effective only for a very narrow range of new technologies. Hence, a "balanced portfolio" spread across many sectors may not succeed.
- Successful private funds are both flexible and active. They are involved with the inevitable shifts in direction and personnel associated with fast-moving companies. Public funds require the same involvement.
- Public funds have to be ruthless in jettisoning underperforming companies. Performance has to be judged according to the criteria of private venture funds.
- Public venture capital funds can be used to demonstrate to financial institutions the presence of a potential market.
- Public funds should be more "patient" than private funds in performing their role.
- The experience of OECD countries is inconclusive as to whether public venture capital funds supplement or lead the provision of venture capital. Such funds are generally young, and assisted firms have had insufficient time to grow.

Financial Support to Innovative SMEs. In the United Kingdom, capital investment grants are provided in the form of financial support to encourage sustainable development and job creation in most disadvantaged areas. Other measures include export credit guarantees (that is, government-backed guarantees, insurance, and reinsurance against the risk of nonpayment) to help exporters secure overseas contracts. To provide enterprises with the skills and expertise to secure private sector funding, governments offer a combination of specialist information and expertise.

In central Europe, Nordic countries such as Sweden, southern European countries, Japan, and Korea, the financing and incentive structures of the national innovation system have been geared primarily toward stimulating productivity improvements and growth in large manufacturing groups. Incentives for starting firms and generating growth of SMEs have been weak. These countries are not particularly well equipped with public preseed financing mechanisms.

The financial environment for supporting innovation is a great barrier in many emerging economies. In China, for example, when regional actors speak of venture capital, that funding usually comes from public sources. For small firms, access to bank loans is repeatedly cited as a major barrier for investment in innovation and overall development. Nevertheless, Shanghai reports an active private venture capital community for the biotechnology industry, for example. Weaknesses in the financial environment also help explain the lesser economic impact of certain innovation investments.

In the countries of the Middle East and North Africa, if private equity fund-raising is limited (see table 3.1), individuals with high net worth can provide a base for venture capital finance, thereby acting as a substitute for institutional investors or bank finance. In addition, private equity and venture capital firms based in Europe and United States may seek opportunities to invest abroad and put money into emerging markets. Since risk capital can involve equity participation, private equity and venture capital are also well suited to Islamic models of finance (which prohibit interest) as exemplified by the success of the Bahrain-based venture capital bank launched in 2005.[9]

Countries such as Egypt, Israel, Jordan, and Morocco have introduced public guarantee instruments in cooperation with the banking sector to meet the borrowing requirements of young firms. But these efforts are not sufficient to meet the region's entrepreneurship financing needs, especially since start-ups require funding for the period during which they do not generate enough

Table 3.1 Private Equity Fund-Raising in Emerging Markets, 2003–05

US$ millions

Region	2003	2004	2005
Africa and the Middle East	350	545	962
Asia[a]	2,200	2,800	15,446
Central and Eastern Europe[b]	406	1,777	2,711
Latin America	400	1,020	2,067
Total	3,356	6,142	21,186

Source: OECD 2006b.
a. Excluding Australia, Japan, and New Zealand.
b. Including the Russian Federation.

revenue to cover expenses. Despite its recent growth, private equity and venture capital are still at an early stage of development in this part of the world. Unlike their U.S. and European counterparts, Middle Eastern and North African countries have yet to develop strong venture capital markets or associations of private equity and venture capital. The establishment of the Gulf Venture Capital Association, however, is a step in that direction.

The Gulf Venture Capital Association was formed to disseminate knowledge about venture capital and best practices in the region through conferences, technology forums, and workshops. It is not clear how countries in the Middle East and North Africa outside the Gulf will be involved in these activities, however. The association will have to face the diversity of financing requirements and differences in the state of development of venture capital industries in the region (OECD 2006b).

Business Angels

So-called angel investment is an important source of informal equity capital. Angel investors often provide critical know-how as well as capital. Studies suggest that in countries such as the Netherlands and the United States, this source of investment may be significantly larger than the formal venture capital market. Evidence from the United Kingdom suggests that informal investors in small firms would make additional investments if presented with suitable proposals. Under these circumstances, public policy can help develop a supply of relevant information. For example, an initiative has been launched in the United States to create an Angel Capital Electronic Network.

The importance of business angel networks is recognized everywhere, but their strength is quite variable. For example, there are 10 times more in the United Kingdom and 100 times more in the United States than in France. Some networks are financed by the public sector, such as Austria's Federal Ministry of Economic Affairs and Labor, and some by membership fees and donations. And while Austria's i2 network is one of the oldest in continental Europe, according to some evaluations, the number of transactions in this network has not reached critical mass.

Bridging Institutions: Clusters and Networks

In the past two decades, clusters (or local productive systems) have developed in all market economies. They have become an increasingly efficient mode of organization, combining the advantages of competition and cooperation in groups of firms located in a relatively limited physical space. Clusters provide a favorable environment for innovation and technology diffusion. In this context, firms benefit from economic advantages that can be translated into productivity gains and growth opportunities: a larger market for workers with specialized skills, more rapid information flows and knowledge diffusion, and trust between contractual parties, which favors cooperation and specialization. In Italy, for example, both employment growth and productivity are higher in clusters than elsewhere. Firms that are part of industrial districts tend to have rates of return to investment and equity that are 2 percent and 4 percent, respectively, higher than those of isolated firms. They are also more innovative.

Clustering has not only increased in industrialized countries but has also diffused widely in emerging and less developed countries and regions. Often quoted are the metalworking and textile industries in the Punjab, the cotton knitwear industry of Tiruppur, the diamond industry of Surat, and the software and electronics cluster of Bangalore (India); the footwear clusters in Agra in Uttar Pradesh (India; see also box 3.6), in the Sinos Valley (Brazil), and in Trujillo (Peru); the shoe clusters in Leon and Guadalajara (Mexico); the textile cluster in Daegu (Korea); and the sports goods and surgical equipment

---

Box 3.6 SME Clusters in India

India's small-scale industry sector accounts for 40 percent of the country's industrial output and 35 percent of its direct exports. It has achieved significant milestones for India's industrial development. Within that sector, clusters—which have been in existence for decades and sometimes even for centuries—play an important role. According to a UNIDO (United Nations Industrial Development Organization) survey of Indian small-scale industry clusters undertaken in the late 1990s, India has 350 such clusters and approximately 2,000 rural and artisan-based clusters. It is estimated that these clusters account for 60 percent of India's manufacturing exports. Among the larger clusters are the following five:

- Panipat, which accounts for 75 percent of the blankets produced in the country
- Tiruppur, which is responsible for 80 percent of the country's cotton hosiery exports
- Agra, with 800 registered and 6,000 unregistered small-scale units, which makes approximately 150,000 pairs of shoes per day with a daily production value of US$1.3 million and exports worth US$60 million per year
- Ludhiana, known as the Manchester of India, which accounts for 95 percent of the country's woolen knitwear, 85 percent of its sewing machines, and 60 percent of its bicycles and bicycle parts
- Bangalore, which is a world-famous software cluster and deserves special mention.

Despite such achievements, the majority of the Indian small-scale industrial clusters have significant constraints: technological obsolescence, relatively poor product quality, information deficiencies, poor market links, and inadequate management systems. Moreover, with the Indian economy on the path of liberalization, all these clusters (even the best-performing ones) increasingly feel competitive pressures from international markets.

Source: UNIDO 2002.

clusters in Sialkot, the cutlery industry in Wazirabad, and the electrical fan industry in Gujrat (Pakistan). In African clusters, the interfirm division of labor and institutional support tend to be less developed in the metalworking, furniture making, garment, and other clusters in Kenya, Tanzania, and Zimbabwe. Most of these clusters are "low road to competitiveness" clusters (they compete on the basis of low prices, cheap materials, and numerical labor flexibility), but some concentrations of firms also exhibit elements of the "high road" such as innovation, quality, and functional flexibility.

Governments now realize the importance of these clusters, both as a significant share of the economy and as a main driver of innovation performance.[10] Experience in industrialized countries also shows that specialization and cooperation among SMEs can be efficiently promoted through public institutions. Directing policy toward groups of firms lowers transaction cost and facilitates learning. Through collective measures rather than subsidies to individual firms, policies can promote investment in both physical capital and in intangibles (forums for exchange, cluster animation). Groups of local actors with good knowledge of local needs and capacities can provide services. Such initiatives can involve strengthening clusters' demand for technological services and improving the work of intermediaries, linking participating firms with international firms within parks, and enhancing cooperative links, for example, through brokering and related programs.

Improving Access to Know-how and International Markets
Individual SMEs rarely have the resources or connections to take advantage of the global wealth of product and process ideas. One way for them to overcome this barrier is to pool resources and act together. Joint participation in international trade fairs, for example, can allow them not only to sell but also to learn through direct contact with potential customers. Trade fairs were important for the development of Brazil's Sinos Valley shoe cluster, for example. Joint action in the early 1960s led to the institution of a regular trade fair, which attracted buyers from all over the country. Subsequently, groups of producers went to trade fairs in the United States and Europe. Organized by local business associations and subsidized by the government, these groups played a vital role in connecting existing clusters with international buyers and provided a driving force for improving their products. Joint participation in trade fairs was also critical for ceramic producers from the Philippines, who launched themselves internationally in the 1980s. External support allowed them to exhibit a range of products at European fairs.

Technology Institutes and Collective Associations. Another possibility is to rely on a local technology institute, funded by government or foreign donors. Cluster development institutions can encourage firms to take certain kinds of collective action, such as cooperating to acquire new competencies while remaining competitors in other product markets.[11] In Taiwan, China, small firms have been encouraged to cooperate on R&D, with technological guidance provided by a public laboratory. The Ministry of Economic Affairs and trade associations play an important role in this context.

Collaborative institutions, councils, or associations that represent a cluster provide it with a sense of identity and with mechanisms for obtaining contracts and grants. They can combine participating firms' demand for specific types of services (see box 3.7).

They can also organize an advocacy function for clusters and express their interests. They encourage the definition of common standards, rules, and norms that stimulate competition or increase efficiency and set agendas for growth. In addition, they can organize training and the transfer of tacit knowledge among participants in an industrial section. Their role can be important in developing countries (see box 3.8).

Clusters. Policy measures in industrialized countries usually include programs to stimulate cluster development. For example, recent development initiatives in New Zealand through the New Zealand Trade and Enterprise aim to foster growth and innovation in existing clusters. Over 40 cluster development initiatives are underway in biotechnology, optics, pharmaceuticals, organics, software, film, and wool. The Cluster Development Program provides a grant (of up to NZD 50,000), which must be matched by the applicant. A cluster facilitator can thus be engaged to help develop the cluster more rapidly.

Box 3.7 The Role of Trade Associations in Italy

In Italy, the main trade associations representing small firms identify cooperation opportunities, suggest ways in which firms can link complementary skills, create contacts among potential partner firms, and motivate firms to cooperate and mediate critical phases in the establishment of a network.

In Bologna, one of the three major trade associations, the CNA (Confederazione Nazionale Artiglianato) has about 17,000 member firms, 41 local offices, and 500 employees. The CNA prepares 22,000 pay packets every month for 5,000 firms. It keeps the books of 10,000 firms, prepares the income tax declarations for most of its members, and organizes 80 training courses a year on subjects ranging from management and business administration to computing and foreign languages.

In the 1950s, the CNA established a large assessment and guarantee consortium in Bologna, which today has 7,500 member firms and guarantees some US$12 million in loans. So far, it has promoted 41 other consortia dealing with production and joint buying and selling, which now have 8,000 member firms and 42 industrial parks in which 1,030 small firms are located.

Source: OECD 2001.

> **Box 3.8 Sector Associations in Senegal and Cameroon**
>
> In Senegal, textile activities (such as tailoring and dressmaking) are well organized in the informal sector. In 1995, the National Federation of Clothing Professionals was started at the initiative of the clothing sector. The federation has some 10,000 members (including small garment workshops as well as small and medium enterprises) and performs critical activities: research into new commercial channels for national and international markets, creation of a savings and credit union to finance members' production activities, and training of workers in the skills required to produce modern wearing apparel, including those necessary for international subcontracting. In 1999, the Training and Professional Development Centre was established under the federation. It employs 18 part-time instructors and can oversee some 130 trainees. The trainees are workers from small and medium enterprises and apprentices. The center provides both preemployment education, which can last up to 12 months, and skills upgrading sessions that last just a few days. The center is registered with the government and provides its own certificates.[1]
>
> The Groupement Interprofessionnel des Artisans, Cameroon, is an association of over 100 informal sector enterprises in different economic sectors (woodworking, leather products, textiles, and metalworking). The group has been active in the organization of training sessions for its members, regulation of apprenticeships, and production of a newsletter. It has introduced an examination for apprentices (from member workshops), for which it organizes a committee of five members: one from the Ministry of Industrial Development and Commerce, two expatriates from donor agencies, and two local master crafts people. The graduates are presented with a joint certificate from the Groupement Interprofessionnel des Artisans and the Ministry of Industrial Development and Commerce.[2]
>
> *Sources:* 1. Johanson and Adams 2004. 2. Haan and Serriere 2002.

Similar mechanisms exist in emerging countries like Mexico, where state governments have developed ways to assist firms in clusters. For example, in Tamaulipas the government helps its 13 clusters by supplying facilitators. In the electronics and telecommunication clusters, the main task of the facilitator is to identify firms' needs and help build capabilities to meet them (for example, by creating a skills profile to transmit to universities so that they develop the appropriate curriculum). In addition, the Regional Maquiladora Association initiates information sharing and has several committees (human resources, finance, and technology, among others) that can provide expertise. It is expected that firms in the telecommunication cluster will share design practices in the future.

Given that tacit knowledge, which is essential to innovation, is not easy to communicate and its attainment necessitates practice, learning and interaction are widely understood to be basic inputs in technological innovation. Innovation in a firm increasingly requires active acquisition and exploitation of knowledge from other firms and public research organizations. Geographical proximity among learners thus becomes important and demonstrates the advantage of clusters.

Science Parks. Science parks are much used to encourage these agglomeration processes. Because they have not always lived up to expectations, more cautious attitudes now prevail. Revamped and better-organized parks and better-designed technopoles, however, are still expected to create spillover effects. Cross-fertilization and value added are intangibles and difficult to create, maintain, and evaluate. While companies and universities may be close together, for example, cultural barriers may still be difficult to overcome. In addition, particularly in high-technology industries, the technology required may be available only in very few places so that links tend to be global rather than local.

In both emerging and developed economies, many parks seek foreign investment through preferential tax policies and various support services. The proximity of suppliers and subcontractors often facilitates the implantation of these international firms.[12] Foreign investors are assumed to produce significant spillovers to the local business sector, and these are assumed to be faster and stronger when the firms are located in the same facility and involved in organized networking, as is often the case in science parks.

Several mechanisms in Turkey aim to attract FDI to encourage local firms to generate knowledge and thereby increase Istanbul's innovation capacity and the country's economic internationalization. Those mechanisms involve creative forms of joint ventures, acquisition of foreign technology licenses, and turnkey projects. Technology parks, which provide an environment for catalyzing strategic alliances, offer a suitable environment for technological start-ups. After an incubation phase, the firms can be relocated in technoparks, which house a more mature group of firms.[13]

In China, an important objective is to replicate the success of "clusters" in industrialized countries by promoting industrial and science parks, albeit on a larger scale and involving a complex set of overlapping structures. China's science parks have evolved over time from focusing on high-technology manufacturing exports to including entities that support endogenous innovation.[14] Firms that locate in a science park hope that this placement will help leverage government support, among the other benefits of participation such as preferential tax policies. The number of actors and the degree of government control are in any case greater than what would be found in industrialized countries.

China has also provincial and local initiatives for such parks, in addition to those designated as national parks, although in view of their proliferation, they are now prohibited from offering certain tax incentives. It has been estimated that there are approximately 12,300 "clusters" across China.

It has also been estimated that there are approximately 6,741 development zones (presumably also a form of park) (Sigurdson 2004). There are, for example, 120 regional high-technology zones in addition to the national ones, although they do not benefit from the same degree of tax exemption as the national zones.

Supporting Innovation in Networks

Unlike clusters, which do not require membership in an association or a collective entity, firm networks work in cooperation, though not necessarily in the same place or linked by some type of agreements. In "hard" networks, small groups of companies come together to achieve shared objectives through formal agreements. "Soft" networks are larger groups with more flexible internal relationships. In most industrialized countries, programs are limited to hard networks. In the United States, however, soft networks predominate because they are easier to form, involve less risk, and seek short-term results. Examples include hosiery companies in western North Carolina, metalworking in Arkansas, and the Berkshire Plastic Networks in western Massachusetts.

An important policy strategy to stimulate networks was the 1989 initiative of the Danish government, which launched a scheme for training and mobilizing brokers to create networks (see box 3.9). While the program was temporary, it served as a prototype for replication in Australia, Canada, France, New Zealand, Norway, Portugal, Spain, the United Kingdom, and the United States.

Box 3.9 Denmark's Network Program: Brokers and Scouts

Denmark's Network Program, implemented in the early 1990s, offered monetary incentives to promote cooperation among groups of at least three independent firms that committed themselves contractually to a long-term relationship. Grants were provided for three phases of network creation: feasibility studies to evaluate the potential for cooperation, planning grants to prepare an action plan or budget for a network, and start-up grants for operating costs in the first year. Eligible activities included R&D, production, joint marketing, and problem solving.

- *Network brokers:* The network broker was the key to the program and served as an external facilitator or systems integrator for network functions. In some instances, brokers were consultants expecting to earn a living in this way, but most already worked for agencies that served SMEs. Because the idea of working with groups of firms was uncommon, Denmark designed a training and certification program.
- *Network multipliers:* These people were very familiar with the companies and able to detect and assess opportunities for collaboration that could be passed on to brokers. Sometimes referred to as "scouts," they included staff members

continued

> **Box 3.9 continued**
>
> of chambers of commerce, trade associations, banks, accounting firms, law offices, trade centers, technical colleges, and technology extension services that serve SMEs.
> - *Incentives for rural networks:* Denmark offered sequenced incentives to compensate small firms for some of the costs of participating in activities with uncertain returns. The Danish program was based on the U.S. Small Business Innovation Research program, with a small 100 percent concept grant (up to US$10,000), larger planning grant (up to US$50,000) and still larger implementation grant (up to US$500,000).
> - *Information campaign:* Denmark also distributed information widely through the media, brochures, and newsletters on the potential value of networks and funding opportunities. The distribution venues ranged from conferences to pubs.
>
> While not formally assessed, the Danish Network Program, which terminated in 1993 after three years of operation, was considered a success on a number of grounds: (a) 5,000 enterprises were involved in forming networks out of a target group of 10,000–12,000; (b) the idea, and often the practice as well, has disseminated widely, and networking has become a natural option to consider in the face of new business challenges; and (c) in the interim survey, 75 percent of participating enterprises felt that networking was raising their ability to compete, and 90 percent of respondents expected that they would continue the practice of networking beyond the subsidy period.
>
> *Source:* Rosenfeld 2005.

In the United States, network programs are state based and modest. They are not viewed as subsidies but as incentives to change attitudes toward cooperation. They are designed for a finite period of time. In recent years, the importance of a network environment has become clear. Overlaps with cluster policies have been emphasized, and networking programs have been included in cluster initiatives, with goals such as creating skills alliances, technical assistance (in MEP programs), building social capital, and fighting poverty. Networks have become "the conventional wisdom of business practices as a result of exhortation to cooperate" by managers, business schools leaders, and policy makers (Rosenfeld 2001).

The Danish model has also been followed in a number of emerging countries. For example in Chile, SERCOTEC introduced an initiative to encourage networking between groups of firms and to provide a focus for channeling support to small firms. It established a series of subprojects, each involving three stages: *preparation*, in which officials work to identify firms in a particular locality, diagnose their problems, and establish the credibility of SERCOTEC; *consolidation*, in which a manager is appointed to coordinate the network, act as an interface with various government and marketing agencies, facilitate the take-up of training and other support services, and develop better interfirm relations; and *independence*, when, after three years, participating firms are expected to take on responsibility for the manager's salary. The idea is that the participants in the network benefit enough that private initiatives will sustain it. Although the program is small, results have been encouraging. Most participating firms succeeded in gaining access to new domestic or international markets. The majority of networks also showed the capacity to be self-sustaining. Government officials have been sufficiently encouraged to develop a new initiative designed specifically for exporting firms.

Networking is increasingly understood in a wider sense, that is, in the context of productive chains that include small firms, large companies, and multinationals.[15] This shift of emphasis is mirrored in the new innovation policies increasingly implemented in industrialized counties (see box 3.10). In Italy, in the wake of the new law on transfer of technology and innovation, which emphasizes the transfer of power from the national to regional governments, regional agencies such as ERVET (Territorial Development Agency of the Emilia Romagna Region) have refocused their strategies from subsidizing services to sectoral districts toward focusing on territorial approaches, productive chains, promotion of public-private partnerships, and investment funds (Dall'Olio 2007). This new trend

---

Box 3.10 Networking Programs: The International Experience

Governments concerned about economic development have frequently encouraged large local employers to engage SMEs more actively in value chains. Supplier development programs involving SMEs in industrialized countries reflect the increasing recognition that the delivery of a final product or service to the customer involves the linking of often numerous suppliers in a "value chain." SMEs rarely initiate value chains or deliver the final products and services.

Incentives for creating value chains can stem from adversity. The United Kingdom's Accelerate program was implemented in the West Midlands, which suffered from a continuous decline in the automotive sector dominated by a large company (MG Rover). Many local suppliers in the region were dependent on this firm. The Accelerate program encouraged local SMEs to use their wide range of skills, diversify their customer portfolio, and improve their modes of production and organization. This goal was achieved through the provision of subsidized consultants who worked closely with firms. Over seven years, Accelerate worked with more than 1,000 companies and safeguarded more than 16,000 jobs.

A recent review of SMEs in global value chains concludes that they are likely to grow in importance. It argues that in addition to facilitating SME financing, protecting intellectual property, and helping SMEs comply with international standards, governments seeking to increase the role played by SMEs should seek to raise awareness of the roles SMEs can play in this respect.

Source: OECD 2006d.

is replicated in many lower-income countries. In Mexico, for example, the objective is clearly to encourage the creation of centers to initiate production, to strengthen integrators (hubs for infrastructure investment projects), and to trigger new supply chains through the cofinancing of projects and the national program of suppliers. In 2006, the federal government of Mexico devoted almost half its assistance to SMEs to these networking programs.

Conclusions

In most countries, support to innovators has become an important policy task. This support is ensured by various institutions that provide specific and relevant services for entrepreneurs and firms. In industrialized countries, the business services infrastructure has been in place for several decades and has considerably improved its offer, with a focus on professional, mature, and highly segmented services. At the same time, while technical centers formerly tended to link with traditional and medium-technology firms, more high-technology ventures have sought sources of expertise in universities or leading public laboratories. Nevertheless, low-technology and incremental innovation continue to account for a considerable share of GDP and employment in all industrialized countries. They remain a basic factor in the innovative performance of countries.

Delivering business services through networks is acknowledged as a very efficient way to maximize their contribution to regional development and innovation. Cooperation is obviously favored by geographical proximity, institutional coordination, and physical opportunities (shared space and facilities), but international communication and cooperation substantially broaden the scope of possible links. At present, these are only partially exploited. To be sustained, the networking trend needs active encouragement from public policies.

In developing countries, support institutions have often copied those of industrialized countries. The spectrum of performance is extremely wide, not only between countries but also within them. Many of these institutions do not function effectively and tend to be of poor quality, with inadequate equipment and a poorly remunerated staff. They also often exacerbate the initial pitfalls of their model. First, they overemphasize the supply side and are often out of touch with industry needs; in particular, they pay insufficient attention to the need to enhance firms' absorptive capacity. Second, unrealistic strategies that place an exaggerated emphasis on leading-edge technologies are commonplace.

To increase policy efficiency, many industrialized countries tend to put the firm at the center of their strategies. Policies are then designed to support small firms and start-ups. In the wake of these policies, comprehensive sets of public initiatives and support are being implemented to cover the life cycle of new products, from design to internationalization. In this context, incubators play a major role in the survival rate of young companies. In developing and especially emerging economies, incubators have also proliferated. The concept is not well established, however, and policies to encourage their professionalization are not easy to design.

The lack of finance available for the early stages of the innovation process has been underlined not only by business circles but also by policy makers. Industrialized countries have improved stock market regulations and intensified support for venture capital through supplementary budget allocations in favor of venture capital, especially for SMEs or technology-based start-ups or the formation of partnerships with private venture capitalists. In continental Europe, the absence of an efficient secondary financial market explains part of the lag in venture capital and business creation vis-à-vis the United States. In most developing countries, these financial markets are embryonic.

Finally, governments everywhere are more aware of the need to support clusters and networks because of their innovative potential, their collective efficiency, and their increasing share in business activities. Physical proximity and a shared "regional culture" (shared practices, attitudes, expectations, norms, and values) that facilitate the flow and sharing of tacit and other forms of proprietary knowledge are the cornerstones of clusters. In industrialized countries, the most popular initiatives for enhancing productivity and capabilities for innovation include the funding of facilitators, efforts to stimulate spillovers, and greater effort to strengthen productive chains.

As a last remark, some countries are clearly concerned about the proliferation of innovation support measures and the need for rationalization and simplification. Not all innovation schemes are cost-effective, and they may be confusing for business. The United Kingdom considers it necessary to merge and simplify these schemes and to expand the number and role of one-stop shops. Most countries are now taking a systems approach that emphasizes the need to optimize the combination of supports and implement structural reform. This approach may be useful as well to many developing countries.

Notes

1. Although these data are relatively old (see Bellini 1998), they probably stand as minimums.

2. The regional investment contract simplifies procedures and helps advance the installation of start-ups and develop new investment projects in the region. It addresses all types of companies that manage operations in the production of goods and services, R&D, and innovation. It gives priority to highly innovative sectors such as new sources of energy.

3. Brazil is the Latin American country with the highest number of quality certification approvals (about 3,000) and one of the world's leaders in the relative increase in approvals.

4. The idea is to ensure early revenue and a route to market for firms that face barriers for their early development.

5. A pilot scheme for grants for investigating an innovative idea also helps U.K. SMEs obtain practical advice when exploring ideas for innovative products, services, and processes by covering 75 percent of the cost of outside experts.

6. The business support services typically provided by incubator management include business planning, advice on accessing capital, marketing, identification of suitable business partners, and general strategic advice. Other business support services—such as specialist legal services, accounting, and market research—tend to be provided by external providers with which the incubator management has established relations. Business incubator managers are often experienced former businesspersons and play an essential role in supporting and nurturing early-stage businesses. Case study evidence and survey work suggest that incubators can help address the traditional market failure in the provision of business support services to small businesses. Larger private sector business support organizations and management consultancies often do not deal with SMEs.

7. As a consequence, from 2000 to 2004 the number of incubators more than tripled from 131 to 464, and the number of firms rose from 7,693 to 33,213.

8. This includes a scheme such as Austria's High Tech Double Equity program.

9. This bank is the first Sharia-compliant venture capital bank in the region. It focuses on SMEs and uses a rigorous system to ensure Sharia certification of investments (see OECD 2006b).

10. Clusters account for a significant and growing share of industries and services in a wide variety of sectors. In Italy, for example, output of the industrial districts account for more than 40 percent of manufacturing production and more than half of industry exports. In the Netherlands, they represent 30 percent of output, and in Norway they employ 22 percent of the workforce. According to Porter (1999), high-technology clusters account for only 8 percent of employment and 2.5 percent of total U.S. employment. The most populated clusters in the United States include business services, financial services, tourism, education and knowledge, distribution, construction, and logistics.

11. A more detailed discussion of policy measures for cluster promotion is to be found in chapter 10, which discusses building innovative sites.

12. In the case of proximity, they tend to form some type of hub-and-spoke clusters.

13. Turkey has 17 technological incubators, which provide the infrastructure for technological start-ups.

14. The Zhongguancun Science Park in Beijing, approved in 1988, is one of the first examples. In this science park, there are 71 institutions of higher education with 300,000 students, including Peking and Tsinghua universities, 213 research institutes, 65 multinational firms, and 54 multinational R&D centers as well as other intermediaries. The Shenzhen High-Tech Industrial Park in the Guangdong province takes advantage of the Shenzhen Special Economic Zone, multiple incubators, and the Shenzhen Software Park, which serves as a base for the national Torch Plan Software Industry program.

15. While districts are characterized by flexible specialization and dense networks of local centers, productive chains integrate elements of productivity, knowledge, and social capital.

References and Other Resources

Bellini, N. 1998. "Services to Industry in the Framework of Regional and Local Industrial Policy." Paper presented at the OECD international conference, "Building Competitive Regional Economies: Upgrading Knowledge and Diffusing Technology to Small Firms," Modena, Italy, May.

Cardullo, Mario, W. 1999. *Technological Entrepreneurism: Enterprise Formation, Financing and Growth.* Baldock, UK: Research Studies Press Ltd.

Dall'Olio, R. 2007. ERVET Presentation, Bologna, Italy, October 19, 2007.

Haan, Hans, and Nicholas Serriere. 2002. *Training for Work in the Informal Sector: Fresh Evidence from West and Central Africa.* Turin, Italy: International Training Center, International Labour Organization.

Izushi, H. 2005. "Creation of Relational Assets through the Library of Equipment Model; An Industrial Modernization Approach of Japan's Local Technology Centers." *Entrepreneurship and Regional Development* 17 (3): 183–204.

Johanson, Richard, and Arvil Adams. 2004. *Skills Development in Sub-Saharan Africa.* Washington, DC: World Bank.

NIST (National Institute for Standards and Technology). 2007. *Strategic Planning and Economic Analysis: FY 2007 Annual Performance Plan.* Washington, DC: U.S. Department of Commerce.

OECD (Organisation for Economic Co-operation and Development). 2001. *Territorial Review: Italy.* Paris: OECD.

———. 2005. *Innovation Policy and Performance: A Cross-Country Comparison.* Paris: OECD.

———. 2006a. *Economic Policy Reforms: Going for Growth.* Paris: OECD.

———. 2006b. *MENA Investment Policy Brief: Venture Capital in MENA Countries: Taking Advantage of Current Opportunity.* Paris: OECD.

———. 2006c. *SME Financing Gap, Theory and Evidence.* Vol. 1. Paris: OECD.

———. 2006d. "Enhancing the Role of SMEs in Global Value Chains: Conceptual Issues Draft Report." Paris: OECD.

———. 2007a. *Higher Education and Regions: Globally Competitive and Locally Engaged.* Paris: OECD.

———. 2007b. *SMEs in Mexico, Issues and Policies.* Paris: OECD.

Porter, Michael. 1999. "The Economic Performance of Regions." *Regional Studies* 37: 549–78.

Rosenfeld, S. 2001. *Networks and Clusters: The Yin and the Yang of Rural Development.* Heidelberg: Springer Berlin.

———. 2005. "Industry Clusters: Business Choice, Policy Outcome or Branding Strategy." Paper presented at the Centre for Regional Innovation and Competitiveness conference, "Beyond Clusters: Current Practices and Future Strategies," University of Ballarat, Victoria, Australia, June 30–July 1.

Shanghai Municipality Science and Technology Commission. 2006. "The Innovation System of Shanghai." Presentation to a delegation from the Organisation for Economic Co-operation and Development, Shanghai, China, October 9.

Shapira, Philip. 2007. "Putting Innovation in Place." Presentation given at the Center for Japanese Research, University of British Colombia, workshop, "Japanese Approaches to Local Development, Clusters, Industry-University Linkages and Implication for British Columbia," Vancouver, University of British Columbia. March 7–8.

Sigurdson, Jon. 2004. "Regional Innovation Systems (RIS) in China." Working Paper 95, European Institute of Japanese Studies, Stockholm School of Economics, Stockholm, Sweden.

Singapore Economic Development Board. 2006. Singapore.

UNIDO (United Nations Industrial Development Organization). 2002. *Industrial Report 2002/2003.* Vienna: UNIDO.

4

Improving the Regulatory Framework for Innovation

Innovation in developing countries is based mostly on adoption, recombination, and adaptation of existing technologies rather than on development of new technology. Innovation is therefore more "new to the market" or "new to the firm" than "new to the world" (WBI 2007; World Bank 2008). In consequence, the capacity of developing countries to innovate depends, on the one hand, on foreign sources of knowledge and technology and, on the other, on the country's capacity to absorb, adapt, and diffuse innovation. International trade rules and practices and intellectual property agreements strongly influence countries' ability to attract partners and foreign investments, benefit from technology transfer through increased trade opportunities, and stimulate local innovation.

The international context aside, building an enabling environment that is both attractive to foreign investment and locally supportive of innovation, adaptation of technology, and dissemination of knowledge requires an adequate institutional framework. Recent studies agree that government policies to support innovation should embark on reforms that update the regulatory and institutional framework for innovation and remove bureaucratic, legislative, and regulatory obstacles to innovation (Chandra 2006; WBI 2007; World Bank 2008). These obstacles affect competition laws, licenses to operate, government authorizations, technical norms and standards, customs procedures, and many other regulations and processes.

This chapter begins by exploring the international context as it relates to knowledge dissemination, technology transfer, and innovation. While trade and

---

This chapter was prepared by Thais Leray.

foreign direct investment (FDI) are well-recognized channels for technology transfer, trade regimes and conditions in which prices are fixed, particularly for agriculture goods, distort innovative efforts and achievements in a number of low- and medium-income countries. Tariff structures, and in particular tariff peaks and tariff escalation, prevent developing countries from diversifying their exports and moving up value chains. Intellectual property rights (IPR) regimes are also not favorable to innovators and innovation in low-income countries.

The chapter then turns to issues that depend primarily on domestic policies such as competition, customs practices, aspects of land property rights, organization of commercial and distribution networks, and infrastructure weaknesses and goes on to highlight reform processes that have been used successfully in various countries. Finally, the chapter draws attention to an essential proactive measure, public procurement, that can further stimulate innovation.

International Trade and Investment Framework

Two critical issues in the international trade and investment framework are technology transfer and the intellectual property rights regime. This section considers impediments to trade, such as tariffs and other barriers to trade and intellectual property rights regimes in developing countries.

Technology Transfer and Trade

Long recognized as an engine for wealth creation, growth, and poverty reduction, trade contributes to technology and knowledge transfer in at least three ways: through embodied technology in the form of goods and services; through knowledge, practices, and processes linked to the use of technological goods and to contacts with foreign suppliers and customers; and through capital and investment (notably FDI).

Channels for Technology Transfer. Imports enhance the technological knowledge of developing countries in various ways. The technological know-how embodied in goods and services, for example, enables developing countries to employ more efficient production processes and thus raise the quality of their own products and processes. Licensing also typically involves the purchase of production or distribution rights for a product and the underlying technical information and expertise for producing it. At the same time, trade openness and competition from technologically superior imports may produce large technology spillovers and boost domestic productivity (Keller 2004; World Bank 2008).

Export activities with foreign countries may also generate technological spillovers through interaction with foreign buyers and customers, for example, when exporters have to meet new specifications or higher standards. These can also support technological progress by increasing product consistency and improving product performance. Foreign buyers also provide information about foreign markets and can assist with process improvements (Schiff and Wang 2006; World Bank 2008) as well as generate additional demand that may lead to economies of scale.

Finally, FDI constitutes a major channel for technology and knowledge transfer. Foreign firms can offer a package of mobile, tangible, and intangible assets that include capital, technology, know-how, skills, brand names, organizational and managerial practices, access to markets, competitive pressures that stimulate innovation, and environmentally sound technologies (UNCTAD 1999). Transnational corporations (TNCs) may promote local innovation in other ways, as follows: by acting as role models and enhancing competition; by developing local capabilities of workers through training and experience that can then spread locally through worker mobility; by encouraging efficiency and technical change in local firms and suppliers, especially when strong backward and forward links are established; and by engaging in collaborative research and innovation activities (UNCTAD 1999; World Bank 2008). The extent of spillovers depends on domestic absorptive capacity and may be greater when the difference in technological levels between host and home countries is not too large. Finally, rules and regulations affecting the investment climate determine not only how attractive the country is to FDI but also the degree to which TNCs are encouraged to upgrade the transfer of technology and skills and raise local capabilities and links.

Tariffs, Tariff Peaks, and Tariff Dispersion. Trade barriers have fallen in many countries, following unilateral efforts and bilateral, regional, or multilateral agreements. Indeed, over the past decade, most-favored nation (MFN) average tariffs have fallen dramatically, and a substantial amount of trade is conducted at a zero MFN tariff rate or through preferential trade agreements, free trade agreements, or customs unions (Islam and Zanini 2008; Portugal-Perez and Wilson 2008). But low average tariffs do not reflect the whole picture. Access to markets is often still restricted because of either tariff or nontariff barriers or a combination of the two.

While the average tariff at which international trade is conducted has been dramatically lowered in recent times, tariff barriers and tariff peaks still prevail in certain sectors and subsectors owing to developed countries' interest in protecting these sectors or subsectors.[1] These tariff peaks often apply to products in which developing countries have a comparative advantage: agriculture and the food industry, textiles and clothing, footwear, leather, and travel goods, as well as the automotive sector and a few other transport and high-technology goods, such as consumer electronics and watches (UNCTAD and WTO 2000; Watkins 2003; World Bank 2007; IAASTD 2008; Islam and Zanini 2008).

Contrary to perceptions, developing countries generally have higher tariff rates than developed ones. In agriculture, for example, the South Asian and

East Asia Pacific regions have the most restrictive policies, followed by high-income countries of the Organisation for Economic Co-operation and Development (OECD) (Islam and Zanini 2008). These policies limit both North-South and South-South trade. Although developing countries' tariff rates are generally higher, tariff dispersion and maximum tariffs applied are generally much higher in developed countries. Canada, the European Union, Japan, and the United States maintain tariff peaks as high as 350–900 percent for important export products of developing countries, and the European Union's food industry accounts for about 30 percent of all tariff peaks (UNCTAD and WTO 2000; Islam and Zanini 2008).

Tariff Escalation. It is mainly the tariff structure itself that constitutes a serious impediment to innovation, technological transfer, and upgrading. It is well known that developing countries have difficulty moving up the value chain in certain markets, as tariffs escalate with the degree of product processing.[2] According to one researcher, this escalation "has the effect of reducing the demand for processed imports from low-income countries, preventing appropriate structural adjustment in developed countries and frustrating the diversification of low-income countries into high value-added exports" (Oyejide 2003). For example,

> although food processing is a major export industry of developing countries, their exports are largely concentrated in the first stage of processing. More advanced food industry products make up only 5 percent of the agricultural exports for LDCs [less-developed countries] and 16.6 percent of those of developing countries as a whole, against 32.5 percent for developed countries. (UNCTAD and WTO 2000)

This tariff structure is a serious problem for exporting countries wishing to diversify their exports and develop their industrial and manufacturing capabilities, especially since tariff escalation occurs precisely in those activities that would otherwise offer a chance for industrialization: food, textiles, clothing and shoes, and wood industry products (UNCTAD and WTO 2000).

Quotas, Subsidies, and Other Nontariff Barriers. Nontariff barriers—including quotas, antidumping measures, countervailing duties, and safeguard measures—significantly affect trading opportunities of developing countries. Such measures generally tend to restrict the volume of traded goods and are often used in combination with high tariffs. For example, prohibitive tariffs of up to 220 percent apply to above-quota imports of bananas into the European Union (UNCTAD and WTO 2000).

Subsidies, which also significantly distort international trade, are most prevalent and controversial in the agricultural sector. Overall trade policies depress prices of agricultural products in international markets by an average of 5 percent (World Bank 2007). Cotton is one example. It is estimated that with a fully liberalized market, the European and U.S. share of cotton production would decrease dramatically (by 70.5 and 60.7 percent, respectively) and that the lowest-cost producers (Benin, Burkina Faso, Mali, Tanzania, and Uganda) could expand their share by 12.6 percent (Baffes 2004; Watkins 2003). Developed countries are shifting policies toward "decoupled" payments—that is, payments not directly linked to the type, volume, and price of products. These measures are considered less distorting than output-linked forms of support, but the subsidies are still substantial.[3] Nor are they always neutral for production as they reduce aversion to risk (wealth effect), reduce variability in farm income (insurance effect), and allow banks to make loans to farmers that they otherwise would not make (World Bank 2007).

Preferences along with special and differential treatment aim to facilitate developing countries' access to developed country markets. A certain number of African countries benefit from trade preferences with both the United States and the European Union. Restrictive rules of origin, bureaucratic and administrative barriers, lack of institutional capacity, and volume limits on exports, however, make it difficult for these countries to benefit from the advantages of trade. These constraints impede further development and diversification and thus slow technological improvement and innovation. Moreover, while preferences are similar in different export markets, the rules of origin differ. Portugal-Perez and Wilson (2008) find econometric evidence that relaxing rules of origin by allowing the use of fabric from anywhere under the African Growth and Opportunity Act (AGOA) for less developed countries increased exports of apparel by about 300 percent for the top seven beneficiaries of AGOA's "special rule" (for less developed countries), while broadening the varieties of apparel exported by these countries.

Technical Barriers to Trade, Standards, and Norms. Finally, norms, standards, and technical regulations applying to products and processes seek to address concerns and mitigate risks relating to health and safety, quality, environmental threats, and social conditions of production. On the one hand, they may encourage exporters to upgrade technology and improve the consistency and quality of their products and processes. By conveying valuable information relating to quality, safety, good practices, and the like, they also reduce transaction and information costs in the importing country. On the other hand, they may also restrict international trade and limit developing countries' participation by raising the costs of compliance, so that it becomes necessary to alter production processes to adapt products to the importing country's standards and regulations. Moreover, certification aiming to demonstrate compliance can generate additional costs for the exporter (Portugal-Perez and Wilson 2008).

Such costs are especially onerous when exporters face a range of constraints for exporting similar products to different countries. Processes that can be useful in mitigating those adverse effects include mutual recognition agreements (which are used extensively within the European Common Market), unilateral recognition of equivalence (clearly defined criteria for accepting foreign standards, measures, and qualifications as equivalent to domestic ones when they pursue the same regulatory objective), promotion of supplier's declarations of conformity, and the like (OECD 2005). As a priority, governments should seek to reduce regulatory barriers to trade and investment arising from divergent and duplicative or outdated requirements, notably by developing standards and norms that build on international standards and seeking harmonization with them.

Intellectual Property Rights Regime: Rationale and Controversies

Intellectual property rights are often seen as having an important impact on stimulating innovation and encouraging technology dissemination.[4] These are enforceable legal measures that confer monopoly rights on innovators for a specified period of time, after which they fall into the public domain and can be freely used by others. The underlying assumption and motivation behind IPRs are that they foster innovation by ensuring that innovators are sufficiently rewarded for their investments, including both their creative energy and financial capital.

Rationale. The last decades have witnessed an unprecedented increase in the scope and level of protection of intellectual property rights. New rights are created, and standards are being harmonized throughout the world (WBI 2007). Opinions about the impact of stronger IP policies on developing countries vary widely. Supporters argue that developing countries that wish to stimulate knowledge generation and diffusion and to attract and benefit from technologically rich investments need to establish strong IPR regimes. Opponents argue that strong IP protection can reinforce economic concentration and, by restricting competition, enable owners to maintain high prices and stifle innovation (see box 4.1).

Developing countries remain largely dependent on foreign technology and products, and effective technology transfer is paramount for their innovation strategy. The challenge is thus to devise IPR policies that strike an appropriate balance between effective generation of creativity and innovation, on the one hand, and diffusion of innovation in various ways and in a wide range of economic and technological contexts, at the lowest possible cost, on the other. But as the degree and scope of protection have increased in the past decades, controversies have arisen over the availability of knowledge and technologies, most notably in the fields of pharmaceuticals, traditional knowledge and folklore, and education.

Controversies over IPRs. First, strong IPRs are believed to impede knowledge diffusion and research in developing countries by depriving educational

> **Box 4.1 Brazil's Policy on HIV/AIDS**
>
> Committed to the policy of free universal access to diagnosis, prevention, and treatment of HIV/AIDS, the Brazilian government began mobilizing local manufacturers at the end of the 1990s to produce 10 low-cost generic versions of antiretroviral drugs (ARVs) within the national therapeutic guidelines. This initiative was possible because the 1970 reform of the IP law refused to recognize patents on processes or molecules and thus permitted the legal copying of molecules. As a result of these reforms, 56 percent of all ARV drugs consumed in Brazil in 2001 were produced nationally, with a price reduction of 82 percent over the period 1996–2001. In addition, the Brazilian government promoted intense price negotiations with multinational pharmaceutical companies to achieve consistent price reductions of patented ARVs. During these negotiations, the state used the threat of compulsory licensing as an argument. Since Brazil had produced ARVs on its own soon before the opening of these negotiations, it had demonstrated its capabilities in the field. However, its commitment to the TRIPS (Trade-Related Aspects of Intellectual Property Rights) agreement led the government in 1996 to immediately amend its IP legislation to recognize pharmaceutical products and processes and renounce its 10-year transition period.
>
> In the end, the Brazilian negotiations provided short-term benefits—notably a reduction of 46 percent in the unit price of capsules and the prompt introduction of a new, reduced daily-dose formulation. According to many, however, the agreement contains a number of restrictive provisions: the Ministry of Health is prohibited from allowing flexibility on any formulation that includes patented molecules until the agreement expires in 2011; the prices on ARVs are fixed until the expiration of the agreement; and some formulations of medications are barred from use in the first stages of treatment, which raises the price of the overall treatment. Brazil thus now faces a dilemma in striking a balance between the financial sustainability of its national anti-AIDS program and access to newer (and more efficient) ARVs.
>
> *Source:* Coriat, Orsi, and d'Almeida 2006.

systems of access to valuable copyrighted material. Because academic journals tend to be very expensive, for example, the availability of educational materials for developing country schools and university students is limited. Access to inventions for research use (biotechnologies) or for further improvement or adaptation (software) has reportedly been hampered by patents in a number of cases.[5] The extension of digital rights management systems is also meeting with resistance, as IPRs put tight restrictions on the rights of users, thereby reducing de facto the scope of "fair use" of copyright law (OECD 2007a). An inadequate balance between diffusion and protection may encourage copying of such material or turning a blind eye to these practices.

Second, appropriating knowledge through IPRs poses several problems. For one thing, it is seen as creating monopolies that maintain high prices in specific goods and services. The pharmaceuticals industry is a case in point.

Appropriating knowledge through IPRs raises questions about access to drugs, local manufacturing capacity, and the development of new drugs, even though developing countries may use compulsory licenses, parallel imports of patented products, or exceptions to grant authorization to a third party to exploit a patented invention for the domestic market under "national emergency" or "extreme urgency."[6] For another, knowledge appropriation generates ethical concerns relating to plants, animals, genes, and gene fragment patenting, but also (mis)appropriation of indigenous knowledge (UNCTAD and ICTSD 2003; WBI 2007; IAASTD 2008). In fact, indigenous knowledge is increasingly recognized as a valuable asset in industrialized and developing countries alike, as it provides input into many modern industries (pharmaceuticals, cosmetics, agriculture, food additives, industrial enzymes, biopesticides, and personal care). Yet firms in industrialized countries appropriate most of the value added in such cases, with their advanced scientific and technological capabilities that make appropriation possible without the prior informed consent of the holders of that knowledge (Commission on Intellectual Property Rights 2002; cited in WBI 2007). Beyond ethical considerations, patents on plants, animals, genes, and gene fragments may stifle innovation as they raise costs and restrict experimentation by the individual farmer or public researcher and potentially undermine local practices that enhance food security and economic sustainability.

Finally, strengthened IPRs are increasingly seen as limiting the development of local capabilities and retarding developing countries' future innovation capacities. Copying and counterfeiting affect various constituencies: consumers, whose health and safety may be put at risk; rights holders, whose sales decline; governments, which suffer lost tax revenues while facing the costs associated with fighting counterfeiting and piracy; and the innovative environment, in which copying and counterfeiting divert creativity, entrepreneurship, and incentives away from genuine innovation (OECD 2007a). While basic copying (such as compact discs or misappropriation of trademarks) provides little avenue for learning, the situation may be quite different for the manufacture of products that require the application of complex processes whose operation and adaptation to local conditions may require high levels of knowledge and skill (UNCTAD and ICTSD 2003). Imitation often serves as a learning process and as informal technology transfer by making it possible to establish basic competence on which to build innovations. History shows that becoming good at imitating, for example through reverse engineering, is a vital stage in the process of becoming innovative (see box 4.2).

The Use of IPRs. The third and last set of problems relates to the use of IPRs. In developed countries, few small and medium enterprises (SMEs) have the knowledge or capacity to take advantage of IPR systems effectively and efficiently,

> **Box 4.2 From Duplicative Imitation to Creative Imitation and Innovation**
>
> Recent studies seem to suggest that imitation with local improvements is a precursor of innovative capabilities. Examples from Indonesia and the Republic of Korea seem to support this view.
>
> Small businesses in East Java, for example, now produce good quality leatherware. At one time, they simply copied Western designs. Employees would watch the carousels at the airport, waiting for examples of the latest designs from the most fashionable designers. They made exact imitations, including the brand name. After warnings from the Indonesian government, they changed the name simply to resemble the fashionable name. They have now begun to adapt the designs as well, and, with changes in design, they have also begun to use their own brand names.
>
> In Korea, most large local pharmaceutical and cosmetic companies and some paper and chemical firms have evolved from small ventures, developing their own primitive production processes through imitation to become significantly large, innovative operations. Leading local pharmaceutical firms first started as importers and dealers of packaged finished drugs and later entered the drug manufacturing business by packaging imported bulk drugs. They gradually moved into more intricate operations by formulating imported raw materials and then, through backward integration, by producing the chemical components. Through this process, they grew in size and in technological capabilities. As a result, in the early 1980s local firms accounted for almost 90 percent of the domestic drug market in Korea as compared to 22 percent in Brazil, 47 percent in Argentina, and 30 percent in India. During this period, Korea respected process patents but not product patents in the chemical, cosmetics, and pharmaceutical industries. Local producers could therefore work around patented processes to produce relatively well-known chemical and pharmaceutical products. Were it not for the lax IPRs, the local pharmaceutical firms could not have achieved so much. Some local firms have advanced to the point that they can undertake serious research and development activities and discover new drug compounds.
>
> *Source:* Kim 2003; Macdonald, Turpin, and Ancog 2005.

because they lack the information, understanding, and resources (human, time, and capital) to research the field and to make sure their IPRs are enforced. What is true of developed countries is even truer of developing countries, where SMEs dominate the economy.[7] Developing countries that strengthen their IP laws, however, often lack qualified examiners to handle the volume of patent applications. They therefore accumulate large backlogs of unexamined applications, create legal uncertainties, and generate concerns about the quality of the patents awarded (OECD 2007a).

The literature on the links between stronger IPRs, investment flows, research and development (R&D) spending, and technology transfer is inconclusive. While some studies find a positive influence on FDI or licensing decisions by multinational corporations, others find no relation between the level of IP protection and FDI or R&D spending.[8] Between these extremes, recent studies seem to suggest that the effects of a stronger IPR regime depend on the country's level of development, the technological nature of the economic activities involved, and the absorptive capacity of individual firms (UNCTAD and ICTSD 2003; World Bank 2008).

Such findings suggest that the strength of IPR protection should evolve in line with local technological capabilities. This is not to say that developing countries should not protect intellectual property, but they should perhaps focus on stimulating adaptation by domestic enterprises, for example, through the use of utility models, industrial design, and compulsory licensing.[9] Soft IPR protection alone will not suffice to raise technological abilities without complementary policies in education and R&D. Encouraging technology transfer on generous terms, rather than trying to promote domestic innovation by making strong legal rights available to all, might best achieve technological capacity building in the early stages of development. In the evolving international regulatory regime, however, emerging economies appear to have little opportunity for instituting IP policies that support their development goals (UNCTAD and ICTSD 2003).

Domestic Institutional and Regulatory Framework

Although innovation in developing countries comes mainly from technology transfer, the general business environment can also foster a climate conducive to innovation. In the first place, it determines a country's attractiveness for foreign investment in comparison with other potential locations. And in the second, it influences the country's ability to benefit from technology transfer; to learn, adapt, and disseminate innovations; and to maximize technological spillovers.

Rules and regulations that apply specifically to foreign companies can either attract or discourage FDI. Many countries still require transnational corporations to obtain a number of permits and licenses to invest and operate. This requirement lengthens the approval process (UNCTAD 1999). Most governments have by now, however, gradually made entry, establishment, and operations of foreign companies easier, notably by reducing sectoral restrictions on FDI and opening up privatization programs; removing foreign equity participation restrictions, compulsory joint ventures, or local-content requirements; replacing screening and authorization requirements by simple registration; loosening restrictions on foreign ownership and rules governing the nationality of board membership and management; relaxing some types of operational restrictions (such as limits on the entry of professional and managerial personnel); guaranteeing legal protection, national treatment, fair and equitable treatment, and most-favored nation status; and establishing bilateral treaties for the promotion and protection of FDI and treaties for the avoidance of double taxation (UNCTAD 1999; OECD 2006).

More generally, the investment climate affects both local and foreign firms and their ability to generate knowledge transfer and innovation. The quality of regulation and its enforcement are recognized as critical determinants of the capacity of new and innovative firms to grow and expand. Restrictions on firm entry, exit, and activities can impede technological progress by propping up inefficient firms and limiting the expansion and creation of innovative ones. An inadequate regulatory environment inhibits business development in general but affects smaller firms even more. A review of the regulatory burden in Australia, for example, indicated that compliance matters can consume up to 25 percent of the time of senior management and boards of large companies (World Bank 2004; Regulation Taskforce 2006). Such regulations can stifle innovation and crowd out productive activity, especially since small businesses have to spread the fixed costs of compliance over a smaller revenue base and often lack the necessary in-house resources or expertise.

Many issues affect a country's investment climate. They range from firm start-up to business closure, from competition to access to land and credit, from customs practices to business setup procedures (table 4.1).[10] This section highlights some of the main regulatory obstacles linked to the innovation agenda: competition policies as well as trade-related issues as they affect business creation, the movement of goods, and technology transfer. It gives some examples of how countries have successfully overcome them.

Table 4.1 Example of Transaction Costs Related to the Legal and Regulatory Environment

Area of operation	Transaction	Enterprise exposure	Effects on
Business entry	Registration, licensing property rights, rules, clarity, predictability, enforcement, conflict resolution	Monetary costs to firm, time costs (including compliance and delays), facilitation costs, expert evaluations of rules and their functioning, number of rules and formalities	Rate of new business entry, distribution of firms by size, age, activity, size of shadow economy, rate of domestic investment, FDI inflows, quantity and quality, investment in R&D
Business operation	Taxation, trade-related regulation, labor hiring/firing, contracting, logistics, rules, clarity, predictability, enforcement, conflict resolution	Cost of compliance, higher costs of operation, costs of conflicts and conflict resolution, search costs and delays, insufficient managerial control, "nuisance" value, problems in making contracts, problems in delivery	Business productivity, export growth, size of shadow economy, growth of industries with specific assets or long-term contracting, rate of innovation and R&D, rate of business expansion, rate of investment in new equipment, subcontracting
Business exit	Bankruptcy, liquidation, severance/layoffs, rules, clarity, predictability, enforcement, conflict resolution	Rate of change of rules, changes in costs and number of rules, availability of rules and documents to firms, rates of compliance or evasion, use of alternatives to formal institutions	Rate of exit (and entry), prevalence of credit, distribution of profitability of corporations

Source: World Bank 2006.

Barriers to Entry and Competition Policy

The general environment for competition influences both the intensity of innovation efforts and the pace at which innovations spread to the market. Low levels of competition and regulations restraining competition in the product market have an adverse effect on productivity growth, while sluggish competition among suppliers may increase the cost of inputs, slow the adoption of best-practice production techniques, retard the diffusion of new technology by discouraging investment in equipment that embodies the latest technology, and reduce the diffusion of technology from abroad through FDI. It may also hinder the competitiveness of other companies or industries when they provide intermediate outputs (OECD 2007b).

Benefits of Competition. Policies that encourage or intensify competition in product markets may instead have positive effects on innovation as firms strive to adapt to competition, changing situations, and new market opportunities to stay ahead of competitors or to differentiate their products as they target different market segments. The competition policy regime may also encourage enterprises, local and foreign, to invest in developing local capabilities. In general, the more competitive and outward looking the regime, the more dynamic this process will be. It may prompt companies to move toward international standards while providing them with access to new markets. According to the *World Development Report 2005* (World Bank 2004), "Firms facing strong competitive pressures are at least 50 percent more likely to innovate than those reporting no such pressure."

Greater competition may be achieved in various ways: elimination of state-owned and legal monopolies, barriers to entry and exit such as unnecessary licenses (see box 4.3), and other interventions into commercial decisions such as price controls (Jacobs and Astrakhan 2006). Areas requiring close attention include abuse of dominant market positions, mergers (to assess effects on competition and potential market dominance[11]), horizontal price-fixing agreements (cartels), vertical agreements on resale prices, and restrictions such as exclusive dealing or territorial assignments (OECD 2007b).

Crowding out may result from heightened competition. It can be positive if it increases the efficiency of local firms and forces inefficient ones to exit. It may be negative if it affects potentially efficient domestic enterprises (infant industry considerations). Distinguishing between sound competition and crowding out is not easy, however, and inappropriate restrictions can result in technological lags. A highly protected regime, or a regime with stringent constraints on local entry and exit, discourages technological upgrading and isolates the economy from international trends. In India for example, the production of more than 600 manufacturing products is still reserved to small-scale companies in the ill-founded belief that it is good for employment. In fact, this regime has cost India many jobs, for example, by preventing it from

> **Box 4.3 Kenya's Radical Licensing Reform, 2005–07**
>
> In 2005, the government of Kenya launched a reform to reduce the growing number of business licenses and fees and the related corruption. Moving beyond previous strategies based on reforming licenses one at a time, the government adopted a broad "guillotine approach" to rapidly identify, review, and streamline all business licenses and associated fees. A central reform committee was created under the authority of the Ministry of Finance, and a government-wide program began. The first task was to assemble Kenya's first complete inventory of licenses and fees. Ultimately, 1,325 business licenses and fees imposed by more than 60 government agencies and 175 local governments were identified, far more than originally expected. Moreover, regulators continually imposed new licenses. Many were found to be unneeded, illegal, or unnecessarily costly. One reason for the growing problem was that the ministries and regulatory bodies, including local agencies, had a direct financial interest in creating new licenses and business fees because these revenues support staff salaries and expand opportunities for corruption.
>
> Once identified, licenses were rapidly reviewed against clear criteria by a neutral body to ensure consistency and quality across the government. The burden of proof was on the regulators to show why a license had to be maintained. As a condition of maintaining their requirements, regulators had to demonstrate that they were acting in the public interest.
>
> At the end of the process, any license that was not successfully justified as legal and needed for future policy needs for market-led development was eliminated, and any license that was needed but not business friendly was to be simplified to the extent possible. As of October 2007, 315 licenses had been eliminated and 379 simplified. A total of 294 were retained. Of the remaining licenses, approximately 300 have been deferred because new bills were under preparation or new laws had already passed; 25 were reclassified and not counted as a license.
>
> *Source:* Jacobs and Astrakhan 2006; Jacobs, Ladegaard, and Musau 2007.

being competitive with China in the apparel sector (Palmade 2005). To date, only a handful of countries have managed to support and strengthen indigenous technologies by sheltering them from competition.

Detrimental Effects of Competition. According to some, competition may be detrimental to innovation, owing to the reduction of monopoly profits that would reward successful innovators (the idea being that the prospect of high profits may stimulate entry). Others claim that competitive pressures enhance efforts to innovate and to diffuse innovation.[12] Striking the right balance between protecting innovators' efforts too much and protecting them too little creates incentives to innovate and ensures competition.

Network Industries. Some sectors and state-owned companies, such as network industries (telecommunications, electricity, air and rail transport), are typically excluded from the competition regime on the grounds of consumer protection, security of supply, or universal service provision. Yet striking the right balance between regulation and competition in these sectors is also important. Indeed, it is crucial for the successful diffusion and implementation of technologies and for developing domestic competencies (World Bank 2008). For example, because of electricity's importance as an intermediate input, the reliability of electrical supply may be even more important for technology diffusion than its availability, as many machines are sensitive to the quality of electrical power and many processes are intolerant of interruptions.[13] Likewise, well-developed air transport and road networks are essential for linking producers to markets and thus for the diffusion and widespread adoption of technologies.[14] Information and communication networks are also positively correlated with the uptake and diffusion of innovation (introduction of new products, services, and business processes and applications) (OECD 2007a). Other recent studies suggest that the removal of anticompetitive regulations that impede the unbundling of information and communications technology software from hardware, the breakup of telecommunications monopolies, and the removal of restrictions on entry in parcel delivery or air transportation have often spurred major waves of innovation (OECD 2007b).

Countries in which older technologies have yet to penetrate deeply may also face limitations on the extent to which other technologies diffuse. Authorities therefore need to focus on ensuring that publicly supplied technological services are available as widely, reliably, and economically as possible, whether they are delivered directly by the state or by private firms (World Bank 2008).

Regulation of some segments of network industries is necessary to prevent monopoly abuse, but competition should be possible in others. For example, securing nondiscriminatory third-party access to the network is crucial to inducing competition in the competitive segments of network industries (OECD 2007b). The challenge is to ensure a level playing field between state-controlled enterprises and private firms, on the one hand, and between domestic and foreign firms, on the other. In addition, the right incentives should be in place for investment in network industries in a more market-based environment, especially as capacity expansion may not be in the interest of a network owner if expansion undermines its capacity to charge high prices, if parts of the network are franchised, or if the franchising period is relatively short (see box 4.4). As universal service obligations in network industries in more competitive markets can no longer be financed through traditional cross-subsidization from profitable market segments, appropriate price regulation can, in principle, help stimulate investment in new capacity by ensuring adequate rewards.

Changes in the regulatory environment and in the nature of technologies partly explain the acceleration in the rate at which they penetrate developing

> **Box 4.4 Railways and Competition**
>
> A certain degree of unbundling of vertically integrated railway companies is desirable to encourage competition, but reform must be carefully designed to take account of country-specific characteristics (such as possibilities for competition on parallel tracks and competition from other modes of long-distance transportation) to avoid regulatory failure. Efficiency gains in the sector have been achieved in Australia, Denmark, Italy, and Switzerland, for example, through reduced regulatory restrictions, notably by lowering entry barriers, or in Denmark, Germany, Italy, and the Netherlands by improving market structures, especially in the freight business. Entry of alternative providers was made possible in Denmark, Finland, France, Italy, Germany, Hungary, Norway, and Sweden through accounting or legal separation of the network. Deregulation of the railway industry, however, is controversial owing to the unresolved question of how to provide market-based investment incentives in the network segment of the industry. In particular, regulatory authorities in the United Kingdom faced this problem after privatizing the rail sector, because of the lack of clear assignment of responsibility for investing in tracks and the lack of incentives to invest in rolling stock, partly owing to the short duration of franchise contracts.
>
> *Source:* OECD 2007b.

countries. Many old infrastructure technologies, such as roads, railroads, sanitation, and fixed-line telephone systems, are often provided by the government and are subject to public sector budget constraints and the risk of government failure. By contrast, the most common new technologies, such as the Internet, mobile phones, and computers, are delivered in a regulatory environment that encourages competition and harnesses private capital (domestic and foreign) to provide basic infrastructure. The example of the diffusion of telecommunications technology in Africa illustrates this point:

> About one-half of all low-income countries have opened their telecommunications markets to competition, leading to growing markets, lower costs, greater innovation, and customized services for different groups of users.... Ten years ago one million phones were available in all of Africa; now there are well over 100 million, mainly mobile. In addition Internet use has also grown rapidly; the number of users increased by more than four-fold between 2000 and 2005. (WBI 2007)

Moreover, the past 10 years have been more politically stable than the 1980s and 1990s, which has likely boosted the diffusion of newer technologies (World Bank 2008).

Movement of Goods across Borders

Cross-border trade is also a significant conduit for knowledge and technology transfer. For most businesses, speed of delivery of goods, predictability, and transparency throughout the process are of paramount importance. The ease of trading across borders thus affects decisions on whether to operate in a given country. Bureaucratic processes, corruption, and unofficial payments prevent the smooth movement of goods across borders and keep businesses from efficiently trading in international markets.

Failure to meet the requirements of government agencies frequently causes delays, while the regulatory prerogatives of the border control agencies that deal with agricultural, veterinary, health, phytosanitary, and standards requirements, in addition to basic customs procedures, often lead to duplication of requirements and controls. These overlaps increase compliance costs, risks of error, and delays (IFC 2006). Governments can take several specific measures to minimize the incidence of customs interventions and speed up control processes:

- Eliminate, simplify, and streamline complex data and documentary requirements, work and paper flows, procedures, and controls
- Minimize and rationalize nontariff regulations[15]
- Ensure that proposed reforms are in full compliance with international customs conventions, recommended practices, and agreed standards.[16]

One way to reduce such delays is to authorize the release of goods before all controls have been imposed, while ensuring that the release may take place at the facility at which the goods are stored. Another is to implement "single windows" or "one-stop shops" for import and export formalities involving all border agencies. This arrangement minimizes reporting and clearance processes by eliminating or combining procedural steps from all border agencies involved. In the same vein, conducting joint inspections helps reduce delays, while mutual recognition of inspections from the exporting and the importing countries helps ensure that a single inspection suffices. The use of risk-management techniques can also reduce the number of physical inspections and delays. Authorizing prefiling of customs documents before arrival and the use of information and communications technology systems so that data requirements can be exchanged wherever possible in advance of cargo arrival are yet other ways to smooth the process (IFC 2006). Finally, ensuring that customs laws, regulations, and requirements are easily accessible and applied uniformly and consistently helps fight corruption.

Other initiatives may help modernize and transform customs administration into more efficient service providers. For example, customs services may offer highly compliant importers and exporters payment deferral regimes, release of goods upon presentation of a simplified declaration (with the full declaration presented at a later time), and a lower level of physical examinations of consignments.[17] Other measures enable manufacturers to import materials without paying the applicable duty or tax until such materials are re-exported as components of finished goods. In addition, the tariff burden on certain imports may be removed or reduced so that exporters gain access to industrial inputs at world prices, thus making their exports more competitive (IFC 2006).

Removal of Obstacles

Reform can remove regulatory, bureaucratic, and legal obstacles to innovation in various ways:

- Identification and sequencing of important reforms that are credible and feasible yet achieve substantial results
- Strategies to mobilize support and get reform on the policy agenda and to mitigate and eventually overcome opposition from interest groups
- Creation of incentives and capacity for implementation and institutional mechanisms to ensure implementation and sustain reform (Kikeri, Kenyon, and Palmade 2006; Jacobs and Astrakhan 2006).

Steps in Reform. Some reforms require little political negotiation or legislative change. Most of the constraints linked to bureaucracy and red tape can be overcome by simplifying procedures. Modern technologies such as the Internet also help simplify procedures and speed up processes. They can also increase transparency and limit the potential for corruption. For example, publishing rules and regulations may help limit options for corruption through the imposition of unofficial requirements. Cases of corrupt judges being caught and punished can be publicized.

Successful regulatory reform processes generally include the following:

- Strong political leadership seems to be chief among the factors explaining successful reform processes. Once high-level political commitment and leadership are ensured, a number of factors—building on previous successful reforms,[18] spillovers from trade (for example, by becoming a WTO member), new information (such as international benchmarking, indicators, and cross-country comparisons),[19] times of crisis, or pilots[20]—can put reform on the agenda.

- Successful reform processes can also benefit from an independent cross-jurisdictional unit to ensure that the process is inclusive and ongoing and that reforms are seen as independent from entrenched interests. If located at arm's length from the president, prime minister, or ministry of finance, for example, the unit should have clear authority and be able to provide leverage for ensuring the cooperation of other parts of the administration, as well as coherence with the budget cycle.

- Identification of priorities for reforms and the appropriate sequencing of them are critical yet challenging steps in the reform process. Reformers, with a long list of constraints and potential reforms, are faced with the arduous task of identifying the reforms that will trigger support and momentum.

Fortunately, a growing set of diagnostic tools and information can help identify priorities. Benchmarking indicators,[21] country rankings, business surveys,[22] industry-specific analyses, and consultations with stakeholders can help identify key constraints in a country's investment climate and therefore help target the priorities for reform (Kikeri, Kenyon, and Palmade 2006; Ladegaard, Djankov, and McLiesh 2007).

- Transparency, communication, and extensive consultation with stakeholders not only help identify top priority areas or regulations but also trigger interest, generate support, and reduce resistance to change. Building coalitions to support reform is crucial. Stakeholder engagement and public participation should help identify supporters and then leverage and empower them to become "champions of reform." Opposition can be reduced through dialogue, consultation, and, where appropriate, compensation.

- The reform process should include a provision to ensure that the underlying causes of regulatory problems are dealt with and that reregulation does not annihilate its achievements. Sound (re-)regulation may be achieved by putting in place regulatory impact assessment of new regulation, cost-benefit analysis of options for assessing new laws, cost of compliance assessments, or consultations in the process of developing regulations.

Pace of Reform. There are two prevailing views on the pace of reform. According to the incremental approach, governments should proceed by targeting a few regulatory constraints at a time, hoping to achieve quick wins and thus build gradual reforms and momentum on the basis of these first successes (OECD 2007b). Proponents of this view believe that broader and bolder reforms are not possible, given the resources available and the strength of resistance to change. Yet, to critics, "small reforms to big and expanding regulatory systems will not substantially or sustainably improve the business environment. Reforms aimed at single processes and rules will never catch up with the productive capacities and incentives of governments to create regulations and controls" (Jacobs, Ladegaard, and Musau 2007). Incremental or partial reforms can be risky if they produce little in the way of results—or even produce adverse effects and thus undermine the credibility of the entire reform process.[23]

Supporters of the alternative view, therefore, believe that radical solutions to improving the regulatory environment, like the guillotine approach,[24] work better than small reforms. While for tactical reasons the government might start with small, manageable reforms that can be accomplished rapidly, the end result should remain in focus to keep reform moving in the right direction and to reassure investors (Jacobs, Ladegaard, and Musau 2007). Finally, initiatives that obtain visible results quickly can help, especially in removing regulatory and legal obstacles to innovation.

Procurement Policies for Innovation

Aside from reforms to remove legal and regulatory obstacles, governments can also take proactive steps to encourage innovation. Because innovation is traditionally believed to come from the supply side, proactive innovation policies generally aim at supporting product or service providers through targeted grants, fiscal incentives, or equity support. While demand was overlooked until recently, it is also a major potential source of innovation.

Indeed, in a recent survey of more than 1,000 firms and 125 federations, over 50 percent of respondents indicated that new requirements and demand are the main source of innovation. Illustrations of demand-driven innovations come from a variety of sources, from firms targeting bottom-of-the-pyramid consumers to public authorities using procurement policies to stimulate innovation.[25] While supply-side measures frequently support innovation, demand-side policies can also generate innovations by increasing demand, defining new functional requirements for products and services, and articulating needs more clearly (Edler and Georghiou 2007).

Innovation-Friendly Procurement Policies

Public procurement is one way to drive the demand for innovative solutions, goods, or services, while improving the delivery of public services. It is demand-side policy that is now gaining momentum among policy makers: "An analysis of the Sfinno database collecting all innovations commercialized in Finland during 1984 and 1998," write Edler and Georghiou, "shows that 48 percent of the projects leading to successful innovation were triggered by public procurement or regulation" (2007). Recent reports commissioned by the European Union also emphasize the importance that public procurement policies can have for encouraging innovation (European Commission 2008). They specifically identify several application areas: e-health, pharmaceuticals, energy, environment, transportation and logistics, security, and digital content. As Edquist, Hommen, and Tsipouri (2000) observe:

> A public agency acts to purchase, or place an order for, a product—service, good or system—that does not yet exist, but which could probably be developed within a reasonable period of time, based on additional or new innovative work by the organizations(s) undertaking to produce, supply, and sell the product being purchased.

Since procurement is spread over a wide range of actors and contracting authorities, figures are not easily calculated. It is estimated, however, that the U.S. public sector spends US$50 billion per year on R&D procurement (European Commission 2007b); that public procurement in Europe represents 17 percent of EU-25[26] GDP and 35 percent of EU-25 public expenditure (European Commission 2007c); and that the magnitude of central government purchases ranges from 9 percent to 13 percent of GDP for the Middle East and Africa. Such figures suggest that public procurement can offer substantial market potential for innovation, first, because the state is frequently more willing or more able to pay the higher prices typically asked at the introduction of innovations and, second, because state demand often rapidly achieves critical mass, in particular by bundling the demand generated by various government agencies and bodies. The concentration of public demand brought about by such coordination creates clear incentives for suppliers and reduces their market risk (Fraunhofer 2005).

Three Types of Innovation Procurement Policies
Public authorities stimulate innovation in three main ways: (a) the first occurs through the public procurement of innovative goods and services when government purchasers specifically look for innovative or alternative solutions to meet their needs and thus enhance public service delivery; (b) the second takes place when public entities procure for goods or services for which R&D still needs to be done and is referred to as *precommercial procurement* or *technology procurement*; and (c) the third, *catalytic procurement*, occurs when the government acts as launch customer for goods intended to be diffused more widely (Georghiou 2007).

Procurement of Innovative Goods and Services. In the first instance, innovative solutions can be promoted by using clear and robust output specifications and by setting functional or performance criteria, thereby leaving tendering companies room to propose solutions. Another way to encourage innovative solutions is to hold project-based competition and design contests. According to a European Commission study (2007a),

> A design contest can be a powerful means of developing and testing new ideas. It gives firms room to come up with solutions, making optimum use of the market's creativity. Contracting authorities can award the contract directly to whoever comes up with the best idea. This makes it attractive for companies to bring their innovative ideas forward.

The advantages of such an approach are manifold: it helps improve the quality and performance of public services by ensuring that they are dynamically updated and upgraded; it stimulates private innovation by creating strong incentives to maximize the efficiency and performance of the products and services offered; it creates a market for innovative solutions and products that may otherwise not exist; and, finally, this one-time market, by example, can then trigger new demand by the private sector and eventually open up additional market opportunities (see box 4.5).

Precommercial Procurement. The objective of precommercial procurement is to create innovative solutions in areas for which solutions are not currently

> **Box 4.5 Variable Message Signs for British Highways**
>
> The English Highway Agency tendered for the development and installation of new variable message signs on motorways in 2001. The signs were to provide information to drivers on advisable speed, lane availability, and the like. The existing signs had very limited flexibility in the messages they could display.
>
> Contrary to earlier tenders, the agency used an output specification and allowed for the application of new technology in the proposed solutions. The use of an output specification allowed suppliers to continue to develop their product. The result was a sign of a type not previously seen, capable of generating graphics as well as text. As a result, the Highway Agency acquired a good and innovative product. The company went on to win a Queen's Award for Enterprise in Innovation and sold to new markets in the Netherlands and the Russian Federation.
>
> *Source:* European Commission 2007a.

available.[27] According to a report by VINNOVA (2007), the Swedish innovation agency, precommercial procurement

> requires the contracting authority to be aware of its long-term needs. The authority also needs the skills to conduct a development process that involves several possible suppliers, to ensure that one or more of the finished solutions can match the functional requirements of the authority.

Technological innovations such as the Internet Protocol or the Global Positioning System (GPS) were developed in this way (European Commission 2007b). The United States, Japan, and the Republic of Korea use precommercial procurement as a strategic tool for creating a strong domestic economy for domestic suppliers in areas of national strength (VINNOVA 2007). For example, the United States and Japan have significantly reduced the cost of fuel cell stations through R&D procurement, enabling buses powered by fuel cells to become a viable energy-efficient public transportation option. China's last national long-range science and technology plan officially introduced public technology procurement in China as a way to encourage innovation (European Commission 2007c).

In practical terms precommercial procurement is in fact an R&D service contract, given to a future supplier in a multistage process, from exploration and feasibility to R&D up to prototyping, field tests with first batches, and finally commercialization (Edler and Georghiou 2007). Because the product or service does not yet exist, the risks of procuring such innovations is inherently higher. To reduce the R&D risks and costs associated with precommercial procurement, one can split the process into different phases and spread it over time, with constant competition to create a range of options (figure 4.1). In an exploratory phase, a selection is made among competing suppliers that have submitted proposals for possible solutions. A prototype phase follows, in which

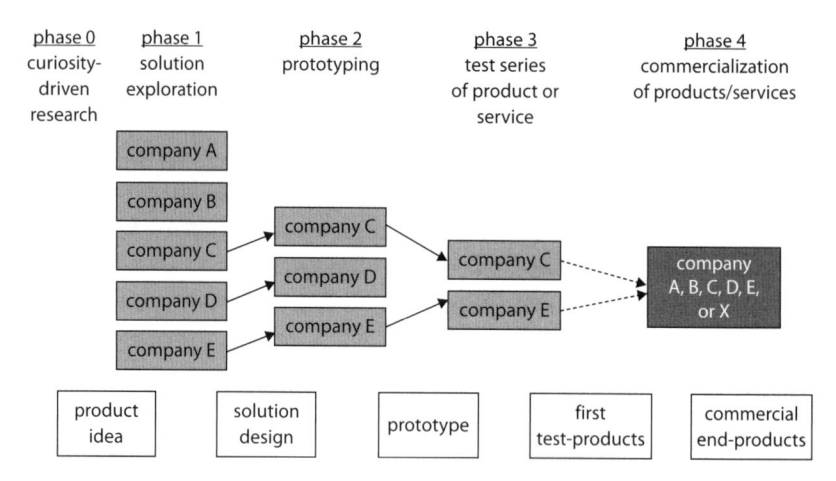

Figure 4.1 Example of a Phased Precommercial Procurement Process

phase 0	phase 1	phase 2	phase 3	phase 4
curiosity-driven research	solution exploration	prototyping	test series of product or service	commercialization of products/services

Phase 1: company A, company B, company C, company D, company E
Phase 2: company C, company D, company E
Phase 3: company C, company E
Phase 4: company A, B, C, D, E, or X

product idea → solution design → prototype → first test-products → commercial end-products

Source: European Commission 2007c.

selected suppliers are offered the opportunity to develop their prototypes (see box 4.6). These are evaluated step by step, and the number of competing suppliers is reduced. In the final phase, at least two suppliers should remain to secure future competition in the market (European Commission 2006).

This type of procurement enables public purchasers to filter out technological R&D risks and to identify the best possible solution the market has to offer before committing to a large-scale commercial rollout. For developing countries, it could be a way to test adaptation of solutions to the local context and conditions rather than adopting an "off-the-shelf" solution that may have been developed for a different context. Precommercial procurement, for example, may increase the chances of success for provision of e-government services or for a railway construction and maintenance system adapted to a Sub-Saharan context.

The advantages of precommercial procurement include sharing the risks and benefits of designing, prototyping, and testing new products and services with suppliers, without involving state aid. In addition, testing prototype products in an operational customer environment enables public purchasers to align product development with customer priorities and to select progressively the solutions that best fit public sector needs. Better anticipation of demand for new solutions shortens time to market for suppliers and helps public authorities introduce new solutions faster. It also enables public authorities to detect potential policy and regulatory issues that need to be addressed earlier to ensure timely introduction of the new solutions into public services and other markets (European Commission 2007a, 2007c).

Catalytic Procurement. Finally, in catalytic procurement, procurement is conducted on behalf of end users other than the public authority, as in the case of

> **Box 4.6 The Swedish Energy Agency's Procurement Procedures**
>
> The Energy Agency has developed a systematic procedure for technology procurement in seven phases. Initially, a feasibility study is conducted to investigate the market and determine the potential for improvement. Then, user and buyer groups are formed. These groups formulate the requirements for the product or system, which are developed into specifications. The tendering phase follows, in which manufacturers that seem to meet the requirements are allowed a period to develop a prototype, which is then evaluated and tested. One or several manufacturers can be named as winners. In certain cases, the Energy Agency pays an investment subsidy to the first buyers to stimulate interest. The group of users and buyers and the manufacturers pass information on the technology procurement to others to create demand for the new technology from more buyers. Many products and systems will continue to need further development after procurement, and those manufacturers in particular that did not fulfill all requirements may need to improve their products to keep up with developments. In the great majority of cases, the technology procurement process results in more efficient solutions.
>
> *Source:* VINNOVA 2007.
> *Note:* While this example is a good illustration of the three-step precommercial procurement process, according to EU rules, this is not considered procurement, but support to individual enterprises.

market transformation programs in the energy sector in the 1990s. Such programs involved, for instance, the procurement of energy-efficient home appliances, the main end users of which would not be public sector organizations but private individuals and households. Such policy schemes may aim, for example, to accelerate the diffusion of energy-efficient technologies by aggregating demand and initiating a technology procurement process.

If procurement is to permit innovative solutions, the evaluation criteria should shift from the traditional focus on price (the lowest-price bid) to one on solutions that offer the greatest advantage to users over the whole life of the purchase. Innovations are sometimes more costly, especially initially. To encourage innovative procurement, policies need to take into account the full life-cycle costs of the products or services and adopt most economically advantageous tender criteria rather than lowest-cost criteria for the awarding of bids (Edler and Georghiou 2007). Tendering for new lighting equipment and equipping a whole building with new low-energy light bulbs, for example, would create higher upfront cost but much lower running costs. Apart from price considerations, public purchasers may include a range of other criteria, such as running costs, lifetime maintenance costs, patterns and intensity, and potential downtime, among others.

Demand-driven procurement should not be seen as a replacement but as a complement to supply-side innovation policies. The role that public procurement can play with regard to innovation relates to its importance in public spending, its ability to provide incentives to innovate while seeking to improve public services, its potential for tailoring solutions to the local context, and, more generally, as a way to stimulate a culture of research and innovation.

Conclusions

All bureaucratic, legislative, and regulatory rules that directly or indirectly support or impede trade, investment (foreign and domestic), and business setup, running, and closure may subsequently support or impede innovation. It is especially important to improve the business climate for innovation, given that business is the principal impetus behind it. The OECD sums up critical elements of improving the conditions for innovation: "More innovation-friendly regulation, combined with lower barriers to trade and foreign direct investment would enhance competition and would foster the flow of technology and knowledge across borders" (OECD 2007a). It is widely recognized, however, that a supportive regulatory framework will not in itself suffice to promote innovation if science education and other policies are not well designed.

Finally, when an innovation-friendly regulatory strategy has been devised, implementing and enforcing reform to sustain it will be crucial but difficult. As an OECD study sums it up,

> Some of the required reforms may affect vested interests, such as in universities and scientific institutions, as well as business sheltered from competition, benefiting from public support or confronted by technology-induced structural change. Strong political leadership and efforts to develop a clear understanding by the various stakeholders of the problems and of the solutions—including the costs they involve—can all help to communicate the need for reform and foster acceptance. (2007a)

Thus, tackling such obstacles requires systematic audits, inspired, for instance, by the Investment Climate Surveys of the World Bank. Such audits should then be followed by sustained actions to ensure that the obstacles identified are duly reduced or removed, which in turn implies a somewhat functional and independent judiciary system.

Notes

1. Tariff peaks are defined as tariff rates above 12 percent ad valorem (UNCTAD and WTO 2000).

2. According to Watkins (2003), average EU tariffs on fully processed foods are twice as high as on products in the first stage of processing.

3. According to World Bank (2007), producer support in member countries of the OECD still represents 30 percent of the gross value of farm receipts in 2003–05.

4. The issue raised by current international patent regimes and their impact on R&D in developing countries is discussed more in depth in chapter 5 ("Strengthening the R&D Base").

5. The cost of software is a major problem for developing countries and the reason for the high level of illicit copying. Copyright can also be a barrier to the further development of software to meet local needs and requirements (WBI 2007).

6. Developing country governments may use compulsory licenses to grant authorization to a third party to exploit a patented invention, generally against remuneration to the patent holder or parallel imports of patented products when they are obtainable in a foreign country (where a patent also exists) at lower prices. They may also establish *exceptions* to the exclusive rights, such as the *early working exception* (also known as the "Bolar exception"), which allows generic firms to initiate and obtain marketing approval of a patented drug before the expiration of the patent (UNCTAD and ICTSD 2003; WBI 2007). In addition, the "use of a patent's subject matter under compulsory licensing is permitted under TRIPS (Trade-Related Aspects of Intellectual Property Rights) agreement even without prior negotiation 'in the case of a national emergency or other circumstances of extreme urgency' or in cases of public non-commercial use, and should be 'predominantly for the supply of the domestic market.'"

7. A survey of SMEs in the United Kingdom found that about half did not apply for patents even on inventions they thought were patentable. And of those that did patent an invention, 87 percent would have developed the invention even without a patent. See Macdonald, Turpin, and Ancog (2005).

8. For detailed references, see World Bank (2008), UNCTAD and ICTSD (2003), and OECD (2006).

9. Japan permitted compulsory licensing when the patent had not been worked continuously in Japan for more than three years or for public interest reasons (Kumar 2002).

10. See the *Doing Business* publications, www.doingbusiness.org.

11. While the European Union, for example, sets thresholds for market shares and concentration ratios of the merged entity above which competition is potentially at risk and therefore further investigation is needed, Canada, New Zealand, and the United States accept mergers that strengthen a dominant position as long as there are no barriers to entry and the merger results in efficiency gains (OECD 2007b).

12. The empirical evidence tends to favor the positive effect of competition on innovation. However, the impact of competition may depend on how far a country or an industry is from the technology frontier. Competition may be more important at the technological frontier both because it stimulates entry and forces firms to innovate to survive. According to OECD (2007b), however, competition has particularly powerful effects on productivity in countries far from the technological frontier, owing to stronger incentives to adopt new technologies.

13. In Bangladesh, for example, where transmission and distribution losses represent only 9 percent of produced power, some 70 percent of managers indicate that unreliable power is a serious constraint on business (see World Bank 2008).

14. A recent study estimates that trade among West African countries could expand by up to 400 percent on average if the road network were upgraded. Similar investment could increase trade in southern Africa by up to 300 percent, and several times more for some countries (World Bank 2006). Likewise, investment in transport infrastructure has allowed Brazil's interior states to enter global markets for soybeans and other crops, whereas rice and maize, usually tradable commodities, are effectively nontradable in rural areas of Madagascar and Ethiopia, respectively, because of high transportation costs (World Bank 2007).

15. For example, using a single, standardized document format and content for multiple agency reporting purposes and customs regimes may help facilitate and simplify preparation and minimize opportunities for errors during transcription.

16. For example, the World Customs Organization's *Revised Kyoto Convention on the Simplification and Harmonization of Customs Procedures* sets out internationally accepted best practices, recommendations, and standards governing customs import and export procedures and controls.

17. See, for example, IFC (2006) and Regulation Taskforce (2006).

18. In Kenya, for example, the approach, competencies, and support developed in the licensing reform have helped expand efforts to improve the capacities of regulatory institutions (building up skills for

regulatory impact analysis, regulatory quality control) and to reduce red tape costs by a further 25 percent by 2010 (Jacobs, Ladegaard, and Musau, 2007).

19. See Kikeri, Kenyon, and Palmade 2006.

20. Pilots may provide important learning, a testing ground, and demonstration for larger reforms, especially when there is uncertainty or strong opposition. China put pilots at the center of its reform strategy, using special economic zones to test market-oriented policies such as land use rights before extending them nationwide. Jordan, Peru, and South Africa also used pilots to learn about potential difficulties and to assess the feasibility and effectiveness of reform programs in land registration and customs (Kikeri, Kenyon, and Palmade 2006).

21. *Doing Business* surveys benchmark and rank the cost and quality of business regulations for key cross-cutting investment climate issues.

22. Annual business surveys that ask entrepreneurs to identify the top 10–20 regulatory burdens they face help reveal annoyance factors.

23. Before adopting a bold land reform program, Mozambique first took an incremental approach; Korea and the Slovak Republic did the same for regulatory reform. In these cases, the incremental reforms were unsuccessful and costly to taxpayers. Similarly, piecemeal inspections reforms in the Philippines and the Russian Federation were no more than short-term palliatives, and they quickly became victims of backtracking and reversals (see Kikeri, Kenyon, and Palmade 2006, 29–30).

24. "The guillotine . . . is a means of rapidly reviewing a large number of regulations, and eliminating those that are no longer needed without the need for lengthy and costly legal action on each regulation. . . . It is a quick scan process, and does not replace the more detailed reviews and revision that are needed for many regulations, and that can occur in later phases . . . . the guillotine should be seen as an entry point to implementation of reforms within a sustained strategy" (Jacobs and Astrakhan 2006). From the mid-1980s onward, the guillotine approach and variants have been used by countries as diverse as Hungary, Kenya, Korea, Mexico, Moldova, and Ukraine (see OECD 1999, 2001). Jacobs and Associates used these countries' experiences to develop a systematic, practical guillotine process that can be widely applied in different countries. The guillotine approach is a trademark of Jacobs and Associates.

25. For example, Prahalad (2004) shows that small innovations and adaptations to products, packaging, or the like sometimes are enough to help spread new products, services, or technologies among the poor and thus contribute to the diffusion of innovation.

26. The "EU-25" are the 25 countries that constituted the European Union in 2006.

27. This form of R&D procurement is called "precommercial" because it applies to areas in which there is no commercial offer (see European Commission 2007b).

References and Other Resources

Baffes, John. 2004. "Cotton, Market Setting, Trade Policies and Issues." World Bank Policy Research Working Paper 3218, World Bank, Washington, DC.

Chandra, Vandana. 2006. *Technology, Adaptation, and Exports: How Some Developing Countries Got It Right*. Washington, DC: World Bank.

Coriat, Benjamin, Fabienne Orsi, and Cristina d'Almeida. 2006. "TRIPS and the International Public Health Controversies: Issues and Challenges." *Industrial and Corporate Change* 15 (6): 1033–62.

Edler, Jakob, and Luke Georghiou. 2007. "Public Procurement and Innovation: Resurrecting the Demand Side." *Research Policy* 36: 949–63.

Edquist, Charles, Leif Hommen, and Lena J. Tsipouri, eds. 2000. *Public Technology Procurement and Innovation*. Boston, MA: Kluwer Academic Publishers.

European Commission. 2006. "Pre-Commercial Procurement: Public Sector Needs as a Driver of Innovation," Paper prepared by Research Directors Forum Working Group.

———. 2007a. "Guide on Dealing with Innovative Solutions in Public Procurement: 10 Elements of Good Practice," Commission Staff Working Document SEC (2007) 280.

———. 2007b. "Commission Advocates New Approach to Investing Public Money in Risky High-Tech Research." Press Release IP/07/1931. December 14. http://europa.eu/rapid/pressReleasesAction.do?reference=IP/07/1931&format=HTML&aged=0&language=EN&guiLanguage.

———. 2007c. "Pre-commercial Procurement: Driving Innovation to Ensure Sustainable High Quality Public Services in Europe." Communication from the Commission to the European Parliament, the Council, the European Economic and Social Committee and the Committee of the Regions. COM (2007) 799 Final. Brussels.

———. 2008. Information Society and Media, "Encouraging Innovation-Friendly Procurement in eHealth." http://ec.europa.eu/information_society/activities/health/docs/policy/encouraging-innov-friendly-procurement200804.pdf.

Fraunhofer Institute Systems and Innovation Research. 2005. "Innovation and Public Procurement: Review of Issues at Stake." Final Report for the European Commission, ENTR/03/24.

Georghiou, Luke. 2007. "Innovative Procurement: Why, What and How?" Presentation at the Malta Council for Science and Technology, "Innovative Procurement in Government Workshop," Malta, May 14.

IAASTD (International Assessment of Agricultural Knowledge, Science and Technology for Development). 2008. *Synthesis Report, Executive Summary.* http://www.agassessment.org/index.cfm?Page=IAASTD%20Reports& ItemID=2713.

IFC (International Finance Corporation). 2006. *Reforming the Regulatory Procedures for Import and Export: Guide for Practitioners.* Washington, DC: World Bank.

Islam, Roumeen, and Gianni Zanini. 2008. *World Trade Indicators: 2008 Benchmarking Policy and Performance.* Washington, DC: World Bank.

Jacobs, Scott, and Irina Astrakhan. 2006. "Effective and Sustainable Regulatory Reform: The Regulatory Guillotine in Three Transition and Developing Countries." Jacobs and Associates, Washington, DC.

Jacobs, Scott, Peter Ladegaard, and Ben Musau. 2007. "Kenya's Radical Licensing Reforms, 2005–2007: Design, Results, and Lessons Learned." Paper prepared for the Africa Regional Consultative Conference, "Creating Better Business Environments for Enterprise Development: African and Global Lessons for More Effective Donor Practices," Accra, November 5–7.

Keller, Wolfgang. 2004. "International Technology Diffusion." *Journal of Economic Literature* 42 (September): 752–82.

Kikeri, Sunita, Thomas Kenyon, and Vincent Palmade. 2006. *Reforming the Investment Climate, Lessons for Practitioners.* Washington, DC: World Bank.

Kim, Linsu. 2003. "Technology Transfer and Intellectual Property Rights: The Korean Experience." UNCTAD-ICTSD Project on IPRs and Sustainable Development, Issue Paper 2, International Centre for Trade and Sustainable Development, Geneva.

Kumar, Nagesh. 2002. "Intellectual Property Rights, Technology and Economic Development: Experiences of Asian Countries." Background Paper 1b, Commission on Intellectual Property Rights, London.

Ladegaard, Peter, Simeon Djankov, and Caralee McLiesh. 2007. "Review of the Dutch Administrative Burden Reduction Programme." Investment Climate Occasional Paper, World Bank, Washington, DC.

Lund, Francie, and Caroline Skinner. 2003. "The Investment Climate for the Informal Economy: A Case of Durban, South Africa." World Development Report 2005 Background Paper, World Bank, Washington, DC.

Macdonald, Stuart, Tim Turpin, and Amelia Ancog. 2005. "Maximizing the Contribution of IP Rights (IPRs) to SME Growth and Competitiveness." Final report, REPSF Project 03005.

OECD (Organisation for Economic Co-operation and Development). 1999. *Regulatory Reform in Hungary.* Paris: OECD.

———. 2001. *Regulatory Reform in Korea.* Paris: OECD.

———. 2005. *OECD Guiding Principles for Regulatory Quality and Performance.* Paris: OECD.

———. 2006. *Going for Growth 2006. Economic Policy Reforms.* Paris: OECD.

———. 2007a. *Innovation and Growth: Rationale for an Innovation Strategy.* Paris: OECD.

———. 2007b. *Going for Growth 2007: Economic Policy Reforms.* Paris: OECD.

Oyejide, T. Ademola. 2003. "Trade Reform for Economic Growth and Poverty Reduction." *Development Outreach* 5 (2): 4–5.

Palmade, Vincent. 2005. "Industry Level Analysis: The Way to Identify Binding Constraints to Growth." Policy Working Paper 3551, World Bank, Washington, DC.

Portugal-Perez, Alberto, and John S. Wilson. 2008. "Lowering Trade Costs for Development in Africa: A Summary Overview." Development Research Group Working Draft, World Bank, Washington, DC.

Prahalad, C. K. 2004. *The Fortune at the Bottom of the Pyramid: Eradicating Poverty through Profits.* Philadelphia: Wharton School Publishing.

Regulation Taskforce. 2006. "Rethinking Regulation: Report of the Taskforce on Reducing Regulatory Burdens on Business." Canberra, Australia.

Schiff, Maurice, and Yanling Wang. 2006. "North-South and South-South Trade-Related Technology Diffusion: An Industry-Level Analysis of Direct and Indirect Effects." *Canadian Journal of Economics* 39 (3): 831–44.

UNCTAD (United Nations Conference on Trade and Development). 1999. *World Investment Report 1999: Foreign Direct Investment and the Challenge of Development.* New York: United Nations.

UNCTAD and ICTSD (International Centre for Trade and Sustainable Development). 2003. "Intellectual Property Rights: Implications for Development." Intellectual Property Rights and Sustainable Development Series Policy Discussion Paper, ICTSD, Geneva.

UNCTAD and WTO (World Trade Organization). 2000. *The Post-Uruguay Round Tariff Environment for Developing Country Exports: Tariff Peaks and Tariff Escalation.* Geneva: UNCTAD and WTO.

VINNOVA. 2007. "Public Procurement as a Driver for Innovation and Change." Report on government commission to Nutek and VINNOVA. VINNOVA Policy VP 2007:03.

Watkins, Kevin. 2003. "Farm Fallacies That Hurt the Poor." *Development Outreach* 5 (2).

World Bank. 2004. *World Development Report 2005: A Better Investment Climate for Everyone.* New York: Oxford University Press.

———. 2006. *Doing Business 2006: How to Reform.* Washington, DC: World Bank.

———. 2007. *World Development Report 2008: Agriculture for Development.* Washington, DC: World Bank.

———. 2008. *Global Economic Prospects 2008: Technology Diffusion in the Developing World.* Washington, DC: World Bank.

WBI (World Bank Institute). 2007. *Building Knowledge Economies: Advanced Strategies for Development.* WBI Development Studies. Washington, DC: World Bank.

5

Strengthening the Research and Development Base

Research and development (R&D) are important not just for pushing back the frontiers of knowledge but also for keeping up with global trends, acquiring knowledge, adapting knowledge to local circumstances, and advancing knowledge. To put the R&D effort of developing countries into context, this chapter first looks at data on global R&D spending and the main actors. A recent trend, with important implications for developing countries, is the increasing internationalization of R&D in general and of R&D activities of multinational corporations (MNCs) in particular.

The chapter then turns to the broader context for the R&D effort in developing countries and the importance of competitive pressure, both domestic and foreign, for encouraging firms to focus on improving their technology. It also raises the complex issue of intellectual property rights. The following sections discuss in turn R&D by the public sector, the private sector, and universities, before turning to international R&D efforts. A brief summary concludes the chapter.

Global Overview of R&D

The global R&D effort in 2006 is estimated to be on the order of US$1 trillion in current purchasing power parity,[1] or almost 2 percent of world gross domestic product (GDP). Overall R&D spending has been increasing slightly faster

---

This chapter was prepared by Carl Dahlman.

than world GDP. In 2006, for example, the Organisation for Economic Co-operation and Development (OECD) total was US$817.8 billion, up from US$468.2 billion in 1995. The United States accounts for 41 percent of the OECD total, the European Union for 30 percent, and Japan for 17 percent. Japan spends more on R&D as a share of GDP than the United States or the EU-27 (that is, the the 27 countries that constitute the European Union).[2]

Developing countries have been increasing their R&D expenditures faster than OECD countries. Their share of the total increased from 11.7 percent in 1995 to 18.4 percent in 2005. This increase is due largely to the very rapid rise in China's expenditures, which grew at an annual average rate of 19 percent in real terms between 2001 and 2006. India and the Russian Federation have also increased their R&D spending. Figure 5.1 shows the evolution of R&D spending for the three main OECD regions and China, in absolute value and as a share of their respective GDP (OECD 2008a, 20–21). Although China started from a much lower base, it is approaching the EU-27 average and is becoming a major player in global R&D in both expenditures and R&D output.

Main R&D Actors

In the past, government was the main funder of R&D, largely because of the very large role of defense spending in the United States. With the end of the Cold War, however, U.S. government spending has declined, and the business sector is now the largest performer of R&D, accounting for 69 percent in the

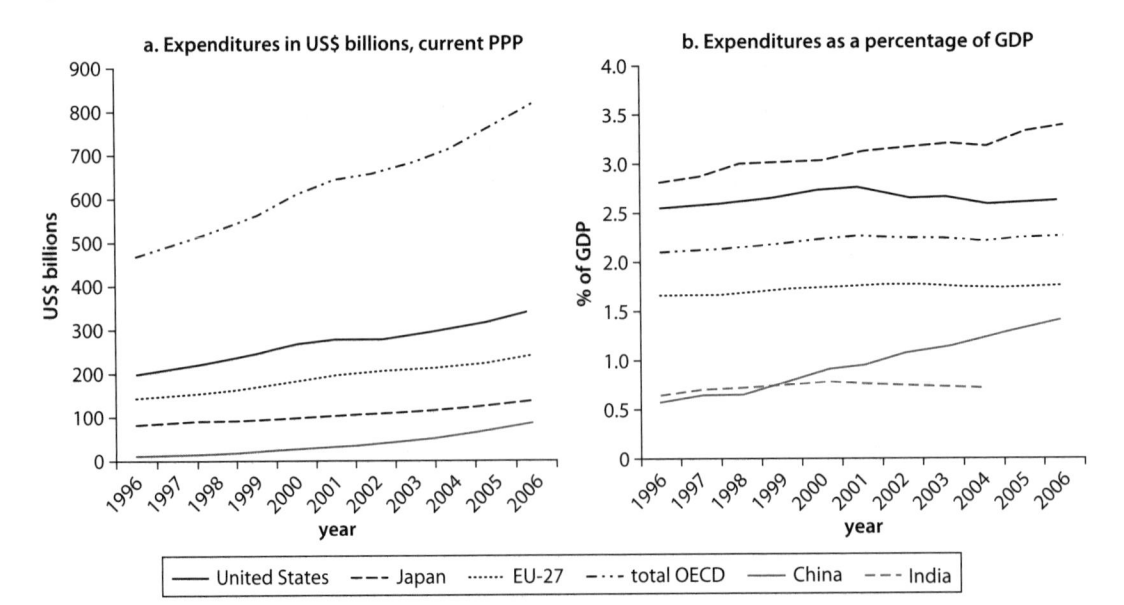

Figure 5.1 Gross Domestic Expenditures on R&D by Area, 1996–2006

Source: OECD 2008a, 31.
Note: EU = European Union; OECD = Organisation for Economic Co-operation and Development; PPP = purchasing power parity.

OECD area. The share spent by business is higher in Japan and in the United States than in the EU-27. China, which started in 1996 with very low business sector R&D, has almost caught up with the EU-27 average, although in absolute value it is still far behind.

Total global R&D for 2007 is estimated at US$982 billion in current dollars, of which 62 percent was carried out by business. The top 1,000 innovating firms were responsible for 50.1 percent of the global total and the next 1,000 for an additional 3.7 percent. Smaller companies carried out another 8.7 percent, while government and not-for-profit institutions (presumably including universities) account for the remaining 37.6 percent.[3]

In 2007, R&D by the 20 largest firms was US$128.493 billion (current dollars), or 13.1 percent of the global total. The leaders were Toyota (US$8.4 billion), followed by General Motors and Pfizer (US$8.1 billion each), and Nokia (US$7.7 billion) (Jaruzelski and Dehoff 2008). Spending by any of these individual multinationals was larger than the total R&D expenditures of any developing country except China, Brazil, and Russia.

Government financing of R&D has been falling in OECD countries. In many, government has reduced its direct support for business R&D in favor of indirect support through tax incentives, which amount to some US$5 billion in the United States, US$4.5 billion in Japan, US$2 billion in Canada, over US$800 million in the United Kingdom and France, and lesser amounts in other countries. The value of these incentives is not counted in reported totals of R&D spending (OECD 2008a, 27).

Governments also fund R&D carried out in universities. In the OECD area, these institutions perform more R&D than the government (table 5.1). Furthermore, R&D conducted in government laboratories continues to decline, while that performed by universities is increasing. In addition to government funding, universities receive funding from the business sector.

Table 5.1 R&D Performed in Government and Universities as a Percentage of GDP, 1996–2006

Research	1996	1997	1998	1999	2000	2001	2002	2003	2004	2005	2006
Government											
Japan	0.26	0.25	0.28	0.30	0.30	0.30	0.30	0.30	0.30	0.28	0.28
United States	0.33	0.31	0.30	0.29	0.28	0.31	0.32	0.33	0.32	0.31	0.29
Total OECD	0.29	0.27	0.27	0.27	0.26	0.27	0.27	0.27	0.27	0.27	0.26
EU-27	0.27	0.26	0.26	0.25	0.25	0.24	0.24	0.24	0.24	0.24	0.24
Higher education											
Japan	0.41	0.41	0.45	0.45	0.44	0.45	0.44	0.44	0.43	0.45	0.43
United States	0.31	0.30	0.30	0.31	0.31	0.33	0.36	0.37	0.37	0.37	0.37
Total OECD	0.34	0.34	0.35	0.35	0.36	0.37	0.39	0.40	0.39	0.40	0.39
EU-27	0.35	0.35	0.35	0.36	0.36	0.38	0.39	0.39	0.38	0.39	0.39

Source: OECD 2008a, 31.
Note: EU-27 = the 27 countries of the European Union; OECD = Organisation for Economic Co-operation and Development.

In OECD countries, the share of university research financed by the business sector has averaged between 6 and 7 percent since 1990. In China, it has been about 36 percent since 2001 (OECD 2007b, 31).

Globalization of R&D

The globalization of innovation and R&D is an important trend, driven by a number of factors: economic activity, the increase in scientific and technical human capital around the world, the greater ease of managing global R&D projects because of advances in information technology, the strengthening of intellectual property rights, and the increasingly favorable tax treatment for R&D in foreign countries. This trend manifests itself in many ways: first, through international co-authorship of scientific publications, which increased more than threefold between 1985 and 2005 to 20.6 percent; and, second, through the share of patents with co-inventors in two or more countries, which nearly doubled from 4 percent in the early 1990s to more than 7 percent in the early 2000s (OECD 2008a, 33).

In addition, MNCs are carrying out more of their R&D abroad. Of the top 1,000 companies doing R&D, 91 percent conducted R&D outside their home country and spent on average 55 percent of their R&D abroad. A detailed analysis of the top 100 spenders and the top 50 spenders in each of the three main sectors (electronics, pharmaceuticals, and autos)—a total of 184 companies—found that they spent US$350 billion (roughly one-third of global R&D spending or 57 percent of all private sector spending). They had 3,400 labs in 47 countries and spent only 47 percent of R&D in their home countries.

U.S.-based firms were the top "exporters" of R&D spending, followed by Japan and Switzerland. The top "net importer" of R&D by these companies was China, where companies spent US$24.7 billion for R&D. The second largest net importer of R&D was India, where companies spent US$12.9 billion. Other large net importers were Canada, Israel, and the United Kingdom (Jaruzelski and Dehoff 2008, 55). These findings are corroborated by macro data for OECD countries, which show that on average 11 percent of business R&D is financed from abroad, but as much as 26 percent in Austria and 25 percent in the United Kingdom. Furthermore, among the larger European countries, the share of R&D performed by foreign affiliates ranged from 39 percent in the United Kingdom to 26 percent in Italy (OECD 2008a, 32).

Until relatively recently, cross-border R&D aimed at adapting products and processes to the needs of the host countries. However, MNCs now also seek to source technology internationally and tap into technical human capital and other knowledge resources abroad. This effort increasingly includes lower-cost scientists and engineers in developing countries (OECD 2008a, 31). Jaruzelsky and Dehoff (2008), however, found that lower costs explained only one-third of the move of R&D facilities to developing countries. The search for specific talent is equally important. India, for example, is known for information and communication technology (ICT) and automotive engineering and China for electronics.[4] Also important is the need for proximity to the market and the capacity to deploy R&D to respond to specific needs and opportunities (Jaruzelski and Dehoff 2008, 56).

Increasing R&D in New Areas

Another global trend is increased research in three newer areas: biotechnology and genetic engineering, nanotechnology, and the environment. The first two reflect rapid advances in the science base. As noted in chapter 1, advances in science make it possible to generate new life forms as well as create new materials. At the same time, greater awareness of natural resource and environmental constraints is leading to increased effort in those areas.

R&D in Developing Countries

Although the role of R&D in developing countries is somewhat different from that in developed countries, developing countries need research capability to know what knowledge is relevant, and to acquire that knowledge. They also need to be able to adapt technology to local conditions. In agriculture, for example, developing country researchers need to understand various soils, climates, weather, pests, and tastes. For industry, they need to understand various raw materials, climates, and local preferences. For services, they must understand various forms of social organization, cultural norms, and customs.

At early stages, R&D focuses mainly on the search for and acquisition of existing technology and on its adaptation to local conditions. As countries catch up with the world frontier and increase their R&D capability, they begin to push back that frontier. They may have done so earlier when trying to develop technologies more appropriate to their specific circumstances, as part of the green revolution in agriculture, for instance. Eventually, though, these countries also conduct more basic research. Some countries, however—even a country as advanced as Japan—still do relatively little basic research and continue to concentrate primarily on applied R&D. Although the United States formerly did more basic research than any other country, its share of basic research has declined with the cutbacks in government spending. In fact, some are concerned that the country is now doing too little basic R&D (National Academy of Sciences 2007).

For the largest spenders on R&D, figure 5.2 compares R&D expenditure and the relative intensity of scientists and engineers. The data relate to 2006, and both China and India have considerably increased their R&D spending since then. The Chinese government has an explicit strategy to go beyond acquiring global knowledge through copying, reverse engineering, foreign direct investment (FDI), and technology licensing and to invest in innovation on its own account. In 2006, it announced a 15-year plan to increase expenditures on

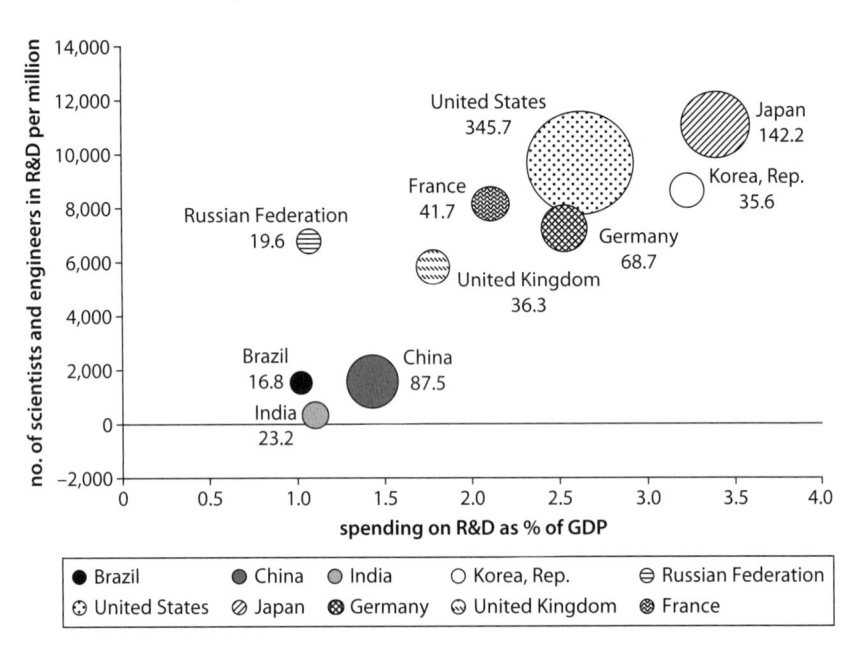

Figure 5.2 Relative R&D Expenditures and Number of Scientists and Engineers in G5 Countries and BRICs, 2006 PPP

[Bubble chart showing no. of scientists and engineers in R&D per million (y-axis) vs spending on R&D as % of GDP (x-axis):
- United States: 345.7
- Japan: 142.2
- France: 41.7
- Korea, Rep.: 35.6
- Russian Federation: 19.6
- Germany: 68.7
- United Kingdom: 36.3
- Brazil: 16.8
- China: 87.5
- India: 23.2

Legend: Brazil, China, India, Korea, Rep., Russian Federation, United States, Japan, Germany, United Kingdom, France]

Source: Author's calculation based WDI 2008, with some adjustments for India based on Dutz 2007.
Note: BRICs = Brazil, Russia, India, and China; G5 = France, Germany, Japan, United Kingdom, and United States; PPP = purchasing power parity.

R&D to 2.0 percent by 2010 and to 2.5 percent (the average level of developed countries) by 2025. In addition, as part of the global outsourcing phenomenon described above, many MNCs are increasing their R&D in developing countries, particularly in China and India. By 2006, MNCs maintained more than 750 R&D labs in China and over 250 in India.

In India, additional R&D investment by MNCs, as well as increased investment by the domestic private sector (particularly in pharmaceuticals, ICT, electronics, and auto parts) raised Indian R&D expenditure from a 20-year average of 0.88 percent of GDP to 1.1 percent in 2005 (Dutz 2007).

The efficiency of domestic R&D spending in India, and particularly in China, is still very low, however. Of other developing countries, only Brazil and Russia also have the necessary critical mass for R&D.[5] Most developing countries will get more immediate returns by putting their efforts into acquiring and making effective use of existing knowledge. Even the BRICs (Brazil, Russia, India, and China) can still get more mileage from active efforts to acquire and use global knowledge. Nonetheless, they and other developing countries need to do more—and do it more efficiently—to develop their own R&D (Figure 5.3).

The Main R&D Actors

In OECD countries, the business sector finances on average 63 percent of R&D, the government finances 30 percent, and others (including universities and

Figure 5.3 R&D Expenditures as a Percentage of GDP for Selected Economies, 2005 PPP
current US$ billions

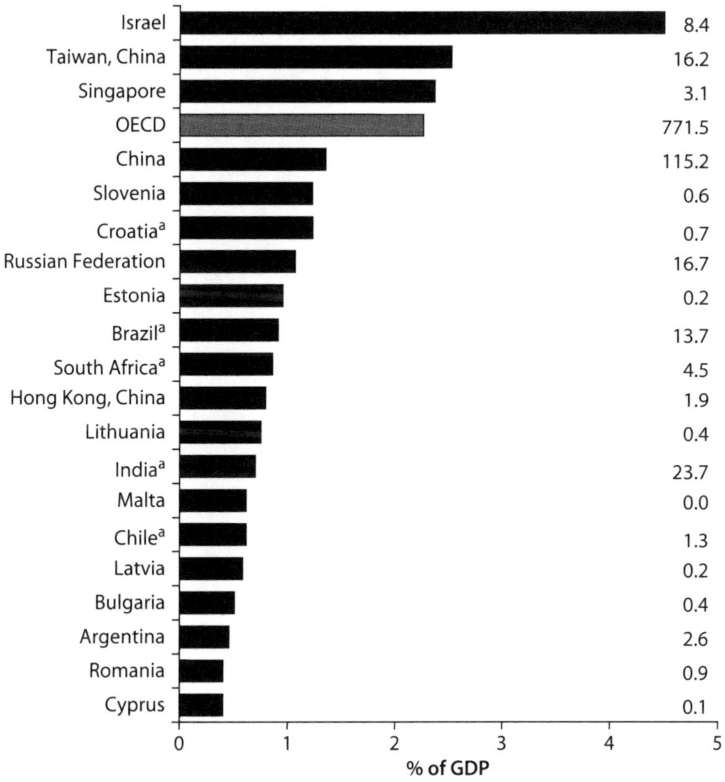

Economy	% of GDP (approx.)	Value
Israel	~4.5	8.4
Taiwan, China	~2.5	16.2
Singapore	~2.4	3.1
OECD	~2.3	771.5
China	~1.4	115.2
Slovenia	~1.2	0.6
Croatia[a]	~1.2	0.7
Russian Federation	~1.1	16.7
Estonia	~1.0	0.2
Brazil[a]	~0.9	13.7
South Africa[a]	~0.9	4.5
Hong Kong, China	~0.8	1.9
Lithuania	~0.8	0.4
India[a]	~0.7	23.7
Malta	~0.6	0.0
Chile[a]	~0.6	1.3
Latvia	~0.6	0.2
Bulgaria	~0.5	0.4
Argentina	~0.5	2.6
Romania	~0.4	0.9
Cyprus	~0.4	0.1

Source: OECD 2007.
Note: OECD = Organisation for Economic Co-operation and Development; PPP = purchasing power parity. Expenditures for China and India are overstated because values of purchasing power parity are based on the pre-December 2007 conversion factors that overstated these economies' dollar values by 40 percent.
a. Data are for 2004.

foundations) finance 7 percent. The situation is similar for the performance of R&D, except that the private sector and universities have larger shares since the government finances some R&D undertaken by the business sector and universities. The private sector also finances some university research, thereby increasing the share of R&D conducted by universities. In most developing countries, the government and the business sector play the opposite roles for both financing (table 5.2) and performance of R&D (figure 5.4). The government is the main financier and the main performer of R&D, because the private sector is generally less developed and comprises smaller firms whose limited capabilities still keep them behind the global technological frontier.[6]

These expenditures have been the pattern for some countries recently moved to developed country status. The Republic of Korea is a good example. In the mid-1960s, Korea's per capita income was not much higher than Ghana's, its R&D spending was just 0.5 percent of GDP, and the government financed 80 percent of R&D and the business sector only 20 percent. Because

Table 5.2 R&D Expenditure by Source of Financing: Main OECD and 10 Developing and Emerging Economies, 2005

Country	Business enterprises	Other (other national and foreign sources)	Government
Russian Federation	30.0	8.1	61.9
Poland	33.4	8.9	57.7
Slovak Republic	36.6	6.4	57.0
Turkey[a]	37.9	5.1	57.0
Hungary	39.4	11.1	49.4
Mexico	46.5	8.2	45.3
South Africa (2004)	48.6	15.8	35.6
EU-27[a]	54.0	10.6	35.4
Czech Republic	54.1	5.0	40.9
OECD	62.5	7.8	29.7
United States[b]	64.9	5.8	29.3
China	67.0	6.6	26.3
Korea, Rep.	75.0	2.0	23.0
Japan	76.1	7.1	16.8

Source: Based on OECD 2007, 27.
Note: EU-27 = the 27 countries of the European Union; OECD = Organisation for Economic Co-operation and Development.
a. Data are for 2004.
b. Data are for 2006.

the Korean government was very eager to have the private sector undertake more R&D, it provided incentives such as duty-free imports for research equipment and materials and accelerated depreciation, offered tax incentives, and exempted graduates who opted to go into research from military service. These incentives, however, did not have a major impact. It was only when foreign companies started restricting technology licenses to Korean companies because they were beginning to compete in their global markets that Korean companies began to invest heavily in R&D, which then became an important bargaining tool for access to foreign technology because of the credible threat that Korean companies would develop the technology themselves (Kim 1997). By 2004, the ratio of public to private financing of R&D had been reversed: almost 80 percent private and only 20 percent public, and R&D expenditures had increased to 2.7 percent of GDP.

For R&D output, patenting offers some insight into the strength of different countries and actors. Since patent regimes vary, it is useful to look at patenting in the United States, which is an important market for most countries. Table 5.3 shows utility patents granted in 2008 to U.S. nationals and to 36 foreign countries by number of patents. Of these, 49 percent went to U.S. nationals and 51 percent to foreigners. It is impressive that Korea and Taiwan, China, two latecomers to developed country status, have already become larger patentees than any European country except Germany. Equally significant is China's 11th place among foreign countries, with 1 percent of total patents.

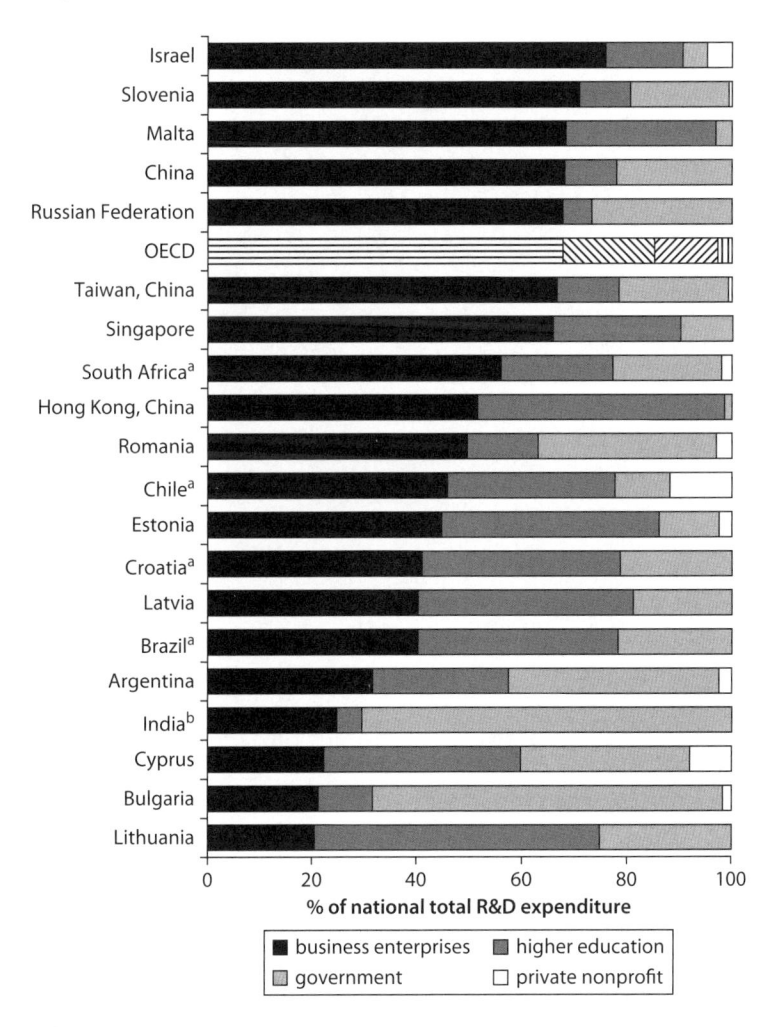

Figure 5.4 R&D Expenditures by Sector as a Percentage of National Total in Selected Economies, 2005

Source: OECD 2007, 29.
Note: OECD = Organisation for Economic Co-operation and Development.
a. Data are for 2004.
b. Data are for 2002.

In terms of who does the patenting, it is significant that both for the United States and for foreign countries overall, government accounts for less than 1 percent of the total. Individuals account for 6 percent of the total in the United States and for 2 percent in foreign countries. The bulk of patenting is done by U.S. corporations (44 percent) and foreign corporations (47 percent), proof of the overwhelming importance of firms in patenting, which basically reflects knowledge thought to have commercial value.

Although generally not based on R&D, much grassroots innovation takes place in developing countries as the result of people's experimentation and practical experience in dealing with their daily challenges. That the efficiency

Table 5.3 Number of Patents Granted by the U.S. Patent and Trademark Office, 2008

Item	Number
Economy	
Japan	33,682
Germany	8,915
Korea, Rep.	7,549
Taiwan, China	6,339
Canada	3,393
France	3,163
United Kingdom	3,094
Italy	1,357
Netherlands	1,329
Australia	1,292
China	1,225
Israel	1,166
Switzerland	1,112
Sweden	1,060
Finland	824
India	634
Belgium	510
Austria	463
Singapore	399
Denmark	391
Hong Kong, China	311
Spain	303
Norway	273
Russian Federation	176
Ireland	164
Malaysia	152
New Zealand	105
Brazil	101
South Africa	91
Hungary	66
Mexico	54
Poland	54
Czech Republic	48
Argentina	32
Saudi Arabia	30
Iceland	26
Others (68)	388
Total	**157,772**
U.S. origin	**77,501**
Foreign origin	**80,271**
Ownership	
U.S. corporations	69,962
U.S. government	676
U.S. individuals	9,021
Foreign corporations	74,465
Foreign government	33
Foreign individuals	3,615

Source: U.S. Patent and Trademark Office, available at www.uspto.gov/web/offices/ac/ido/oeip/taf/topo_08.pdf.

and effectiveness of many of these innovations can be improved with some R&D is acknowledged in countries such as India, which has systematically collected grassroots innovations and has a well-organized grassroots innovation system (see chapter 9).

The Pressure to Innovate and Undertake R&D
The pressure to innovate comes from the degree of competition through trade as well as from domestic competition. Countries with highly protected industries have little incentive to innovate unless domestic competition is strong. Indeed, even if it is strong, the incentive to innovate may be weaker than in a market with competition from imports that embody global technological advances superior to those of domestic competitors. Competition from abroad is thus very important for stimulating domestic R&D even if, at first, it is only to keep up with foreign technology.

Governments should therefore consider the economic context in which firms operate and examine broader policies that may affect firms' incentives to improve performance and their capacity for undertaking R&D, notably policies that affect competitive pressure in the economy. Principal among these is the trade regime, as protected economies offer little incentive for firms to improve their productivity by using better technology already available and even less incentive for developing new technology. Other critical policies involve the degree of openness to FDI, technology licensing, increasing domestic competition, and reducing bureaucracy. Table 5.4 lists the advantages as well as possible shortcomings of these policies.

Macroeconomic conditions also affect not only the degree of R&D but also its nature. Unstable macroeconomic environments are likely to offer less incentive, because R&D is a risky and generally longer-term business. High interest rates and high inflation are also likely to mean less R&D because the longer-term horizon and the inherent risk will make R&D more costly. R&D, however, is sometimes undertaken to develop products or processes that help firms overcome some of the problems of inflation. For example, Brazil developed excellent financial software during periods of high inflation as a way to optimize real-time financial transactions.

Finally, the rule of law, intellectual property protection, and the enforceability of contracts all affect the incentives to undertake R&D and the expected returns.

The Complex Issue of Intellectual Property Rights
Intellectual property rights fall into four basic types: patents, trademarks, trade secrets, and copyright. This section covers only patents, as they are the most relevant for R&D. A patent gives its developer property rights for a fixed period of time over the new, commercially relevant knowledge produced in exchange for public disclosure of that knowledge. Thus, a patent is

Table 5.4 Advantages and Disadvantages of Instruments for Encouraging Innovation and R&D

Instrument	Advantages	Disadvantages
Reducing barriers to imports of goods and services	Freer trade brings in global knowledge at world prices and puts pressure on domestic producers to improve their technologies and perhaps even to undertake R&D. Imports of capital goods are particularly important because they embody technology. Imports of products and services also provide models to copy or reverse engineer.	Imports may kill off domestic infant industry that cannot compete with products and services produced by more experienced and large-scale firms using better technology. Producers of domestic capital goods may be particularly hard hit.
Opening up to foreign direct investment	FDI promotes greater competition in the domestic economy. FDI can also provide technological externalities by putting pressure on domestic firms to improve their technology. Trained local workers and managers may later leave to work in domestic companies. Suppliers and distributors may get technical assistance and also be forced to improve their technology level, including undertaking R&D.	More efficient foreign firms may wipe out domestic firms because of superior technology or scale advantages from international supply and distribution networks. Foreign firms may buy out domestic firms to eliminate local competition. Foreign firms may not be interested in developing local suppliers and distributors because they prefer to use their overseas partners who may also locate domestically.
Liberalizing licensing of foreign technology	Liberalized licensing allows easier access to existing technologies, which increases pressure to produce more efficiently, including perhaps doing adaptive R&D.	Superior foreign technology may wipe out domestic technology that may have improved over time. Easy access to foreign technology may undermine efforts to try to develop technology domestically.
Increasing domestic competition	Increased domestic competition reduces monopoly power. Increased domestic competition facilitates entry and exit of firms, which permits economy to constantly restructure to use more efficient technologies.	Increased domestic competition may undercut firms that lack the scale to compete with large foreign companies that benefit from economies of scale and scope and can export their products or services to domestic markets.
Reducing bureaucracy	Less bureaucracy reduces the transaction costs for setting up and operating businesses.	Too little regulation may lead to problems of safety, predatory actions, or environmental degradation.

Source: Author.

a compromise between the incentive for inventors to produce new knowledge to advance the total knowledge pool and the social welfare benefits of diffusing that knowledge widely. Patent protection is more important for industries such as pharmaceuticals and chemicals, where it is relatively easy to copy a formula, than for other industries, where trade secrecy and first-mover advantage may be more appropriate. Patent protection is usually accorded for 20 years from the filing date. Having a uniform 20-year period across all sectors does not make much sense, though, as the rate of technical change is very fast in sectors such as electronics and communications and much slower in others such as steel or cement.

The importance of patent protection generally increases with economic development (Lerner 2002). When a country is small and very poor, patent protection is not very relevant because the country is not capable of developing new knowledge and its markets are not very attractive to developed country firms able to create technology appropriate to its needs. As a country develops, its ability to copy or imitate technology increases, but it still does not have much capacity for developing frontier technology. Forcing it to adopt and enforce strong IPR means that it will have to pay a rent to protected global knowledge. This burden will constrain its growth (Dutta and Sharma 2008; Maskus 2000). As a country increases its ability to develop new knowledge, the balance becomes more complicated. While patent protection can encourage locals to develop new technology, the country must pay rents to owners of foreign knowledge. Unless the country has great innovative capability, it is likely to lose more from paying rents than it gains from domestic innovation. Although too little patent protection can lead to suboptimal investment, too much can also misallocate resources and reduce the efficiency of innovation. In the United States, for example, some are concerned that innovation is being suffocated by excessive IPR protection (Jaffee and Lerner 2004; Boldrin and Levine 2008; Heller 2008).

There has been much pressure, particularly from the United States, to have developing countries adopt stronger IPR laws and enforcement. This pressure is reflected in WTO agreements, and countries that do not comply face stronger sanctions. In addition, in the rapidly expanding trade treaties it is signing with developing countries, the United States has been pushing for even stronger terms than those in WTO agreements (Fink and Reichenmiller 2005).

Developing countries should resist those pressures and think carefully about what makes the most sense for them at their particular stage of development. The WTO Agreement on Trade-related Aspects of Intellectual Property Rights leaves some room for maneuvering on "novelty," "nonobviousness," and scope of patent protection (Abrahamson 2007). For example, nonobviousness should be interpreted widely to fight blatantly spurious patents. Disclosure can be strengthened to provide additional information spillovers. Competition laws can also be used to curb many of the adverse effects of IPR. In addition, countries can follow India's lead in actively defending the public use of existing knowledge by fighting attempts to reappropriate the public domain through marginal changes to traditional knowledge. The best examples here are India's challenge to patents for basmati rice and neem (Boldrin and Levine 2008).

Instead of focusing on IPR as the main incentive for innovation, developing countries should promote investing in innovation without creating distortion of monopoly rights. For example, the open source innovation model is proving successful in developing innovations, such as software, through a cumulative and competitive process (Jaffee and Lerner 2004). Other models include public procurement for specific new technologies that meet certain standards, prizes for relevant technology, and international collaboration on public goods for socially important innovation, such as proposals for virtual research networks (Hubbard and Love 2004). Developing countries should also invest more in the broader system in which innovation takes place: education, entrepreneurship, and openness to global knowledge.

Public Sector R&D in Developing Countries

Developing countries need to create and commercialize knowledge because new knowledge is key to competitiveness. This is particularly true for larger countries, even low-income ones, such as India, that have a critical mass of resources and competences for a significant R&D effort. Even smaller poor countries have to have some capacity for creating knowledge. At a minimum, they need R&D capability for assessing relevant global knowledge, helping negotiate and acquire it, and adapting it to local conditions.

Key Policy Issues

The allocation of limited public resources and the effectiveness of their use is a critical policy issue. Unfortunately, most developing countries do not allocate or use these very limited resources very well, and better allocation of public resources should be a priority, including a better definition of what areas the government should support. A second priority is more effective management of these resources, particularly their contribution to the economy. It is difficult to justify pure academic research in countries with pressing social and economic needs when more applied R&D can make a significant contribution. Many developing countries do not monitor public research institutes adequately or impose effective accountability standards. Those institutions that contribute little to meeting the needs of the economy should be restructured. Box 5.1 describes how India restructured one of its premier institutes and made it more relevant by transforming it into a more outward-oriented policy organization.

Poor countries also need to undertake some basic research so that people who understand global scientific and technological trends can help their countries access relevant knowledge, adapt it to their needs, and work with other researchers to solve scientific problems. As has been pointed out, the price of admission to international research networks is local scientists who do basic research (Wagner 2008).

While it makes sense for developing countries to invest in areas in which they already have a comparative advantage to enhance that advantage, not simply maintain it, it is also important for them to invest in new technological areas such as genetic engineering, biotechnology, and nanotechnology. The public sector will have to play a greater role in carrying out this type of riskier

Box 5.1 Becoming a More Internationally Competitive, Market-Driven R&D Organization

The Indian Council on Scientific and Industrial Research (CSIR) was set up in 1942, modeled after the U.K. Department of Scientific and Industrial Research. It predated most other specialized R&D institutes in India and had a wide range of functions, from promoting scientific research to establishing R&D institutions to collecting and disseminating data on research and industry. After India's independence in 1947, CSIR became an independent entity under the prime minister. In the first two decades after independence, it focused on building up an extensive infrastructure, from metrology to R&D for a wide range of industries, with a focus on supporting emerging industry, especially small and medium enterprises.

The global energy shock of the early 1970s coincided with three years of consecutive drought in India. In the pursuit of Indian self-reliance, CSIR concentrated on reverse engineering products and process technology—primarily in pharmaceuticals, chemicals, glass, and other import-substituting industries—and on adding value to technologies using domestic resources such as high-ash coal, small-scale cement plants, and medicinal and aromatic plants.

When India shifted from an inward- to a more outward-oriented and market-driven development strategy as a result of the 1991 economic crisis, CSIR's focus changed as well. With the liberalization of trade and industrial policy, firms began to face more international competition. CSIR was criticized for being unwieldy and ineffective at transforming laboratory results into technologies for industrial production and for spending too much effort "reinventing the wheel" by focusing on known processes. The demands of the crisis led to self-examination and radical change in CSIR's role—from emphasizing technological self-reliance to viewing R&D as a business and generating world-class industrial R&D. More emphasis was placed on outputs and performance and on work relevant to productive and income-earning sectors. Each laboratory became a corporate subsidiary, with rewards for meeting targets. Laboratories were given autonomy in operations based on how well they delivered on committed outputs and deliverables. In addition, continuous efforts to continue streamlining have aimed to improve effectiveness and efficiency.

Although CSIR is still restructuring, the results to date have been quite impressive. It shows what impact a change in the direction and incentive regime of even a very large public research system can have. Between 1997 and 2002, CSIR cut its laboratories from 40 to 38 and its staff from 24,000 to 20,000, and there was a noticeable increase in its output. Technical and scientific publications in internationally recognized journals jumped from 1,576 in 1995 to 2,900 in 2005, and their average impact factor increased from 1.5 to 2.2. Patent filings in India rose from 264 in 1997–98 to 418 in 2004–05. Patent filings abroad quintupled from 94 in 1997–98 to 500 in 2004–05, and CSIR accounted for 50–60 percent of U.S. patents granted to Indian inventors. In addition, CSIR increased earnings from outside income from 1.8 billion rupees in 1995–96 to 3.1 billion rupees in 2005–06 (about US$65 million). Today it has 4,700 active scientists and technologists supported by 8,500 scientific and technical personnel. Its government grant budget has roughly doubled since 1997 and is now 15 billion rupees (US$325 million); its earnings are about 20 percent of its grant budget.

Source: Based on Bhojwani 2006.

and more uncertain research as part of a strategy of exploring new areas with potentially high returns. Such investments are needed so that countries can move rapidly into areas that show promising results.

Therefore, countries need to put in place not only appropriate policies but also public and private supporting institutions to create new knowledge and to facilitate the acquisition and dissemination of that knowledge. In addition, a key problem in most developing countries is that even when relevant knowledge is created in public labs or universities, it is not commercialized. Therefore, the supportive infrastructure (technology parks, business incubators, technology transfer centers, and venture capital) to commercialize knowledge is essential; East Asia—particularly China; Korea; and Taiwan, China—is a good example of this approach.[7] It is also necessary to make sure that the country develops the necessary human resources ("techno-entrepreneurs") to undertake and manage R&D and to commercialize relevant knowledge.

Obviously, how much a country should invest in its R&D and commercialization infrastructure will depend on its resources and size. The richer and more developed its institutions and human capital are, the more it can do. Even some countries poor in average per capita income, such as China or India, have the critical mass of resources, institutions, and people to create and commercialize knowledge. They will still benefit tremendously, however, from continuing to improve the acquisition, dissemination, and effective use of existing knowledge.

Private Sector R&D in Developing Countries

In developing countries, the productive sector does relatively little research and development, for various reasons:[8]

- Because most firms are behind the global technological frontier, it makes more sense for them to buy or copy existing foreign technology, which is generally cheaper than undertaking risky R&D.
- Because domestic markets are generally less competitive and more segmented than those in developed countries, they face less pressure to develop new technology and must overcome more barriers to entry and to exit.
- Most firms do not have the scientists and engineers to undertake formal R&D.
- The very large majority of firms are too small to have the resources to invest in R&D.
- The cost of capital is also generally higher than in developed economies.
- The macroeconomic environment is often more unstable and not conducive to undertaking lengthy R&D.

- Because intellectual property regimes are generally less developed, firms face a greater risk that any technology they develop will leak out or be appropriated by others.
- Transactions costs are higher for setting up, operating, and expanding firms than in developed countries (IFC 2009).

Main Firms Doing R&D in Developing Countries

The firms that undertake R&D tend to be large public enterprises in natural resources (such as oil or minerals) or large conglomerates in electronics, telecommunications, auto and engineering, domestic appliances, and basic commodities, such as paper, mining, iron and steel, food products, or other products based on natural resources (table 5.5). The exceptions include aircraft in Brazil (Embraer) and pharmaceuticals in India (Ranbaxy and Dr Reddy).

Only 93 developing countries are among the 1,000 companies that spend the most on R&D worldwide (table 5.6). Almost three-fifths are concentrated in Korea and Taiwan, China, followed by China and India. Companies in East Asia specialize mostly in computing and electronics; in India and Eastern Europe, in health; and in the rest, mostly in natural resources, health, and some industrials.

Not surprisingly, more or less the same countries account for the bulk of patenting by developing countries in the United States (table 5.7). A comparison of cumulative patents through 2008 with the total number of patents in 2008 clearly shows that India and, in particular, China are increasing patenting in the United States as they invest more in R&D.

Table 5.8 lists some of the better-known frontier-level innovations resulting from R&D in companies in developing countries. They include everything

Table 5.5 Top-10 R&D Companies from Developing and Emerging Economies, 2007

Company	Country	Industry	R&D expenditures (US$ millions)
Samsung (9)	Korea, Rep.	Computing, electronics	6,536
Hyundai Motor (62)	Korea, Rep.	Auto	1,197
LG Corporation (63)	Korea, Rep.	Other	1,952
Petrobras (117)	Brazil	Chemicals, energy	879
Cia Vale do Rio Doce (140)	Brazil	Minerals	717
Petrochina (142)	China	Chemicals, energy	699
Kia Motors (148)	Korea, Rep.	Auto	649
Korea Electric Power (149)	Korea, Rep.	Other	649
Hynix Semiconductor (150)	Taiwan, China	Computing, electronics	635
Gazprom (159)	Russian Federation	Chemicals, energy	605

Source: Jaruzelski, Dehoff, and Bordia 2005.
Note: The figure in parentheses is position among global 1,000.

Table 5.6 Number of Developing Economy Companies among the Global 1,000, 2007

Economy	Number of companies and main areas
Taiwan, China	30: computing and electronics, software, industrials
Korea, Rep.	24: electronics, software, telecom, auto, chemicals, energy, industrials
China	10: petrochemicals, auto, industrials
India	6: auto, health, industrials, other
Israel	5: software, health
Brazil	4: natural resources, aerospace, power
Hong Kong, China	4: consumer goods, industrials, chemicals, energy
Singapore	3: computing, electronics
South Africa	2: industrials, chemicals, energy
Hungary	1: health
Russian Federation	1: chemicals, energy
Slovenia	1: health
Turkey	1: other

Source: Jaruzelski, Dehoff, and Bordia 2005.

Table 5.7 Utility Patents Granted by the U.S. Patent and Trademark Office to the Top-15 Developing and Emerging Economies, 2008

Economy	Cumulative patents through 2008	Patents granted in 2008
Taiwan, China	70,643	6,339
Korea, Rep.	59,958	7,549
Israel	16,805	1,166
USSR[a]	6,994	0
China	5,162	1,225
Singapore	4,097	399
India	4,080	634
Hong Kong, China	3,805	311
South Africa	3,976	91
Hungary	2,871	66
Mexico	2,509	54
Russian Federation[a]	2,409	176
Czechoslovakia[b]	2,121	0
Brazil	2,094	101
Argentina	1,249	32
Malaysia	947	152

Source: U.S. Patent and Trademark Office 2008, available at http://www.upto.gov.
a. USSR was patenting around 175 per year in 1990–91. Patenting as the USSR ceased in 2000 and patenting by the Russian Federation started in 1993.
b. Patenting by Czechoslovakia ceased in 2000; patenting by the Czech Republic started in 1994 and totaled an additional 360 by the end of 2008. Patenting by Slovakia started in 1996 and totaled an additional 49 by the end of 2008.

from new processes in agriculture and industry to new products such as pharmaceuticals, cars, and airplanes, as well as new forms of business services. Moreover, some also come from smaller companies in lower-income countries and include new ways to deliver social services such as education and sanitation.

Table 5.8 Illustrative Examples of Innovations by Developing Economy Firms

Economy	Company	Innovation
Brazil	Embraer	Airplanes
	Petrobras	Deep sea oil exploration platforms and processes
India	Ranbaxy, Dr Reddy	New pharmaceutical products
	Tata	Nano car for US$2,500
Mexico	HYLSA	Direct reduction technology for producing steel
	TELMEX	Prepaid phone card for low-income users
Taiwan, China	Acer	Small, high-capacity network computers

Source: Author.

Government Support for Business R&D

Government can support R&D in various ways. Direct support instruments include tax incentives, grants, accelerated depreciation on R&D equipment, duty exemptions on imported equipment and other research inputs, and venture capital to support high-technology start-ups. Table 5.9 summarizes the advantages and disadvantages of each. The two most important instruments are tax incentives and grants. As noted above, OECD governments are moving away from grants toward tax incentives, largely because they prefer automatic neutral support to targeted interventions. Developing countries with scarce resources, poor tax systems, and limited R&D capability in enterprises, however, should carefully consider the trade-offs between neutral and more targeted support.

Government can also support business R&D more generally by investing more in public R&D, developing technical human capital, and promoting links between firms and public R&D labs and university research. Table 5.10 summarizes the principal advantages and disadvantages of each. While these measures may be helpful, they may not work well if public R&D or the technical human capital produced is of poor quality or if the productive sector has little incentive or capacity for undertaking R&D or for exploiting public investments in R&D.

Multinational Corporation Labs in Developing Countries

As noted earlier, MNCs increasingly undertake R&D in developing countries that have a critical mass of high-quality R&D personnel, primarily, Brazil; China; the Czech Republic; Hungary; India; Israel; Malaysia; Russia; Singapore; Taiwan, China; and Thailand (see figure 5.2). From the perspective of the host countries, there are positive and negative sides to multinational R&D. On the positive side, local scientists and engineers acquire training in R&D management and methods when working in the MNC labs. They also connect into the companies' international research networks, which offer valuable opportunities for researchers' professional growth, practical experience, and contacts. The country can also take advantage of any R&D that meets its specific needs

Table 5.9 Direct Instruments for Supporting Business R&D

Instrument	Advantages	Disadvantages
Tax incentives for R&D	Provides functional intervention, not picking winners Offers less distortion, more automatic Generally requires less bureaucracy to implement, although advisable to have monitoring and spot checks	Has unclear fiscal costs in advance, could be high Is difficult to ensure that R&D increase is induced by tax incentives (additionality) Is not very relevant for start-up firms that do not yet have taxable revenue streams Is blunt instrument, cannot target specific companies, although it can target specific sectors
Grants for R&D projects	Allows specific targeting on case-by-case basis Can control amount of subsidy granted Can be given in tranches against defined goals Can be structured as matching grants that may help improve quality or efficiency	Requires large bureaucracy to administer May not select the best project Is also difficult to ensure additionality
Accelerated depreciation for R&D equipment	Reduces the capital costs of R&D projects	Does not provide incentive for noncapital costs such as personnel and material inputs
Duty exemption on imported inputs into R&D	Reduces cost of world class inputs if country otherwise has high import duties	Results in loss of tariff revenue Is distortionary to extent that it favors R&D over other activities
Venture capital to facilitate commercialization of research results	Helps overcome financial market failure in making capital available to start-ups with no collateral or track record	Requires detailed knowledge of sectors to evaluate technical and commercial prospects Is often not successful because of limited deal flow and shortage of techno-entrepreneurs Also requires developed stock market so investors can sell off shares and reinvest in new projects

Source: Author.

Table 5.10 General Science and Technology Instruments for Supporting Business R&D

Instrument	Advantages	Disadvantages
More public R&D	Is supposed to fund basic research, which provides public good that can be used as input into more applied commercial development	Public sector may be very inefficient in undertaking R&D. Productive sector may not exploit publicly financed R&D.
Development of technical human capital	Is supposed to prepare human capital to manage and undertake research	Often there is no uptake by the productive sector if it does not see a need to undertake research.
Promotion of links with universities and public research institutes	Is supposed to facilitate complementarity of basic research capability of universities and public research institutes with more applied research and commercial needs of industry	The productive sector is often uninterested in undertaking R&D, may not have high enough regard for capability of domestic university or public R&D institute to want to work with them, or may be concerned about intellectual property leaking out to competitors.

Source: Author.

and enjoy the benefits from links and interactions between the corporations' R&D and domestic research at universities, public labs, and perhaps domestic firms. In addition, the country benefits when national personnel with valuable R&D experience gained from working for the multinational leave to work in national R&D institutions (public or private labs and universities) or to set up their own high-technology companies.

On the negative side, MNCs may appropriate valuable domestic human resources for their own use, as an increasing amount of the R&D appears geared to MNCs' global research projects, without much value for the host country. In addition, absorption of domestic R&D personnel by MNCs may force up salaries for national scientists and engineers. Although the individuals themselves benefit and the higher salaries may lead to expanded university training for scientists and engineers, they also increase personnel costs for national R&D institutions and firms.

Thus, there are clearly trade-offs. No detailed studies of positive or negative effects of MNC operations in host countries have been conducted, except for data on rapidly rising salaries of scientists and engineers in some countries. The net benefits to the host country will depend on its situation. If the country can increase the supply of its R&D personnel, however, the positive effects are likely to outweigh the negative ones.

University R&D in Developing Countries

Universities are key institutions for research and development in two respects.[9] First, they train scientists and engineers, the principal input into R&D, as well as managers and other technical support personnel. Although domestic universities are not the only suppliers of scientists and engineers (many students go abroad to study at all levels, from undergraduate to PhD), they are an important source of talent. Second, domestic universities carry out R&D in their research labs. Therefore, the number and quality of universities in developing countries are an important part of domestic R&D capacity. One quick way to assess the strength of developing country universities in R&D is to look at global rankings of the best universities. The most comprehensive ranking, with a strong focus on research capacity and quality, is that of Shanghai's Jiao Tong University (table 5.11).[10]

The United States dominates, with 54 percent of the top 100 universities and 31.6 percent of the top 500, which is greater than its share in global GDP. The only two developing countries among the top 100 are Israel and Russia, with just one university each. Among the top 500, China is the top developing country with 30 universities, followed by Korea with 8, Brazil with 6, and South Africa with 3. Chile, Hungary, India, and Poland have 2 each. The other five developing countries among the top 500 have only one.[11] Overall, the countries on this list are very similar to those with the most patents or those in which MNCs undertake R&D.

Table 5.11 Selected Research Universities from the World's Top 100 and 500, by Country

Country	Percent of top 100	Percent of top 500	Percent of world GDP	Percent of world population
United States (1)	54.0	31.6	27.2	4.6
United Kingdom (2)	11.0	8.3	4.9	0.9
Germany (3)	6.0	8.0	6.0	1.3
Japan (4)	4.0	6.2	9.0	2.0
Israel (13)	1.0	1.2	0.3	0.1
Russian Federation (15)	1.0	0.4	2.0	2.2
China (16)	0	6.0	6.6	20.5
Korea, Rep. (19)	0	1.6	1.8	0.7
Brazil (22)	0	1.2	2.2	2.9
South Africa (25)	0	0.6	0.5	0.7
Chile (26)	0	0.4	0.3	0.3
Hungary (28)	0	0.4	0.2	0.2
India (29)	0	0.4	1.9	17.0
Poland (30)	0	0.4	0.7	0.6
Singapore (32)	0	0.4	0.3	0.1
Argentina (33)	0	0.2	0.4	0.6
Czech Republic (34)	0	0.2	0.3	0.2
Mexico (35)	0	0.2	1.7	1.6
Slovenia (36)	0	0.2	0.1	0.0
Turkey (37)	0	0.2	0.8	1.1

Source: Academic Ranking of World Universities, available at http://www.arwu.org/.
Note: Figures in parentheses represent country rank among the top 100.

In most developing countries, university scientists and engineers do not interact very much with enterprises or even with public research labs, and universities tend to have the largest proportion of the countries' scientists and research engineers. Universities also tend to produce scientific and technical publications but few patents.[12] For this reason, many have sought to promote greater interaction among university researchers, enterprises, and public research institutes (known as the "triple helix").[13]

In the United States, the lack of interaction between universities and the productive sector and the lack of a commercial focus in universities and public research institutes led to the Bayh-Dole Act in 1980. It gave recipients of federally funded research at universities, public labs, and small and medium enterprises intellectual property rights over the inventions they developed as a result of that funding. The objective was to get them to patent and commercialize their inventions.[14] The perceived success of the Bayh-Dole Act in stimulating the patenting and commercialization of publicly funded research has led many to recommend that developing countries provide a similar incentive for researchers at universities and public labs to make their research more commercially relevant.

Some economists, however, have disputed the merits of the Bayh-Dole Act. They argue that it did not lead to a strong increase in patenting by universities and note that the increase in technology licensing revenues came from a few very profitable licenses (Mowery and others 1999, 2004; Mowery 2007; Su and others 2008). Moreover, some theorists have argued that the act actually distorts the role of public research and the vocation of the university, which are to advance basic knowledge that should be available to all (Nelson 1959; Dasgupta and David 1994). Many large U.S. multinationals also claim that the overly commercial orientation of U.S. universities has made it very difficult for them to negotiate joint research and technology licensing agreements with them and that they are therefore doing more R&D with universities abroad (Thursby and Thursby 2006).

China offers an instructive example of such distortions. Before 1990, Chinese universities were academic "ivory towers" isolated from the needs of the economy and the productive sector. The government then passed a series of reforms that drastically cut public funding to universities and public research institutes to force them to seek contract research for their funding and become more responsive to the needs of the productive sector. The policy was all too successful. Universities, particularly major prestigious research universities such as Tsinghua, Beijing, Fudan, and Jiao Tong, became very commercially oriented. They spun off hundreds of commercial enterprises, a few of which (such as Legend Computers, which later bought IBM's PC business and became Lenovo) became major companies on the Chinese stock market. The commercial focus, however, distracted universities from their function of educating highly skilled workers to run the economy and push back the frontiers of knowledge. Therefore, after the year 2000 universities refocused on their education and public research roles (Xue 2006, 2007).

Other mechanisms for stimulating universities to do research more relevant to the needs of the country and to commercialize the knowledge they produce include technology transfer offices, science parks, business incubators at or near universities, matching grants or subsidies for cooperative ventures among universities, enterprises, and public research institutes. Table 5.12 summarizes their advantages and disadvantages.

Increasing the involvement of universities in relevant R&D may result in problems: (a) low quality of university research; (b) poor research facilities and equipment; (c) poor technical human capital; (d) poor monitoring and evaluation and accountability systems; and (e) cumbersome university regulations that hamper interaction between university researchers and needs of the productive sector, such as high overheads and the fact that only academic publications count for promotion.

International R&D Cooperation and Research Programs

Besides strengthening their own domestic R&D programs, developing countries can benefit from joining two types of international programs: networks

Table 5.12 Instruments for Promoting Relevant R&D in Universities and Greater Commercialization of Knowledge and Interaction with Enterprises

Instrument	Advantage	Disadvantage
Bayh-Dole–type legislation	Provides an incentive for researchers at universities and public research institutes to produce commercially relevant knowledge and earn income from the licensing or sale of the knowledge produced	May create an excessively commercial orientation in universities or public R&D labs, which compromises the public-good nature of university and public lab R&D Excessive preoccupation by universities and public R&D centers with financial side of contracts may make transactions costs too high for businesses to work with them
Technology transfer offices	Provide economies of scale and experience in patenting applications and technology transfer contracts Create greater incentive to commercialize technology	May put too much pressure on researchers to privatize their knowledge and thus impede the public flow of knowledge Sometimes may not produce enough income to justify cost
Science parks	Provide economies of scale in provision of basic infrastructure May lead to agglomeration economies in interaction between knowledge workers and technology-based firms	May not achieve the economies of scale and agglomeration envisioned because they lack the necessary critical mass May become real estate operations more than knowledge centers
Business incubators at universities	Provide economies of scale in physical and institutional support for start-ups, including help in preparing business plans, matching scientists with business, obtaining permits to set up new businesses, and the like	May not function well because they lack the ability to match business skills with technology skills, or to provide complementary support services May focus too much on real estate rather than on promotion of new technology firms
Matching grants or tax subsidies for cooperation among universities, firms, and public research institutes	Create incentives for potentially mutually beneficial synergies among firms, universities, and public R&D labs	May not be used because of lack of trust between the parties. May subsidize interactions that would have happened anyway

Source: Author.

of researchers working together on topics of mutual relevance and programs focusing on global public goods. A number of countries have participated in these international networks, as follows:

- Chile has drawn on researchers from all over the world, including U.S. universities, to help improve the quality of food exports.
- Canadian, Norwegian, and Scottish companies, universities, and research institutions have linked up to improve their salmon farming.
- Australian, Japanese, and South African researchers have worked together to improve their copper mining processes.

Other possibilities include joining formal and informal research networks that address basic and applied research in specific fields and areas of interest. Participating in such networks allows domestic researchers to keep up with the evolving frontiers of knowledge and to draw on, and contribute to, the evolution of knowledge.

The second type of international program is the large, multicountry program seeking to advance global public goods. One impressive program is the Consultative Group on International Agricultural Research (CGIAR), which grew out of a proposal by the Rockefeller Foundation in the 1970s to create a global research network to help developing countries avoid famines. By 1983, there were 13 research centers in countries ranging from Colombia to Mexico to India and the Philippines; together, they were instrumental in developing high-yield cereal crops. The CGIAR now includes more than 60 governmental and nongovernmental centers and 15 labs studying maize, potatoes, rice, wheat, tropical agriculture, arid agriculture, fish, and forestry.

Health is another important international public good.[15] Multicountry and multipartner initiatives are studying malaria, tuberculosis, and HIV/AIDS; and, while progress has been made, more R&D is needed on these diseases. Developing country researchers should join these international programs to learn about and to contribute to advances in areas that are important in their own countries.

Environmental sustainability in general and efforts to address global warming in particular constitute a third global public good. In this broad and critical area, much more R&D than is currently being performed is urgently needed. For example, no proven technology is available at commercial scale for a process as important but apparently simple as CO_2 sequestration. Different forms of carbon sequestration will be required for specific geological features such as underwater reservoirs, salt mines, or underground storage tanks. Development of appropriate technologies will necessitate research and trials in many different contexts. Even more R&D will be needed to make major breakthroughs in new energy technologies. Again, developing country researchers should be involved in these international programs.

Summary and Conclusions

Although developing countries account for 47 percent of world GDP in purchasing power parity and 85 percent of world population, they perform less than 20 percent of world R&D. While research and development are not the most critical components of the domestic innovation system in developing countries and tapping into existing knowledge is a far more important source of innovation, developing countries need an R&D base. Without it, they cannot follow, assess, acquire, adapt, and use new knowledge to meet their development goals. University, government, and business researchers collaborate with colleagues across national frontiers. Developing countries need to become part of the global R&D research community to keep up to date with the rapid advances in science and technology and to draw on those advances for their specific needs. To do so, they need to perform R&D. This chapter has focused on strengthening R&D in developing countries because of its fundamental importance.

The main R&D performers are firms, governments, and universities. Firms perform the most R&D globally, particularly in developed countries, and are responsible for more than 60 percent of R&D worldwide. In developed countries, universities come next, followed by government. In most developing countries, though, the government conducts most of the R&D, followed by universities and then the productive sector, mainly because the productive sector does not generally operate at the world technology frontier and therefore its greatest need is to acquire and adapt existing knowledge. In addition, many firms in developing countries are too small and lack the financial resources or the human capital to undertake much R&D, even of the adaptive kind.

Developing countries also need to find ways to allocate public R&D resources more effectively; to establish clearer criteria for allocating resources according to their needs among government, universities, and business; and to develop better ways to monitor and evaluate the results of the R&D effort they fund in public laboratories, universities, and the productive sector.

To realize synergies among key actors, governments can do much to encourage more R&D by the productive sector and universities and to promote greater collaboration among these two actors and public R&D labs. In addition, as R&D is increasingly global, governments and researchers in developing countries have to consider how to become part of international R&D networks in general as well as of those promoting international public goods.

This chapter has summarized the advantages and disadvantages of many instruments for encouraging R&D. The balance between advantages and disadvantages will depend on the specifics of each country's situation, which depend in turn on the country's stage of development and its needs, the capabilities of the different actors, the design of the programs, and the broader underlying conditions, including the economic incentives, institutional regime, and quality of human and institutional capital.

Notes

1. Author's estimates based on OECD data and additional data for other developing countries, using the latest purchasing power parity series released in December 2007. All monetary amounts are U.S. dollars unless otherwise indicated.

2. In 2006, the largest relative spender on R&D was Israel at 4.7 percent of GDP, followed by Sweden (3.7 percent), Finland (3.5 percent), Japan (3.4 percent), Korea (3.2 percent), Switzerland (2.9 percent), Iceland (2.7 percent), and the United States (2.6 percent).

3. These estimates were made by a team at Booz Allen Hamilton as part of its fourth annual private innovation survey based on data collected from private firms, estimates of total R&D spending based on OECD data for developed countries, and estimates of R&D spending by developing countries from International Monetary Fund and World Bank data. See Jaruzelski and Dehoff (2008). UNCTAD (2005) also estimated that transnational companies accounted for more than half of global R&D.

4. The increasing importance of innovation as a key element of competitiveness is also leading to intense global competition for talent. See, for example, OECD (2008b).

5. The four so-called BRICs are Brazil, Russia, India, and China.

6. China is an important exception. According to official Chinese data, 65 percent of R&D is done by the productive sector. However, this includes R&D done by state-owned enterprises, and there is some question on what is counted as R&D.

7. See Yusuf and Nabeshima (2008) for an analysis of the successes and failures of science and industrial parks in East Asia.

8. Israel, Korea, and Taiwan, China, are grouped with developing countries because of their still relatively recent transition to higher-income status.

9. For a more detailed treatment of the role of universities, see Santiago and others (2008).

10. The criteria for the ranking are number of alumni and staff winning Nobel prizes and Fields medals, number of highly cited researchers, number of articles published in *Science* and *Nature*, number of articles cited in science and social science citation indexes, and overall weighted score of the above five indicators divided by the number of full-time equivalent academic staff (a quality of staff indicator). See http://www.arwu.org/rank2008/EN2008.htm for more details on methodology and detailed rankings.

11. The major exception is Taiwan, China, which does not appear on the list of universities since China treats it as a province. It is possible that since Academic Ranking of World Universities is based in Shanghai, Taiwanese universities were included in the totals for China.

12. See Rodriguez, Dahlman, and Salmi (2008) for data on Brazil.

13. Etzkowitz (2002) popularized the term in a study of the role of MIT.

14. Prior to the Bayh-Dole Act, the government had accumulated 30,000 patents, only 5 percent of which had been commercialized.

15. Kaul and Faust (2001) discuss why developed countries should put more funding into health as a global public good and how it should be organized.

References and Other Resources

Abramson, Bruce. 2007. "India's Journey toward an Effective Patent System." Working Paper 4301, World Bank, Washington, DC.

Bhojwani H.R. 2006. "Report on the Indian Civilan R&D System." South Asia Finance and Private Sector Development background paper, World Bank, Washington, DC.

Boldrin, Michael, and David Levine. 2008. *Against Intellectual Monopoly*. New York: Cambridge University Press.

Dasgupta, Partha, and Paul David. 1994. "Toward a New Economics of Science." *Policy Research* 23: 487–21.

Dutta, Antara, and Siddharth Sharma. 2008. "Intellectual Property Rights and Innovation in Developing Countries: Evidence from India." http://www.enterprisesurveys.org/ResearchPapers/Intellectual-Property-Rights-India.aspx.

Dutz, Mark, ed. 2007. *Unleashing India's Innovation*. Washington, DC: World Bank.

Etzkowitz, Henry. 2002. *MIT and the Rise of Entrepreneurial Science*. London: Routledge.

Fink, Carsten, and Patrick Reichenmiller. 2005. "Tightening TRIPS: The Intellectual Property Provisions of Recent US Free Trade Agreement." Issue Brief, World Bank, Washington, DC.

Heller, Michael. 2008. *Gridlock Economy: How Too Much Ownership Wrecks Markets, Stops Innovation, and Costs Lives*. Philadelphia: Basic Books.

Hubbard. T, and J. Love. 2004. "A New Trade Framework for Global Healthcare R&D." *PLoS Biol* 2 (2): e52. Published online Feb. 17, 2004. doi:10.1371/journal.pbio.0020052.

IFC (International Finance Corporation). 2009. *Cost of Doing Business*. Washington, DC: Palgrave Macmillan and World Bank.

Jaffee, Adam B., and Josh Lerner. 2004. *Innovation and Its Discontents: How Our Broken Patent System Is Endangering Innovation and Progress, and What to Do about It.* Princeton, NJ: Princeton University Press.

Jaruzelski, Barry, and Kevin Dehoff. 2008. "Beyond Borders: The Global Innovation 1000." *Strategy and Business* Issue 53 (Winter). http://www.strategy-business.com/article/08405.

Jaruzelski, Barry, Kevin Dehoff, and Rakesh Bordia. 2005. "The Booz Allen Hamilton Global Innovation 1000: Money Isn't Everything." Available at www.boozallen.com and www.strategy+business.com (Issue 41, Winter 2005).

Kaul, Inge, and Michael Faust. 2001. "Global Public Goods and Health: Taking the Agenda Forward." *Bulletin of the World Health Organization* 70 (9): 869–74.

Kim, Linsu.1997. *Imitation to Innovation: The Dynamics of Korean Technological Learning.* Cambridge, MA: Harvard Business School Press.

Lerner, J. 2002. "Patent Protection and Innovation over 150 Years. Working Paper 8977, National Bureau of Economic Research, Cambridge, MA.

Maskus, Keith. 2000. *Intellectual Property Rights in the Global Economy.* New York: Institute for International Economics.

Mowery, David C. 2007. "University-Industry Research Collaboration and Technology Transfer in the United States since 1980." In *How Universities Promote Economic Growth*, ed. Shahid Yusuf and Kaoru Nabeshima, 163–82. Washington, DC: World Bank.

Mowery, David C., Richard R. Nelson, N. Bhaven, B. Sampat, and A. Ziedonis. 1999. "The Effects of the Bayh-Dole Act on U.S. University Research and Technology Transfer." In *Industrializing Knowledge: University-Industry Linkages in Japan and the United States*, ed. Lewis M. Branscomb, Fumio Kodama, and Richard Florida, 269–306. Boston: MIT Press.

Mowery, D., R. Nelson, B. Sampat, and A. Ziedonis. 2004. *Ivory Tower and Industrial Innovation: University-Industry Technology Transfer before and after the Bayh-Dole Act in the United States.* Stanford: Stanford University Press.

National Academy of Sciences. 2007. *Rising above the Gathering Storm: Energizing and Employing America for a Brighter Economic Future.* Washington, DC: National Academy Press.

Nelson, Richard. 1959. "The Simple Economics of Basic Scientific Research." *Journal of Political Economy* 49: 297–306.

OECD (Organisation for Economic Co-operation and Develoment). 2007. *Science Technology and Industry Scoreboard 2007.* Paris: OECD.

———. 2008a. *OECD Science, Technology and Industry Outlook.* Paris: OECD.

———. 2008b. *The Global Competition for Talent: Mobility of the Highly Skilled.* Paris: OECD.

———. 2008c. *Review of Innovation Policy: China.* Paris: OECD.

———. 2008d. *The Internationalization of Business R&D: Evidence, Impact and Implications.* Paris: OECD.

Rodríguez, Alberto, Carl Dahlman, and Jamil Salmi. 2008. *Knowledge and Innovation for Competitiveness in Brazil.* Washington, DC: World Bank.

Santiago, Paulo, Karine Tremblay, Ester Basri, and Elena Arnal. 2008. "Enhancing the Role of Tertiary Education in Research and Innovation." In *Tertiary Education for the Knowledge Society.* Vol. 2, ed. Paulo Santiago, Karine Tremblay, Ester Basri, and Elena Arnal, 73–129. Paris: OECD.

Su, Anthony D., Bhaven N. Sampat, Arti K. Rai, Robert Cook-Deegan, Jerome H. Reichman, Robert Weisman, and Amy Kapcyznski. (2008). "Is Bayh-Dole Good for Developing Countries? Lessons from the US Experience." *PloSBiol* 6 (10). *e263.doi:10.1371/journal.pbio.0060262.*

Thursby, Jerry, and Marie Thursby. 2006. *Here or There? A Survey of Factors in Multinational Firm Location of R&D.* Washington, DC: National Academies Press.

UNCTAD (United Nations Conference on Trade and Development). 2005. *World Investment Report 2005.* Geneva: UNCTAD.

Wagner, Caroline. 2008. *The Invisible College.* Washington, DC: Brookings Institution Press.

World Bank. 2008. *World Development Indicators: 2008.* Washington, DC: World Bank.

———. 2009. *World Development Indicators: 2009.* Washington, DC: World Bank.

Xue, Lan. 2006a. "The Changing Roles of Universities in China's National Innovation System." PowerPoint presentation at the World Bank Knowledge Economy Conference, Europe and Central Asia Region, Prague.

———. 2006b. "Universities in China's National Innovation System." Paper prepared for the UNESCO Forum, "Higher Education, Research and Knowledge," Palo Alto, California, November 27–30.

Yusuf, Shahid, and Kaoru Nabeshima. 2008. *Growing Industrial Clusters in Asia.* Washington, DC: World Bank.

6

Fostering Innovation through Education and Training

This chapter discusses the role of education and skills in fostering innovation in a context of structural change and economic development. Recent examples of the promotion of innovation through education and skills development in developed and developing countries provide some insight into how countries can become knowledge-based and innovation-driven economies.

The chapter deals extensively with education and training policies because a good educational and training system is fundamental to building a population receptive to innovation, able to tap into and absorb the sources of global knowledge, and creative in terms of technology and entrepreneurship. For that reason, it is important to discuss such issues as well as the challenges of implementing educational reform. In a final section, we document and discuss the issue of brain drain and the ways it can be turned into a positive force for helping developing countries to respond to trends in the global knowledge economy.

Skills for a Knowledge-Based and Innovation-Driven Economy

Knowledge and human capital accumulation and innovation have become the driving forces of economic and social development around the world. Along with globalization and the rapid dissemination and transfer of knowledge by information and communication technology (ICT), these forces affect all countries and regions in their quest for economic growth and prosperity. In a

---

This chapter was prepared by Kurt Larsen and Florian Theus, with a contribution of Yevgeny Kuznetsov for the section on brain circulation.

knowledge economy, knowledge is created, acquired, transmitted, and used more effectively by individuals, enterprises, organizations, and communities to promote economic and social development. These developments have far-reaching implications for education and training.

The rise of rapidly expanding knowledge-based industries, in particular ICT-related industries and the service industry, has increased the demand for more highly skilled labor. The demand for skills sharply increased in the 1980s and 1990s in middle-income countries, owing more to within-industry skills upgrading than to restructuring from low- to high-skill industries. Evidence suggests that wage differentials between skilled and unskilled workers are rising in many regions, driven by skill-biased technological change (Tan 2008). For example, a study comparing the weekly earnings of Indians and Chinese reveals that median and mean earnings have risen faster in China than in India as education levels in China have risen (Bargain and others 2007). Moreover, growth of highly skilled occupations has been significantly faster than growth of less skilled ones, a trend reinforced by imports of equipment and technology, which can raise demand for skills (Tan 2008). Finally, innovation surveys from Organisation for Economic Co-operation and Development (OECD) countries such as the United Kingdom show that sectors and firms with more highly educated workers are likely to be more innovative (Miles, Green, and Jones 2007).

These trends suggest that education, skills development, and training are key elements of a knowledge-based, innovation-driven economy and affect the supply of and demand for innovation. Human capital and skilled labor complement technological advances: new technologies cannot be adopted in production without sufficient workforce training and education. The demand side is also important, as innovation may not occur if demanding customers and consumers are lacking. This issue applies to both the formal and the informal sectors in developed as well as in developing countries.

Countries able to coordinate policies for education, skills development, and innovation are certainly better positioned to compete in the global economic environment. Indeed, a number of countries are now seeking to do this. Furthermore, today's innovation policies look for new sources of innovation among workers, consumers, and users engaged in formal and informal organizational and learning activities, as innovation is increasingly inspired by social changes and consumers.

This chapter seeks to answer the following questions:

- Which generic skills are needed in an increasingly networked global economy?
- What kinds of skills are important for fostering all types of innovation (product, process, organizational, marketing) across sectors?
- What can international experience and case studies teach about fostering innovation through education, training, and skill development initiatives?

- Finally, how do countries prepare for the knowledge-based economy, and how do they tackle the challenges of education and training reform to create an enabling environment for innovation?

The Demand for Skills in the Age of Innovation

It is difficult to distinguish skills that drive innovation from those required as a result of changes brought about by innovation, an area that calls for further analysis (Miles, Green, and Jones 2007). Nevertheless, it is possible to say something about the nature of skills required for innovation and the implications for policy. Earlier findings show that innovation requires managerial and communication skills in addition to a supply of well-trained scientists and engineers. As innovation is arguably becoming more widely distributed and "democratic," it is also important for the general workforce to be able to engage with, and adapt to, innovation.

Generic Skills for Innovation. The forces at work in a knowledge-based economy clearly indicate the need for a certain set of generic skills across industries, economies, and regions. The ability to innovate will increasingly require individuals to be able to understand the nature of problems and to have the aptitude and creativity to address them. Employees are now expected to move quickly between areas of expertise and to acquire new skills to keep pace with rapidly changing knowledge. Research and development (R&D) is only the tip of the technology development and innovation process, which includes, in addition, such non-R&D activities as the skills for acquiring, using, and operating technologies at rising levels of complexity, productivity, and quality; and the design, engineering, and associated managerial capabilities for acquiring technologies, developing a continuous stream of improvements, and generating innovations. General skills thus become more useful than specialization. As a result, and because skills and knowledge can become quickly outdated, a person's capacity and potential are valued over his or her academic specialization and qualifications.

Since most of the knowledge that companies use for innovation, especially in developing countries, comes from outside, their "absorptive capacity"— that is, their ability to recognize the value of new external information and to assimilate and apply it—becomes essential for innovation (Allinson 2006). This challenge requires a broad set of platform skills provided by a good general education beyond primary education.

The Need for a Set of Key Competencies. The rapidly advancing knowledge frontier warrants a stronger emphasis on generic skills that provide the basis for adaptability and continuous learning. In fact, what is arguably needed are *competencies* that go beyond knowledge and skills to include psychosocial elements such as values, attitudes, and the ability to apply skills in a particular context:

- *Cognitive, academic, and technical.* Possessing skills in language, symbols, text, logic, mathematics, and technology and the ability to use them purposively and interactively are becoming more important with the spread and evolution of ICT and globalization.

- *Problem solving.* Capacity to observe, analyze, think critically, question, challenge, identify parts of a problem, suggest creative solutions, and innovate is increasingly necessary to compete in an innovation-driven economy.

- *Creativity.* A key feature of innovativeness is the ability to combine knowledge across fields—from science to technology to art and design—by "thinking outside the box." It also requires the confidence to take risks (Florida 2004). The importance of creativity for innovation in the laboratory and the factory and the value of creative industries in the economy and society have been highlighted in recent debates on innovation. Florida (2004), for example, shows that the most successful city-regions in the United States are those with a social environment open to creativity. Arguably, creativity is as important in developing countries, where the lack of endowments often puts a premium on finding creative ways to overcome challenges and to be competitive and profitable.

- *Social and interpersonal skills.* The ability to interact and communicate, relate well to others, work in a team both as a member and as a leader, cooperate, negotiate, manage and resolve conflicts, construct arguments, and develop social and professional networks becomes critical in the context of increasingly networked, multidisciplinary, complex, and global innovation processes.

- *Work ethic.* Demonstrating commitment, interest, motivation, and responsibility as well as flexibility and adaptability at work is necessary in a rapidly changing world. An entrepreneurial mindset, which involves a certain degree of risk taking but also goal setting, planning, and initiative, is widely considered a necessary attribute in innovation-friendly societies.

- *Continuous and independent learning.* Motivation to learn, learning to learn, learning independently, concern with one's own development, knowledge of one's capacities, self-confidence, ability to form and conduct life plans and personal projects and to defend and assert one's rights, interests, limits, and needs are becoming essential skills and competencies.

- *A premium on innovation management skills.* Innovation processes are increasingly distributed or "open," requiring clusters of firms and other stakeholders to work together. These processes require managerial skills to form and sustain collaborative arrangements for innovation. An ability to coordinate activities, select appropriate (and appropriately skilled) individuals, assemble teams, motivate and inspire, resolve problems and disputes, generate a creative (and protected) environment, communicate up and down the supply or value chain, and provide focus and leadership are just some of the skills required of managers and innovation leaders in contemporary organizations (Deschamps 2005). Beyond these, management of the innovation process requires an ability to manage and maintain the complex of intra- and interorganizational relationships that frequently characterize both large and more modest innovation projects.

Specific Skills for Innovation. The economic advantage in a knowledge economy comes from the capacity to innovate by producing marketable goods and services. The specific skills needed to nurture this capacity have to be seen in the context of the innovation process. While innovation involves the introduction and sale of new or improved products (product innovation) and the introduction and use of new methods of production (process innovation), it also includes economic and social dimensions and activities that fall under the general heading of business innovation:

- Introducing new forms of business organization, such as franchising, cooperatives, joint ventures, outsourcing agreements, and just-in-time manufacturing
- Finding new uses and applications for existing products
- Developing new markets for existing products and services and new sales and distribution channels (such as market differentiation and Internet-based sale of goods and services).[1]

Different types of innovation may require different kinds of skills and competencies. Through research on innovation (Tether and others 2005), it is possible to highlight the skills needed in specific contexts. The notion of the product cycle helps show how the innovation process triggers changes in the demand for skills and how the evolving skills profile of the organization shapes the direction of subsequent innovation capacity (Tether and others 2005). The results indicate that a single qualification rarely provides all the skills needed for innovation in a person's working life.

The Supply of Human Capital for Innovation

Human capital, the driving force behind the development of a knowledge-based economy, is severely limited in both developed and developing countries. Developed countries offer evidence of skills shortages' affecting innovation performance. For example, the European Innobarometer Survey (2001) showed that the lack of appropriate human resources was the most cited impediment to innovation in the United Kingdom. The Scottish Employers Skill Survey (Futureskills Scotland 2004) found that an inability to fill vacancies with adequately skilled workers caused delays in developing new products in 30 percent of firms and difficulties in introducing new work practices in 24 percent.

Basic mathematical and literacy skills are essential for a knowledge economy. However, in only five OECD countries do more than two-thirds of young people reach or surpass level 3 in reading literacy (comprehension and interpretation of a moderately complex text) in the Program for International Student Assessment (PISA) (OECD 2007). Most OECD countries have a significant minority, even a majority, of students with very low performance in mathematics. Except in Finland and the Republic of Korea, at least 10 percent of students score at PISA level 1 or less in all OECD countries. In 13 OECD countries, they account for one-fifth or more of students.

The situation is even more serious in many developing countries. Evidence from international student assessments suggests that some developing countries and transition economies significantly lag industrial countries in providing people with the skills needed in the knowledge economy. In many developing countries, coverage is insufficient, access is inequitable (especially in tertiary education and employee and adult training), and the quality of education is poor. Adult literacy rates are low, and too few children complete basic education. In transition countries, the quality of education and education relevant to the market are often poor. In terms of the quantity, quality, and relevance of education, the picture of the readiness of many countries for the knowledge economy is quite bleak.

Quantity of Schooling. In a knowledge-based economy, education (especially higher-level) and platform skills are essential for growth (World Bank 2009). However, in developing regions attendance at primary school is typically low, between 69 percent in Sub-Saharan Africa, for example, and around 80 percent in South Asia; and most regions saw very little improvement in their net education enrollments between 2001 and 2006 (EdStats 2008). Secondary education enrollment is a problem in all major regions, with only about half of children in this age group enrolled (EdStats 2008). Although tertiary enrollments have tripled over the past 15 years in Africa, the regional enrollment ratio for tertiary education currently stands at only 5 percent compared to 67 percent for high-income countries (EdStats 2008). Despite recent gains in access at all educational levels, few Africans, for instance, would argue that these enrollment levels are adequate for future development. The quality of education is a further major problem.

Quality of Education. Several indicators can be used to assess the quality of education in developing countries. One is the completion rate. According to the World Bank (2007), 42 percent are still unlikely to reach the goal of universal primary completion by 2015, and 26 percent are unlikely to reach that goal before 2040. Sub-Saharan Africa remains the most challenging region, despite progress made in the past decade: only 12 percent of the region's countries are likely to meet this goal by 2015 (World Bank 2007). Moreover, despite some improvement between 2005 and 2006, the pupil-teacher ratio is also high in Sub-Saharan Africa, which had 9 of the 10 highest student numbers per teacher. Although significant improvement has been made over the last years, the ratios still ranged from 67.4 to 48.1 in 2006—compared to less than 30 percent for the rest, except for South Asia (EdStats 2008).

However, school completion rates do not reveal whether the level of cognitive skills, platform competencies, and the higher skills needed for a knowledge-based economy have been achieved. Students who have completed five or even nine years of schooling in the average developing country have not necessarily mastered the basic cognitive skills. More than half the tested students in many developing countries do not reach the literacy threshold compared to less than 5 percent in leading OECD countries (Hanushek and Woessmann 2007; World Bank 2009).[2] A UNDP study (2008), using slightly different criteria, came to a similar conclusion. In Africa, the countries with the largest shares of functionally illiterate adults are Mali (77 percent), Burkina Faso (72 percent), and Niger (70 percent) and, in Asia, Bangladesh (47 percent) and Pakistan (46 percent). In these countries, more than 70 percent of those in school often do not master the basic cognitive skills. Completion rates, therefore, may camouflage the extent of the lack of quality. Combining the quantitative data on schooling and the qualitative data on cognitive skills makes clear the truly staggering task facing many developing countries.

Relevance of Education. Global competitiveness puts a premium on the relevance of education to the needs of the country and its employers. In the past, educational quality and relevance were often viewed as synonymous; high-quality secondary, vocational, and tertiary education was by definition relevant education. But this is not necessarily the case. Today, high-quality education can be irrelevant to a country's ambitions or to the regional economy around a university campus. Irrelevant education increases the chances of graduate unemployment and brain drain and deprives a nation of an important instrument for development (World Bank 2009). The phenomenon is more pronounced in developing countries. In Africa, for example, mismatches between the education provided and the capabilities required in the job market reportedly contribute to high unemployment among graduates: 35 percent in Mauritania and 17 percent in Nigeria (Teferra and Altbach 2003). In Cameroon, Côte d'Ivoire, Madagascar, Mauritania, Niger, Nigeria, Senegal, and Uganda, the higher the level of education is, the higher the incidence of unemployment (Amelewonou and Brossard 2005).

Developing countries often suffer from a shortage of labor with mid-level craft skills as well as high-level skills. As well-equipped technical and vocational schools beyond the formal primary and secondary education are also in short supply, poor countries will have difficulty moving beyond subsistence agriculture without an adequate supply of personnel trained in a mid-level craft curriculum that prepares them for diverse tasks such as repairing automobiles, repairing and maintaining electrical appliances and electronic equipment, and designing and constructing facilities such as rainwater harvesting systems and schools (World Bank 2007).

The situation is more dire at the tertiary level. Feedback from employers indicates the need for more relevant tertiary education and research. Employer surveys report that tertiary graduates have weak high-level and platform skills such as problem solving, business understanding, computer use, communication, and teamwork skills (Larsen, Kim, and Theus 2009). Employers in Nigeria reported a "total lack of practical skills among technology graduates," and Ghanaian firms voiced similar complaints. This situation lowers competitiveness and the absorptive capacity of companies to innovate.

At the same time, shortages of highly skilled labor for innovation prevail throughout developing countries from Africa (World Bank 2009) to Asia (Froumin and others 2007; Tan 2008). For instance, in India, with average annual growth of over 8 percent (until recently) and strong growth in many sectors, the education and workforce development system is struggling to respond to rapid growth in the demand for skilled labor. Termed the "Bangalore Bug," the skills scarcity faced by the information technology and financial services industries is spilling over to other industries, including those that employ less skilled workers. Recent studies show that the manufacturing sector is losing skilled workers to more knowledge-intensive sectors (Kocchar and others 2006). At the same time, industries that have grown rapidly due to an increased specialization in high-skill services (information technology, finance, and telecommunications) and in skill-based manufacturing (pharmaceuticals, petrochemicals) face shortages of the highly skilled labor that they need to maintain their edge. Similar situations exist throughout the developing world (for Africa, see World Bank 2009).

The array of specific skills needed in a knowledge economy beyond generic skills appears to be underrepresented in developing countries and contributes to the problem. An indication, again using Africa as an example, is the distribution of the disciplines studied (table 6.1). On average in 2005, just 28 percent of students were enrolled in science and technology fields (agriculture, health science, engineering, sciences). Much of Africa's recent growth in enrollments (including in private institutions) has occurred in the "soft" disciplines, a trend that is unlikely to provide the knowledge and specific core skills that African nations need to boost competitiveness and growth (World

Table 6.1 Distribution of African University Graduates by Field of Study, 2005

Field of study	Distribution of graduates (%)
Agriculture	3
Education	22
Health science	7
Engineering	9
Sciences	9
Social sciences and humanities	47
Other	3

Source: EdStats 2008.

Bank 2009). Given the dismal state of the quantity, quality, and relevance of education for the knowledge economy in many developing countries, the question is, What can be done?

Lessons from Developed and Developing Countries

Adaptability to changing circumstances and readiness to obtain new work-related knowledge and skills have become increasingly important for workers and their employers and require lifelong learning. Lifelong learning involves not only continuous upgrading and retraining but also continuous personal and collective learning, growth, and development. To create and leverage lifelong learning for the knowledge economy first requires laying a foundation of basic literacy, then expanding and adjusting formal education from primary to higher education, next providing training and education outside the formal education system, and finally adopting new technologies for distance and networked learning.

A Good General Education and the Development of Platform Skills

The spine of any educational system in the innovation-driven economy is the quality and reach of its primary and secondary education and the competence of its teachers. This is the basis of subsequent learning. Access to primary schooling or basic education in developing countries has improved greatly during the past decade. However, many pupils do not master the competencies, from problem solving to teamwork, that are necessary for adaptation to an innovation-driven economy and for entering a system of lifelong learning (see Filmer, Hasan, and Pritchett 2006). While nearly all countries' educational systems are expanding *quantitatively*, nearly all are failing in their fundamental purpose. Policy makers, educators, and citizens alike need to focus on the real goal of schooling: to equip their nation's youth for full participation as adults in the economic, political, and social activities of the knowledge economy.

To this end, policy needs to promote up-to-date quality and higher standards in learning, teaching, and teacher education. Ways need to be found to adapt educational systems, meet the strong demand for secondary and higher education in developing countries, and deal with the fading frontier between general and vocational education, which is the consequence both of the forces of the knowledge-based economy and of donors' efforts to improve basic education in the 1990s.

Literacy: The Foundation of Lifelong Learning

Lifelong learning starts with the ability to read and write. The International Adult Literacy Survey measures literacy on five levels, with level 3 the minimum required to function in the knowledge economy. The standard for literacy at this level comprises three levels:

- *Prose literacy.* Learners should be able to locate information that requires low-level inferences or that meets specified conditions. They should be able to identify several pieces of information located in different sentences or paragraphs. They should be able to integrate or compare and contrast information across paragraphs or sections of text.
- *Document literacy.* Learners should be able to make literal or synonymous matches. They should be able to take conditional information into account or match up pieces of information that have multiple features. They should be able to integrate information from one or more displays of information and to work through a document to provide multiple responses.
- *Quantitative literacy.* Learners should be able to solve some multiplication and division problems. They should be able to identify two or more numbers from various places in a document. They should be able to determine the appropriate operation to use in an arithmetic problem (OECD and Statistics Canada 2002).

The lack of adequate basic skills directly affects the potential of the knowledge economy. Illiteracy limits the capacity to acquire the basic skills needed for an innovation economy and curbs the productivity potential of the informal and lower-skill sectors. Technological literacy and access to ICT resources are important at the basic level. If individuals do not obtain adequate skills at the basic level, whether through formal or informal education, fewer qualified workers will be available to participate in labor-intensive industries and fewer skilled workers will be available for the innovation system as a whole.

Many children in developing countries face a significant hurdle when they enter formal schooling because the language of instruction is not spoken at home. South Africa's poor performance on the TIMSS,[3] for example, appears to be attributable in part to the high proportion of learners for whom English (the language of the test in South Africa) is a second language (Howie and others 2000). Children are more likely to enroll in school, learn more, and develop positive psychological attitudes—and to be less likely to repeat grades or drop out of school—when initial basic education is offered in their first language, or at least in a language they understand (Klaus, Tesar, and Shore 2002).

To improve the basic skills of their citizens as a foundation for lifelong learning, governments have to focus on two types of reforms. On the one hand, they need to develop innovative approaches to improving the quality of primary and secondary education by modernizing curricula and pedagogy, training teachers accordingly, and creating a more flexible and responsive educational system. ICT literacy must have greater prominence in the early years of education to prepare students for an increasingly ICT-dominated world.

On the other hand, it is equally important to strengthen basic skills, including functional literacy, for the informal sector. Governments should invest in programs that combat illiteracy and help transfer skills to youth and adults in the informal sector by supporting local nongovernmental organizations that provide training that is adequate to meet the needs of the informal economy. This effort should include training instructors, developing curricula, and providing financial incentives to encourage external financing of training programs. In addition, the government should provide regulatory and financial support for informal education through focused, short-term courses and programs, such as training in information technology literacy (Froumin and others 2007).

Adapting the Way Learners Learn to the Knowledge Economy

Traditional educational systems in which the teacher is the sole source of knowledge are ill suited to equipping people to work and live in a knowledge economy. Competencies such as teamwork, problem solving, and motivation for lifelong learning are not acquired in a learning setting in which teachers convey facts to learners whose main task is to learn and repeat them. Providing people with the tools they need to function in the knowledge economy requires a new pedagogical model, which differs from the traditional one in many ways (table 6.2).

The lifelong learning model enables learners to acquire the new skills required by the knowledge economy as well as more traditional academic skills.

Table 6.2 Characteristics of Traditional and Lifelong Learning Models

Traditional learning	Lifelong learning
The teacher is the source of knowledge.	Educators are guides to sources of knowledge.
Learners receive knowledge from the teacher.	People learn by doing.
Learners work by themselves.	People learn in groups and from each other.
Tests are given to ensure that students have mastered a set of skills and to ration access to further learning.	Assessment is used to guide learning strategies and identify pathways for future learning.
All learners do the same thing.	Educators develop individualized learning plans.
Teachers receive initial training plus ad hoc in-service training.	Educators are lifelong learners; initial training and ongoing professional development are linked.
"Good" learners are permitted to continue their education.	People have access to learning opportunities over a lifetime.

Source: World Bank 2003a.

In Guatemala, for example, learners taught through active learning—that is, learning that takes place in collaboration with other learners and teachers, in which learners seek out information for themselves—improved their reading scores more and engaged more in democratic behavior than learners not in the program (see box 6.1).

A lifelong learning system must reach larger segments of the population and address diverse learning needs. It must be competence driven rather than age related. Traditional institutional settings require new curricula and new teaching methods. At the same time, efforts need to be made to reach learners who cannot enroll in programs at traditional institutions.

Aspects of an Effective Learning Environment for the Knowledge Economy

Recent studies commissioned by governmental and international bodies (for instance, Kozma 2003; OECD 2004) have shown that the standard model designed to transmit knowledge to learners is not sufficient to prepare a person as a knowledge worker. Specifically, factual and procedural knowledge is useful only when a person knows how to apply it and how to modify it for new situations, a skill that is essential in the knowledge-based economy. Effective learning environments are based on the way people learn (for a more detailed analysis, see OECD 2008b). According to the findings of education science, this involves several components:

- *Deep conceptual understanding.* When students gain a deeper conceptual understanding, they learn facts in a more useful and profound way and can transfer them to real-world settings.

Box 6.1 What Does a Learner-Centered Classroom Look Like?

Guatemala's Nueva Escuela Unitaria (NEU) program tackles some of the country's poorest and most isolated rural schools. The classrooms in the program reflect the program's learning-centered model:

> One seldom observes any large-group, teacher-dominated instruction. Rather groups of two to six students at a particular grade level can be seen working at a table, a learning corner, the library, or outside working in their self-teaching workbooks. The large chalkboard has been removed from most NEU classrooms, and while these classrooms generally have more instructional materials than a traditional, poor rural school, it is the way materials are used by students rather than their quantity that is exceptional in these classrooms. The library, always under student management, is meant to be used during the school day and books borrowed overnight rather than kept under lock and key. . . . [Evaluations] indicate a very low level of student discipline problems and an extremely high interest level by students "doing their work." (Craig, Kraft, and du Plessis 1998, 89)

Source: World Bank 2003a.

- *Learning by children.* Students think about problems in different ways from educators; teaching content and pedagogy should reflect this difference.
- *Prior knowledge:* Standard schools were developed under the assumption that children have no prior knowledge. This approach does not take into account the individual knowledge structures that students possess when they enter school.
- *Articulation.* Learners learn more effectively when they externalize and articulate their developing knowledge, because the student can think about knowledge and the process of learning. This aspect is an important component of seeking deeper understanding of content.
- *Structuring of articulation and thinking.* Articulation and thinking are more effective if guided in ways most beneficial to learning. Some structuring to tailor and channel knowledge may need to occur so that certain types of knowledge are articulated and lead to reflection.

In the absence of consensus on teaching models, which are often context specific (see, for example, OECD 2008b), certain aspects of learning environments appear to meet the skills demands of a knowledge economy. Further empirical evidence is needed, but a trend in developed countries points toward integrating those aspects into their education practices. Falling under the so-called learner-centered education paradigm, this approach is customized and knowledge rich, involves networking and teamwork, and is assessment driven (see box 6.2) (Desforges 2001).

Box 6.2 Learner-Centered Teaching for the Knowledge Economy

The learner-centered education paradigm is based on learners' active involvement in reflection, interpretation, and self-evaluation. Knowledge and skills are acquired through exploration, drawing from the real world, and applying learning in practice. Learning is social; it occurs in interaction with others and in debating and creatively changing social practices. Learner-centered education supports deep learning and creativity (Hargreaves 2006). A learner-centered environment recognizes that learners acquire new knowledge and skills best if the knowledge and skills are connected to what they already know. Teachers need to know what learners already know and understand before introducing new material. Learner-centered learning allows new knowledge to become available for use in new situations; that is, it allows knowledge transfer and adaptation to a specific context. It includes several specific elements:

- *Customized learning.* Credit hours and time in the classroom may not necessarily be linked in learner-centered education. Although students with background knowledge and experience in a content area may quickly master the course material and required skills, others may need more time and additional help. Consequently, students in learner-centered environments often complete courses at different rates.

continued

> **Box 6.2 continued**
>
> - *Knowledge-rich learning, learning by doing, and learning by using.* Learners' ability to transfer what they learn to new contexts requires a grasp of themes and overarching concepts in addition to factual knowledge as well as its applications. Knowledge-rich learning thus favors teaching fewer subject areas in depth over more subjects in less depth. Learning by doing and learning by using are important ways that learners master the knowledge and concepts being taught. This kind of learning provides learners with a variety of strategies and tools for retrieving and applying or transferring knowledge to new situations.
>
> - *Interconnected networking and teamwork.* In a knowledge economy, it becomes paramount to collaborate with other parties and tap into the global stock of knowledge. Learners also have to be able to learn from one another. Giving learners, both children and adults, the opportunity to work on joint projects is important. Indeed research has shown that learning can accelerate in collaborating student groups (Sawyer 2006). Activities inside the classroom should be linked to what is happening outside the classroom. Working on real-life problems or issues that are relevant to participants increases interest and motivation and promotes knowledge transfer (Appalachian Regional Commission 2004). Moreover, learners need to understand and access sources of information and knowledge outside the classroom.
>
> - *Assessment driven.* Assessment-driven learning is based on defining clear standards, identifying the point from which learners start, determining the progress they are making toward meeting standards, and recognizing whether they have reached them. Assessment-driven learning helps the educational system define the instructional action plan, which needs to reflect learners' points of departure. Education researchers currently experiment with how to reconcile this approach with the accountability standards to which schools are held. However, a consensus is gaining ground that giving learners—even very young learners—a role in tracking their learning achievements and especially engaging them in discussion of the outcomes of these assessments are powerful motivators and tools for improved and independent learning.
>
> *Source:* Author's compilation.

To date, these characteristics are not part of the standard model of schooling, where learning is standardized, knowledge sources are often limited to the teacher and the textbook, most learning is that of solitary learners, and assessment in many developing countries measures the memorization of superficial facts and procedural knowledge. Some of the characteristics of the learner-centered model can be introduced into the standard model; for example, existing classrooms can introduce collaborative learning tasks, as many schools are doing today. But others will be extremely difficult to include in the standard model, such as the notion of customized learning, especially for accountability, which relies on standardized assessments. Efforts are being made to find models of learning that include those characteristics (OECD 2008b).

Adaptation of Curricula to the Forces of the Knowledge-Based Economy
The changing work patterns in the knowledge-based economy are leading to radical new approaches to selecting, organizing, and sequencing the curriculum. In primary and especially in secondary education, greater emphasis on the democratization of access to knowledge, on the formation of social capital, and on a better understanding of youth issues and of how adolescents learn as well as general advances in understanding effective learning environments are all affecting the curriculum. The need to build "creative capital" is reflected in the growing importance of interactive teaching methods and active learning, case-based training, simulations, and team projects—in short, a curriculum based on problem solving.

In many developing and transition countries, however, the secondary education curriculum often remains abstract and unconnected to social and economic needs. It is largely driven by the high-stakes public examinations introduced in many of these countries by the colonial powers that still hold the key to university access and to elite professional jobs (World Bank 2005a). Abstract, fact-centered, and decontextualized narrative knowledge continues to be used as a criterion for selection in a setting of scarce educational and job opportunities and causes high dropout and failure rates among secondary school students (World Bank 2005a). Curriculum relevance not only improves the quality of secondary graduates but also helps retain young students in school.

With progress being made toward mass secondary education and with knowledge becoming the basic economic resource of society, curriculum issues today are less concerned with imparting vocational skills to secondary graduates than with adding basic vocational content to the general curriculum. At the heart of the debate is the question of which school subjects are vocationally relevant. Science, mathematics, English, and philosophy—all traditionally viewed as academic college preparatory subjects—are in increasing demand because of their relevance for careers and work (World Bank 2005a). This trend has blurred the hitherto clear boundary between general and vocational secondary education. Even countries that seek to enhance the labor market relevance of graduates through a secondary education curriculum that strongly emphasizes occupationally oriented skills and competencies tend to ensure at the same time that strong and up-to-date general content remains the central component of the curriculum (World Bank 2005a).

This balance is reflected in the emergence of new subject areas and a continuing reweighting of traditional types of knowledge in school curricula. The topic is strongly debated as more and more subjects, both disciplinary and interdisciplinary, have become socially and economically relevant and seek to occupy a significant place in the curriculum. This is the case, for example, of technology, economics, citizenship education, second foreign language, environmental education, and health (World Bank 2003a).

The growing demand for a scientifically and technologically literate workforce presents further challenges. Efforts are being made to discontinue the practice of using science and technology education to develop future scientists and technologists and instead to retrain science teachers to impart knowledge about science as a process of inquiry rather than as a collection of discoveries.

The importance of rapid and easy communication in labor markets and the increased need to understand people of different cultures and nationalities and to tap into global sources and networks of knowledge create demand for competence in more than one foreign language. Foreign language skill is directly related to a country's overall marketing capabilities. To meet this need, recent curriculum reforms in many developed countries have made the study of at least two foreign languages compulsory.

These trends seem to call for more differentiation of the curriculum to maximize students' potential while keeping a strong general education. As student populations grow and become increasingly diverse, responding to differentiated demand with a more customized curriculum may be the only way to prevent student dropout and achieve high completion rates. Policy choices for curriculum differentiation include tracking students according to their academic ability or achievement or permitting students to choose from a variety of electives, options, or curriculum modules that can be sequenced and accredited in different ways. In many subject areas, common tasks can be set with the expectation of differing levels of achievement depending on students' needs and capabilities.

However, some words of caution are in order. It has been demonstrated, for example, in Eastern Europe that subject overload can inhibit a student's learning (OECD 2006). It also runs counter to the principle of deep conceptual understanding for effective learning. Finland, which excels on the PISA tests, is the OECD country with the smallest number of "intended instruction hours" for 7-to-14-year-olds, less than 70 percent of the total hours in Italy (OECD 2008a). Furthermore, *diversification* and *profiling* are terms with strong political connotations in many countries. They are at the core of the heated political debate that often surrounds equal access to secondary education and recurring waves of reform. Tracking, streaming, banding, and other student grouping arrangements are the practical outcomes of political and ideological stances pertaining to secondary education.

Youth Entrepreneurship Programs. Youth entrepreneurship programs provide an example of how relevant skills are being taught. They prepare young people to be responsible, enterprising individuals who become entrepreneurs or entrepreneurial thinkers and contribute to economic development and sustainable communities. Entrepreneurship education provides opportunities for youth to master core entrepreneurial knowledge, skills, and attitudes, including recognition of opportunities, generation of ideas, creation of ventures, and critical thinking. Entrepreneurship education for youth (see http://www.entre-ed.org) may be described as a process that seeks several specific goals:

- Providing opportunities for youth to start and operate appropriate enterprises
- Reinforcing the concept that successful entrepreneurs take calculated risks based on sound research and relevant information
- Requiring youth to develop a plan for a business that includes its financial, marketing, and operational aspects
- Portraying the relationship between risk and reward and providing opportunities for young people to understand basic economic concepts such as savings, interest, supply, and demand
- Generating an understanding of a variety of industries.

Young people with experience in creating small businesses, either alone or as a team, learn how to apply entrepreneurial skills to many different situations. They learn how to find their best customers, how to manage their finances, and also how to learn from their failures. They learn how to use their resources and start small, reinvesting the profits until the business is as big as they wish it to be. And they learn how to create a business plan. These are experiences that make a difference in the way students identify opportunities, generate ideas, and think critically.

Entrepreneurship courses provide experiential learning opportunities in many ways. School-based businesses have become popular components of youth entrepreneurship courses in recent years and offer excellent opportunities for creative thinking and problem solving. Entrepreneurship education is being developed as a full course in itself or as a unit of study in another class. It can be quite simple in the earlier grades and quite complex in higher education. Adults often use such courses to refine their ideas before actually starting their business or to increase their knowledge when already in business. In the United States, and also in Singapore and Paraguay, entrepreneurial programs in secondary and tertiary education have been pioneered with notable success (see box 6.3).

Implications for Teachers and Teachers' Training

The shift from industry- to knowledge-based school organization and lifelong learning directly affects teacher training and deployment. More flexible arrangements are needed to allow teachers to assume the role of facilitator for learners. The quality of teachers and teaching is perhaps the most important aspect of successful learning. Education decision makers today face the problem of attracting able graduates to the teaching profession and retaining them. In developing countries, especially in Africa, shortages of teachers, particularly in mathematics, science, and technology, pose a major threat to

> **Box 6.3 Entrepreneurship Program at Walhalla High School, South Carolina, United States**
>
> After Walhalla, a county seat in South Carolina, was recognized by the U.S. Department of Education as a New American High School, it used a US$60,000 grant from the Ewing Marion Kauffman Foundation to promote entrepreneurial thinking by students and faculty. The grant was intended to stimulate creativity rather than emphasize business mechanics like accounting and marketing.
>
> In 1998, the school added an entrepreneurship unit to a course taken by all ninth graders and created the Entrepreneurship 1 and 2 courses. In Entrepreneurship 1, students work individually or with partners. The class includes a mix of conventional academic work and hands-on projects. On the academic side, students read and write reports on entrepreneurs whose products, if not their names and history, they already know. On the experiential side, they interview county business owners and learn how entrepreneurs in their own backyards turned ideas into successful enterprises. The students also develop business plans of their own. They make them as realistic as possible, even if they have no intention of putting them into practice. Students submit their plans in writing to EntreBoard, a six-person panel of local entrepreneurs and supporting professionals. The plans judged best earn substantial prizes for their originators. The students, aware that they are being judged by real business people, not "just teachers," work hard on their presentations.
>
> In their second semester, class members work together on a project and are given a chance to make money. The students launch a business that meets two criteria: it must be doable during school hours (or immediately before or after) and be based on a product or service that students and teachers will pay for not just once, but repeatedly. For practical purposes, these considerations point to food. Students can raise start-up capital however they choose. This often involves fundraisers.
>
> The program has been a huge success. It helped attract students from different backgrounds, keep students motivated and in school, and hone a variety of skills from teamwork to taking responsibility and creative thinking; in some cases, businesses were so successful that they made a profit of US$5,000.
>
> *Source:* Appalachian Regional Commission 2004.

the goals of expanding education and enhancing its quality. In Lesotho, for example, it was estimated that almost half the school completion cohort would have to choose teaching to meet predicted demand (Lewin 2002, 229). Teacher absenteeism is another severe problem. A government that cannot ensure that its largest expenditure is yielding even the most basic of returns—that is, that its teachers are actually going to their classrooms—is unlikely to be effective at ensuring that students are learning. Comprehensive policies to attract and retain high-quality teachers need to be designed. They should address the effective and dynamic integration of teachers' professional development and career issues, teacher deployment policies, class size, and monitoring and evaluation practices (for instance, Halsey and others 2006).[4]

A knowledge-based economy requires teachers who are knowledge workers and designers of learning environments and who have the ability to take advantage of the various sources of the production of knowledge. Apart from in-depth knowledge of their discipline, they also need to know how best to teach their specific subject to primary and secondary school students. Furthermore, they have to update their knowledge regularly in all these areas.

Knowledge about teaching is for the most part tacit, difficult to articulate and systematize, and practical and context-based. This complexity makes it difficult to transfer and fully use teachers' knowledge about their teaching. In short, teacher education institutions, schools, and educational systems in general are still very far from meeting the needs of a knowledge-management society (World Bank 2005a). In recent years, however, empirical research and practical experience have pointed to some effective teacher education strategies (for instance, World Bank 2005a; OECD 2005):

- Emphasize quality over quantity.
- Develop teacher profiles to align teacher development and performance with school needs.
- View teacher development as a continuum and a process of lifelong learning.
- Devise more flexible entry and education schemes for teachers.
- Transform teaching into a knowledge-rich and innovative profession.

Beyond Formal General Education

Beyond formal general education, vocational education and training play a crucial role, as well as other informal mechanisms by which youth and adults acquire needed skills.

Vocational Education and Training

In many countries, reforms in the education and training system have tended to concentrate on expanding general education and academic pathways and have given comparatively little attention to vocational education and training (VET) in the process of structural adjustment. Moreover, in many countries VET is part of secondary education and is organized and delivered by colleges or schools, an approach that does not adequately prepare graduates to meet the demands of the world of work. Assessments have shown that employers are often not satisfied with the quality of vocational education and training. In particular, they complain of the low quality of training schemes, trainees' lack of practical skills, and inappropriate training content (see Froumin and others 2007; Larsen, Kim, and Theus 2009). However, economic development depends to a great extent on adapting VET systems to meet social and economic demands. Developing countries' VET systems need to be better aligned to market needs to meet the preservice training requirements of enterprises.

Moreover, many countries need more vocational training. In India, for example, only 3 percent of rural youth and 6 percent of urban youth have received vocational training (Froumin and others 2007).

For this reason, many countries have highlighted the need for greater emphasis on VET in the years to come by taking several specific actions:

- Providing attractive, qualified training programs and continuing training opportunities to enhance employability and occupational mobility
- Designing VET to conform more closely to practice
- Orienting VET to the requirements of the employment system and labor market needs
- Preparing young people for degrees that comply with high standards while opening up forward-looking employment prospects (BIBB 2004).

Updating the Curriculum and Pedagogy. The vocational curriculum should be designed so that graduates are prepared to meet the needs of the market and employers in a knowledge-based economy. It should be updated to reflect modern technologies, with input from the private sector, and should be flexible (Dar 2008). Both the curriculum and the teaching may be context specific and should fit the country's level of development and the structure and characteristics of its economy.

It is important for the curriculum to maintain links with the world of work while keeping the flexibility necessary to explore ideas theoretically. Experience has shown that, in a knowledge-based economy with rapidly changing skill demands, the teaching of practical skills for a certain occupation has to be supplemented with some theoretical training. Those theoretical skills and competencies foster adaptability and absorption of new knowledge. The German dual system of vocational training is a good example, as it combines formal learning at school and on-the-job training in enterprises.

A review by the World Bank also found three pedagogical tools that appear promising, if they are closely and competently supervised: work placement with an entrepreneur as part of the school program, establishment of student enterprises, and compulsory development of a business plan that includes planning specified production, assessing the market, and writing a cost and financing plan (Johanson and Adams 2004). A number of innovations in the design of VET systems have been created over the past years in developing countries to meet the objectives outlined above and to link VET to the demands of a knowledge economy.

Dual Modes of Training. Dual systems link the school and the firm as the two places of learning and focus on work-based learning to acquire vocational competencies. The German dual system can serve as the archetype. Various countries have introduced a dual training system, including Korea as well as

African countries such as Côte d'Ivoire and Namibia. German-speaking countries in Europe have shown dual training to be an effective means of familiarizing trainees early with the work environment. However, the implementation of dual training programs raises several difficulties. First and foremost, local enterprises must be willing and able to provide training. Second, the system requires careful organization, in-company practical training, and supervision. These conditions may be present in emerging economies but are unlikely to be met in most of Sub-Saharan Africa. The lack of an industrial fabric in this region is a major obstacle to the development of dual training. Contacts with enterprises have been more productive when they are part of a policy of opening up training centers, for example, where the centers offer modules of continuing education (Johanson and Adams 2004).

National Qualification Frameworks. National qualification frameworks (NQFs) reflect a conceptual shift from the classic focus on the input process to a more modern focus on outputs and toward adherence to a market-oriented policy agenda. NQFs are based on the assumption that individuals bear the primary responsibility for training. First developed in Australia, New Zealand, and the United Kingdom, the concept and the principles of NQFs are being adopted and implemented in an increasing number of countries, including in Africa (for example, in Botswana and South Africa). NQFs can encourage individuals to continue their education and training by establishing specific steps toward increased qualifications (and incomes). They can lead to cost-effective training by focusing on outcomes regardless of how the skills are obtained, whether in classrooms or out of school. NQFs stress the competencies acquired, not the avenues or the institutions that teach the skills. NQFs can also promote job mobility and therefore increase labor market efficiency (Johanson and Adams 2004). Because NQFs are quite new, little empirical information is available. The application of NQF models in developing country contexts may be problematic because of the different and much weaker educational, economic, and institutional environments, which may not be relevant for countries with low enrollments or for those whose main problem is insufficient access to skills rather than inadequate quality of assessment. The South African experience, for example, has been a complex, bureaucratic, and slow process.

Competence-Based Training. A more realistic objective for less developed countries may be the establishment of competence-based training systems. This objective shifts the emphasis from what courses a trainee has taken and when to what the trainee can do. Competence-based training is usually modular and, in theory, facilitates flexible entry and exit and recognizes different routes for skills acquisition. Tanzania and Zambia, among others, have introduced competence-based training. Implementation is complex and includes the development of standards-based on-the-job analysis, the preparation of new modular curricula, and the design of assessment methods and new performance tests (Johanson and Adams 2004). Competence-based training focuses on the skills needed for performance in a job, and it puts pressure on instructors and center management to deliver these skills. It can lead to a reduction in training duration as well as greater flexibility. One of the lessons of implementation from Tanzania is the need to involve employers in the process and to publicize the concepts widely so that they are understood by enterprises, parents, and trainees (VETA 2002).

External Training Institutes and Programs as Providers of Specific Skills
The most successful reforms appear to be those that combine public financing of training programs with rigorous evaluation of program impact and competition between providers for delivery (Dar 2008). This combination is particularly relevant for skills upgrading. For the development of specific vocational skills, specialized training centers may be more suitable, because they are able to respond to the labor market and have stronger institutional links to that market than secondary schools.

In-Service Training. In-service training is distinguished from vocational education and pre-employment training by the fact that it takes place in the workplace (on the job), is specifically job relevant, and is often relatively informal. Firms in most countries cite in-house training, private institutes, and other firms as the most important sources of formal training. Exceptions are China and Singapore, which rely more heavily on public training institutions (Tan 2008).

Empirical analysis, based mostly on firm surveys in different countries, shows that the share of firms that provide in-service training is significantly correlated with innovation along several dimensions: firm size (bigger companies provide more), industry (technology-intensive industries provide more), R&D activity (those with more R&D activity provide more), export status (exporters provide more training), and foreign ownership (it raises incidence of training) (Tan and Savchenko 2007; Tan 2008). In addition, in India, firms that provide in-service training are 23–28 percent more productive than firms that do not (India Enterprise Survey 2006); other studies have also found evidence that training, especially when repeated, leads to higher productivity growth and wages (for example, Tan and Lopez-Acevedo 2005 for Mexico) especially when knowledge is becoming outdated and skills need to be adapted.

The incidence of in-service training differs broadly by countries' development level, the investment climate for business, and their formal education and training system. In many developing countries, it is still rare, in particular among small and medium companies, those that are less technology intensive, and those that are less exposed to international markets. For example, only 16 percent of Indian manufacturing firms provide in-service training to their employees, as opposed to 60 percent in Brazil and 42 percent in

Korea (Froumin and others 2007). Several major factors affect the often limited in-service training by firms:

- The low level of education of the workforce (illiteracy) makes training ineffective.
- Most firms of all sizes in the formal and informal sectors that did not provide training used "mature" technologies and did not require training or skills upgrading.
- Training was not affordable because of limited funding resources, suggesting a weakness in financial markets.
- High turnover of trained staff that seek opportunities elsewhere prevents companies from recouping the costs of training employees.
- Firm had imperfect knowledge about training (Tan 2008).

In many developed and developing countries, governments have been able to increase employee-targeted training policies by using the following incentives aimed at the private sector:

- *Grant schemes.* Government administrators use earmarked levies to provide grants to employers for state-approved training programs (Singapore and previously in the United Kingdom).
- *Rebate schemes.* Employers are partially reimbursed for approved employee training programs by drawing against their payroll levies (Malaysia, the Netherlands, and Nigeria).
- *Exemption schemes.* Employers are exempted from tax payments if they spend a given percentage of their payroll on training (France, Korea, and Morocco).
- *Tax incentives.* Tax incentives are given to approved employer training programs that are financed with general government revenues (Chile and previously in Malaysia).
- *Entitlement schemes.* Over their lifetime, employees are entitled to government funds (usually vouchers or loans) for additional training to be spent as they determine (Austria, Kenya, Paraguay, and the United Kingdom).
- *Individual learning accounts.* Individual learning accounts provide individuals with discretionary training funds partially financed by the state, employers, and employees (the Netherlands, Spain, and previously the United Kingdom) (Kuznetsov and Dahlman 2008; Tan and Savchenko 2007).

These schemes have been used with success in both industrial and developing countries. Overall experience with training levies suggests several lessons for developing countries (Gill, Fluitman, and Dar 2000). First, employers should be closely involved in the governance of levy funds; Argentina, Brazil, and Chile have vested supervision of levies in industrial bodies. Second, policies should be designed to increase competition for training provision by both public and private providers, including the employer. Third, levy funds should be strictly earmarked for training and not diverted to other government uses as has happened with training levies in several Latin American and African countries. Training levies, however, as experience from Brazil, Chile, and China suggests, do not work particularly well for small and medium enterprises (SMEs), which exhibit the lowest incidence and intensity of in-service training (Tan 2008). Mexico's experience with SME training programs offers some lessons (see box 6.4).

Apart from these incentive programs, some countries use grant-matching schemes to increase the training of their workers and create a training culture. The most successful matching-grant schemes are demand driven, are implemented by the private sector, and aim to create sustained training markets. Programs in Chile and Mauritius, for example, rely on the private sector to administer these initiatives and have achieved positive results (Tan and Savchenko 2007).

Box 6.4 Mexico's Proactive Training Programs for SMEs

The Integral Quality and Modernization Program (CIMO), established in 1988 by the Mexican Secretariat of Labor, has proved effective in serving SMEs. Set up initially to provide subsidized training, CIMO evolved when it became apparent that lack of training was only one of many factors contributing to low productivity among smaller enterprises. By 2000, CIMO was providing an integrated package of training and industrial extension services to over 80,000 SMEs each year and training to 200,000 employees. Private sector interest has grown, and in 2004 more than 300 business associations participated in CIMO, up from 72 in 1988.

All states and the Federal District of Mexico have at least one CIMO unit, each staffed by three or four promoters and housed in business associations that contribute office and support infrastructure. Promoters organize workshops on training and technical assistance services, identify potential local and regional training suppliers and consulting agents, both public and private, and actively seek out SMEs to deliver assistance on a cost-sharing, time-limited basis. They work with interested companies to conduct an initial diagnostic evaluation as the basis for organizing training programs and other consulting and technical assistance. The government does not deliver the training; instead, its role is to identify the most qualified local public and private training providers, which then deliver the training usually on a group or association basis to reduce unit training costs. This strategy is deliberate, since one of the program's objectives is to promote the development of regional training markets able to serve the needs of local enterprises. The CIMO program also targets industrial clusters and works with large firms and their SME suppliers to organize and deliver cluster-specific training programs. Several rigorous evaluations have found CIMO to be a cost-effective way of assisting SMEs.

While CIMO firms tended to have lower preprogram performance than a comparison group with similar attributes, their postprogram outcome indicators tended to show improvements in key areas such as labor productivity, capacity utilization, product quality, wages, and employment.

Source: Tan and Savchenko 2007.

Whatever training policy is eventually adopted, enterprises and employer associations must have meaningful inputs into the design of the policy so that the training system is responsive to their needs and to those of other key stakeholders in the knowledge economy. Where warranted, industry may share responsibility with government for the management and delivery of training, such as the example of Mexico in box 6.4 or the employer-owned and -managed training program for SMEs in Brazil (see Tan and Savchenko 2007). Public-private division of tasks or partnerships are a vital approach for SMEs, which have less financial leeway, in delivering demand-driven, low-cost training that is largely self-financing. Malaysia's Penang Skills Development Center is suggestive of how the private sector in different regions of larger countries can partner with state governments in the reform and management of tertiary professional and technical institutes.

Skills Development for the Informal Sector. In most African countries, 30 percent of total nonfarm employment is in the informal sector, of which two-thirds is in urban areas. Workers in the informal agricultural sector account for about half of total employment in Africa. In view of the low productivity and earnings in many micro and small enterprises, the informal sector is the safety valve for these economies and an increasingly important instrument for poverty alleviation. If the informal sector continues to absorb people and to supply a modest but reasonable return on labor, increasing the skills of its operators is crucial. Improved technical and business skills are of prime importance for enhancing the productivity of informal sector activities as well as the quality of the goods and services produced. Improved skills will strengthen the informal sector's ability to compete and innovate. Technical skills, together with other types of support (for example, access to credit, technology, markets, and information), are imperative. The majority of operators in Africa's informal sector have no formal training: 1.4 percent in Ghana, 2 percent in Tanzania, and 6 percent in Uganda, for example (Johanson and Adams 2004).

The Primacy of Traditional Apprenticeship Training. Traditional apprenticeship training is responsible for more skills development than all other types of training combined in the developing and least-developed countries, especially in Africa. Traditional apprenticeship training is also probably the most important source of technical and business skills for workers in the informal sector. In Ghana, 80–90 percent of all basic skills training comes from traditional or informal apprenticeship, compared with 5–10 percent from public training institutions and 10–15 percent from nongovernmental for-profit and non-profit training providers (Atchoarena and Delluc 2001). One important reason is that traditional apprenticeship training can be the least expensive way to get skills training (for further details, see Johanson and Adams 2004).

Although no single approach prevails, the main characteristics of traditional apprenticeships are as follows. A written or oral agreement is concluded between a "master" and parents or guardians for an apprentice to acquire a set of relevant, practical skills. Sometimes the master receives a training fee. In other situations, the apprentice has to "earn" the training in exchange for work or reduced wages. Training consists primarily of observing and imitating the master. The apprenticeship is usually for a fixed period (three to four years), and, instead of being competence based, it is product specific. Theoretical aspects and basic technical practices (for example, precise measuring) are often ignored.

Traditional apprenticeship training for the informal sector confers substantial advantages over conventional training methods and is a major provider of skills in Sub-Saharan Africa and parts of the world such as Southeast Asia (Adams 2008). The main strengths of traditional apprenticeship are its practical orientation, self-regulation, and self-financing. Apprenticeship also caters to individuals who lack the educational requirements for formal training, serves important target groups (rural populations and the urban poor), and is generally cost effective.

But it also has some disadvantages (see table 6.2 in Johanson and Adams 2004). Traditional apprenticeship is gender biased, screens out applicants from very poor households, perpetuates traditional technologies, and lacks standards and quality assurance. In many countries and business environments, apprenticeship has served the informal sector well but is proving too narrowly focused to cope with the increasing challenges of technical change, skills enhancement, and wider markets that characterize the knowledge economy. Efforts are needed to improve traditional apprenticeship training. Apart from traditional apprenticeships, little training for the informal sector exists because of constraints on the demand and supply sides (Johanson and Adams 2004).

To make traditional apprenticeship more effective, an integrated strategy needs to include the following actions:

- Start with market surveys.
- Assist the poor in financing their apprenticeship training.
- Upgrade the skills of master craftspersons.
- Enhance traditional apprenticeship training.
- Introduce supplementary training for apprentices.
- Evaluate and certify the skills obtained.[5]

From Brain Drain to Brain Circulation

International mobility of talent and its most visible manifestation, brain drain (usually defined as the migration of tertiary educated human capital from less to more developed economies) is central to learning and development.

Size of the Brain Drain

Large stocks of highly skilled (university-educated) expatriates from developing countries can be found in developed countries. Among developing countries in 2000 (the latest year for which data are available), the Philippines had the highest emigration stocks of university-educated expatriates in high-income economies (1,126,260 people), followed by India (1,037,626), Mexico (922,964), and China (816,824) (Schiff and Ozden 2006, 170).

Migration of skills affects both developed and developing countries. Due to the increasing returns to skills, talent seeks an environment with similarly talented peers. A few centers of excellence, such as Silicon Valley, therefore draw skills from developed and developing countries alike. Somewhat counterintuitively, it is the United Kingdom, not China or India, that has the largest stock of tertiary-educated nationals abroad (Schiff and Ozden 2006). Small countries suffer the most from emigration of the highly skilled. More than 85 percent of tertiary-educated individuals emigrate from countries such as Granada, Guyana, and Trinidad and Tobago.

The number of skilled migrants has increased. According to the International Organization for Migration, the international migration of qualified people has increased significantly. In addition, OECD studies indicate that while the total number of migrants between 1990 and 2007 grew by 68.2 percent, qualified migration grew by 111.3 percent. This trend is confirmed by Docquier and Marfouk (2006). The number of foreign-born individuals of working age living in OECD countries increased from 42 million in 1990 to 59 million in 2000. Skilled workers are now much more likely to engage in international migration.

In relative terms, the brain drain remains quite stable. Although the absolute amount of skilled migration has increased significantly, the size of the developing country populations has increased as well, as have educational attainments throughout the world. Hence, somewhat counterintuitively, the relative intensity of the brain drain has not increased over the past decades.

Figure 6.1 presents skilled emigration rates (roughly a ratio of outward migration of the tertiary educated to the overall stock of the tertiary educated) by regions over a quarter of a century. At the world level or at the level of developing countries as a whole, the average skilled migration rate has been stable. Some regions experienced an increase in the intensity of the brain drain (Central America, Eastern Europe, South Central Asia and Sub-Saharan Africa), while significant decreases were observed in others (notably in the Middle East and North Africa).

Heterogeneity of Skilled Diasporas

While the overall dynamics of skills mobility is revealing, diasporas of the highly skilled are heterogeneous in their impact on sending and receiving

Figure 6.1 Long-Run Trends in Skilled Emigration in Developing Countries, 1975–2000

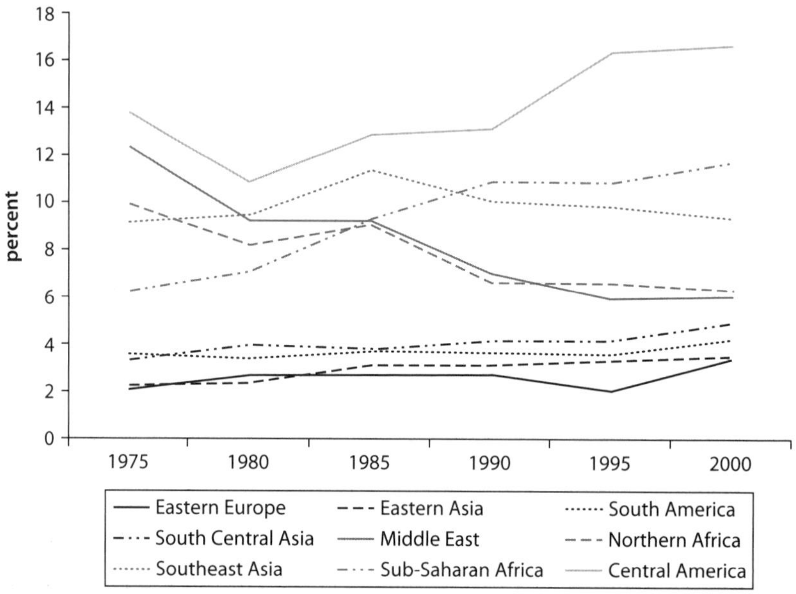

[Line chart showing percent on y-axis (0 to 18) against years 1975, 1980, 1985, 1990, 1995, 2000. Series: Eastern Europe, Eastern Asia, South America, South Central Asia, Middle East, Northern Africa, Southeast Asia, Sub-Saharan Africa, Central America.]

Source: Defoort 2006.
Note: Data represent a rough ratio of outward migration of the population with graduate degrees to the overall stock of the tertiary educated.

countries alike. The so-called "overachievers"—highly skilled professionals who are in a position of influence, the owners of high-technology companies, and those in senior management positions in multinationals—can have the greatest impact. However, their number is not specifically captured by the data. For instance, according to OECD data of the year 2000 (2002), the Philippines boasted the largest skilled diaspora among developing countries. This finding is not surprising, given that country's creation of an industry of skills exports. Yet a relatively small number of these are overachievers, at least in comparison with China and India.

In receiving countries as well, skilled migrants tend to go to particular geographic areas. In Silicon Valley, for example, the population surpassed 1 million in 1970, with only a small community of first-generation immigrants (less than 10 percent), mainly of European and Canadian origin. By 2000, one-third of the region's population and 42 percent of the high-technology workforce were immigrants—largely from Asia. Today, more than half the engineers and scientists working in the region's technology industries (53 percent) are foreign born. This is where foreign-born overachievers tend to reside.

These skilled expatriates can be a significant resource for the development of their home countries. As a well-known example, overseas Chinese contributed 70 percent of China's foreign direct investment over the years

1985–2000. By 1995, 59 percent of the accumulated foreign direct investment in China came from Hong Kong, China, and Macao, China, with a further 9 percent from Taiwan, China (World Bank 2005b).

Expatriates do not need to be investors or make financial contributions to have an impact on their home countries. They can serve as "bridges" by providing access to markets, sources of investment, and expertise. Influential members of diasporas can shape public debate, articulate reform plans, and help implement reforms and new projects. Policy expertise and managerial and marketing knowledge are the most significant resources of diaspora networks. The recent literature emphasizes remittances and their impact on development (see World Bank 2005b, for a summary). However, it appears unlikely that remittances and other financial transfers by migrants can ever have a significant impact on development, although they are certainly an important tool for poverty alleviation.[6]

The most direct and obvious mechanism for transferring the diasporas' knowledge to the home country would be for the highly educated emigrants to return home to work. Yet in spite of aggressive recruitment efforts by home country policy makers (programs to subsidize and encourage scientists and other highly skilled workers to return exist in many middle-income economies) and some information on rising return rates (from a very low base) in places like China and India, no evidence indicates that educated migrants to the United States and other advanced economies are substantially more likely to return permanently to their home economies than they were a decade or two ago. Research also points to a diaspora effect in scientific collaboration by documenting how knowledge, as measured by patent citations and co-authorship, flows disproportionately among members of the same ethnic community, even over long distances (Agrawal, Kapur, and McHale 2004). Yet efforts to demonstrate that scientific collaboration with the diaspora contributes to economic growth in the home country remain unconvincing.

The main contribution of skilled diasporas appears to be to institutional development in home countries through contributions to the transformation of the public and private sectors. It is significant that only a small number of "overachievers" with knowledge, motivation, and institutional resources were involved in such institutional developments. In contrast, while remittances have a visible impact on poverty reduction, their institutional impact is at best negligible and sometimes even negative. In some instances, of course, remittances can result in investments in or by small-scale entrepreneurs. Understanding the hierarchy of diaspora impacts, starting from remittances (subsistence agenda) at the base of the pyramid to institutional reform at the apex, and organizing the transitions is important (see figure 6.2).

Is the globalized knowledge-based economy replacing "brain drain" (of so much concern for developing country policy makers) with "brain circulation"

Figure 6.2 Hierarchy of Diaspora Impact on Institutional Reform in Developing Countries

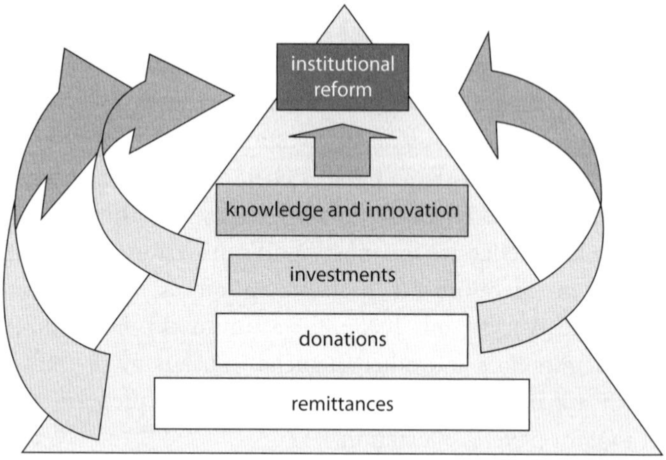

Source: Author.

(the talent engaged both in the receiving and sending country)? The evidence tends to support a relatively limited positive response:

- Significant potential exists for sending countries to benefit from the migration of skills from their economies. Although countries like Chile, China, or India have demonstrated that it indeed can be done, this potential is not automatically realized.
- The capability of the home country economy, its dynamism, and the availability of organizations of excellence with which overseas talent can engage are the main determinants in turning brain drain into brain circulation. The monetary incentives that some countries have instituted to encourage diaspora scientists to return seem to be relatively ineffective. The public sector needs to be creative and experimental in designing programs to engage its talent abroad with the home country.
- The heterogeneity of emigrants' skills plays an important role. Engaging medical professionals is very different from engaging technical professionals, which is again different from engaging scientists. No "one size fits all" approach will encourage brain circulation.

Conclusion

Human resource dimensions in the creation of a receptive innovation climate are numerous and diverse. Governments have to be involved along many fronts, ranging from basic education to informal training and the mobilization of talented diasporas. The shift toward knowledge-based economies makes the renewal of the educational foundations urgent and challenging, particularly for developing countries. Resistance to change is unavoidably strong, however, and dealing with related issues of political economy requires imaginative and bold approaches.

Notes

1. This typology draws heavily on Thornhill (2005).

2. In 11 of the 14 developing countries for which data are available—Albania, Brazil, Colombia, the Arab Republic of Egypt, Ghana, Indonesia, Morocco, Peru, the Philippines, South Africa, and Turkey—less than one-third of the students in recent cohorts are fully literate. In Brazil, Ghana, and South Africa, only 5–8 percent of each cohort achieves literacy (Hanushek and Woessmann 2007, figure 9).

3. That is, the Trends in International Mathematics and Science Study.

4. Chile is probably the best example in this regard.

5. List adapted from Johanson and Adams 2004.

6. For a perspective stressing knowledge and institution building rather than financial flows, see Kapur and McHale (2005).

References and Other Resources

Adams, Arvil V. 2008. "Improving Skills in the Informal Sector: Policies, Providers, and Outcomes." Presentation at the World Bank Labor Market Policy Course, "Jobs for a Globalizing World," World Bank, Washington, DC, March 31–April 11, 2008.

Agrawal, Ajay, Devesh Kapur, and John McHale. 2004. "Defying Distance: Examining the Influence of the Diaspora on Scientific Knowledge Flows." Paper presented at the Fourth Annual Roundtable on Engineering Entrepreneurship Research Conference, Atlanta, GA, December 3–5.

Allinson, Rebecca. 2006. "Shell Step Innovation Programme: Exploring the Absorptive Capacities of Host SMEs." *Cyprus International Journal of Management* 11 (1).

Amelewonou, K., and M. Brossard. 2005. "Développer L'Éducation Secondaire En Afrique: Enjeux, Contraintes Et Marges De Manoeuvre." Paper presented at l'Atelier regional sur l'éducation secondaire en Afrique, Addis Ababa, Ethiopia, November 21.

Appalachian Regional Commission and Ewing Marion Kauffman Foundation. 2004. *Learning by Doing.* Report on Entrepreneurship Education. Kansas City.

Atchoarena, David, and André Marcel Delluc. 2001. *Revisiting Technical and Vocational Education in Sub-Saharan Africa: An Update on Trends, Innovations, and Challenges.* IIEP/Prg.DA/01.320. Paris: International Institute for Educational Planning.

Bargain, Olivier, Sumon Kumar Bhaumik, Manisha Chakrabarty, and Zhong Zhao. 2007. "Earnings Differences between Chinese and Indian Wage Earners, 1987–2004." *Review of Income and Wealth* 55 (1): 562–58.

BIBB (Federal Institute for Vocational Training). 2004. *Modernizing Vocational Education and Training.* Bonn: German Federal Institute for Vocational Training.

Craig, Helen J., Richard J. Kraft, and Joy du Plessis. 1998. *Teacher Development: Making an Impact.* Washington, DC: World Bank and U.S. Agency for International Development.

Dar, Amit. 2008. "Vocational Education and Training Reform: Lessons of Experience." Presentation at the World Bank Labor Market Policy Course, "Jobs for a Globalizing World," World Bank, Washington, DC, March 31–April 11, 2008.

Defoort, Cécily. 2006. "Tendances de long terme en migrations internationales: analyse a partir des six principaux pays receveurs." Doctoral dissertation, Université de Lille, France.

Deschamps, J. P. 2005. "Different Leadership Skills for Different Innovation Strategies." *Strategy and Leadership* 33 (5): 31–38.

Desforges, Charles. 2001. "Knowledge Base for Teaching and Learning." *Teaching and Learning Research Programme Newsletter* 3: 3–4.

Docquier, F., and A. Marfouk. 2006. "International Migration by Education Attainment in 1990–2000." In *International Migration, Remittances, and the Brain Drain,* ed. Maurice Schiff and Caglar Ozden, 151–200. Washington, DC: World Bank.

EdStats. 2008. Washington, DC: World Bank. www.worldbank.org/education/edstats.

European Innobarometer Survey. 2001. Innovation Papers 22. Brussels, Belgium: European Commission.

Filmer, D., A. Hasan, and L. Pritchett. 2006. "A Millennium Learning Goal: Measuring Real Progress in Education." Working Paper 97, Center for Global Development, Washington, DC.

Florida, Richard. 2004. *The Rise of the Creative Class: And How It's Transforming Work, Leisure, Community and Everyday Life.* New York: Perseus Books Group.

Froumin, Isak, Shanthi Divakaran, Hong Tan, and Yevgeniya Savchenko. 2007. "Strengthening Skills and Education for Innovation." In *Unleashing India's Innovation: Toward Sustainable and Inclusive Growth,* ed. Mark Dutz, 129–46. Washington, DC: World Bank.

Futureskills Scotland. 2004. *Employers Skill Survey 2004.* Inverness: Futureskills Scotland.

Gill, Indermit S., Fred Fluitman, and Amit Dar. 2000. *Vocational Education and Training Reform: Matching Skills to Markets and Budgets.* Washington, DC: Oxford University Press.

Halsey, Rogers F., N. Chaudhury, J. S. Hammer, M. Kremer, and K. Muralidharan. 2006. "Missing in Action: Teacher and Health Worker Absence in Developing Countries." *Journal of Economic Perspectives* 20 (1): 91–116.

Hanushek, Eric A., and Ludger Woessmann. 2007. *Education Quality and Economic Growth.* Washington, DC: World Bank.

Hargreaves, Andy. 2006. "The Long and Short of Educational Change." *Education Canada.* Canada Education Association.

Howie, S. J., T. A. March, J. Allummoottil, M. Glencross, C. Diliwe, and C. A. Hughes. 2000. "Middle School Students' Performance in Mathematics in the Third International Mathematics and Science Study: South African Realities." *Studies in Educational Evaluation* 26 (1): 61–77.

India Enterprise Survey. 2006. http://www.enterprisesurveys.org/ExploreEconomies/?economyid=89&year=2006.

Johanson, Richard, and Arvil V. Adams. 2004. *Skills Development in Sub-Saharan Africa.* Washington, DC: World Bank.

Kapur, Devesh, and John McHale. 2005. *Give Us Your Best and Brightest Talent: The Global Hunt for Talent and Its Impact on the Developing World.* Washington, DC: Center for Global Development.

Klaus, David, Charlie Tesar, and Jane Shore. 2002. "Language of Instruction: A Critical Factor in Achieving Education for All." Education Group, Human Development Network Working Paper, World Bank, Washington, DC.

Kocchar, Kalpana, Utsav Kumar, Raghuram Rajan, Arvind Subramian, and Ioannis Tokatlidis. 2006. "India's Pattern of Development: What Happened, What Follows?" Working Paper WP/06/22, International Monetary Fund, Washington, DC.

Kozma, R. 2003. "A Review of the Findings and Their Implications for Practice and Policy." In *Technology, Innovation, and Educational Change: A Global Perspective,* ed. R. Kozma, 217–240. Eugene, OR: International Society for Educational Technology.

Kuznetsov, Yevgeny, and Carl Dahlman. 2008. *Mexico's Transition to a Knowledge-Based Economy: Challenges and Opportunities.* Washington, DC: World Bank.

Larsen, Kurt, Ronald Kim, and Florian Theus. 2009. *Agribusiness and Innovation Systems in Africa.* Washington, DC: World Bank.

Lewin, Keith. 2002. *Options for Post-Primary Education and Training in Uganda: Increasing Access, Equity and Efficiency.* London: U.K. Department for International Development and Government of Uganda.

Miles, Ian, Lawrence Green, and Barbara Jones. 2007. "Mini Study 02: Skills for Innovation." Global Review of Innovation Intelligence and Policy Studies, INNO GRIPS Europe, Brussels.

OECD (Organisation for Economic Co-operation and Development). 2002. *Trends in International Migration.* Paris: OECD.

———. 2004. *Innovation in the Knowledge Economy: Implications for Education and Learning.* Paris: OECD.

———. 2005. "Executive Summary." In *Teachers Matter: Attracting, Developing and Retaining Effective Teachers.* Paris: OECD.

———. 2006. *The Knowledge-Based Economy.* Paris: OECD.

———. 2007. *PISA 2006. Volume 1: Analysis.* Paris: OECD.

———.2008a. *Education at a Glance: OECD Indicators.* Paris: OECD.

———. 2008b. *Innovating to Learn, Learning to Innovate.* Paris: OECD.

OECD and Statistics Canada. 2002. *Literacy in the Information Age: Final Report of the International Adult Literacy Survey.* Paris: OECD. http://www.oecd.org/EN/document/0,,ENdocument-601-5-no-27-21891-601,00.html.

Sawyer, Keith, ed. 2006. *The Cambridge Handbook of the Learning Sciences.* Cambridge: Cambridge University Press.

Schiff, Maurice, and Caglar Ozden, eds. 2006. *International Migration, Remittances, and the Brain Drain.* Washington DC: World Bank.

Tan, Hong. 2008. "Skills, Training Policies and Economic Performance: International Perspectives." Paper presented at the World Bank Labor Market Policy Course, "Jobs for a Globalizing World," World Bank, Washington, DC, March 31–April 11.

Tan, Hong, and Gladys Lopez-Acevedo. 2005. "Evaluating Training Programs for Small and Medium Enterprises: Lessons from Mexico." World Bank Policy Research Working Paper 3760, World Bank, Washington, DC.

Tan, Hong, and Yevgeniya Savchenko. 2007. *Skills Shortages and Training in Russian Enterprises.* World Bank Policy Research Working Paper 4222, World Bank, Washington, DC.

Teferra, Damtew, and Philip G. Altbach, eds. 2003. *African Higher Education: An International Reference Handbook.* Bloomington: Indiana University Press.

Tether, Bruce, Andrea Mina, Davide Consoli, and Dimitri Gagliardi. 2005. *A Literature Review on Skills and Innovation: How Does Successful Innovation Impact on the Demand for Skills and How Do Skills Drive Innovation?* Manchester, United Kingdom: ESRC Centre for Research on Innovation and Competition and University of Manchester.

Thornhill, Don. 2005. "Creativity, Innovation, and Role of Higher Education in Economic Development: Financing of Tertiary Education." Paper presented at the Second Europe and Central Asia World Bank Education Conference, "Tertiary Education: Quality, Financing and Linkages with Innovation and Productivity," Dubrovnik, Croatia, October 2–4.

UNDP (United Nations Development Programme). 2008. *Human Development Indices: A Statistical Update 2008.* New York: UNDP.

VETA (Vocational Education and Training Authority). 2002. "Competence-Based Education and Training: A Practitioner's Guide to VET Competence-Based Education and Training in Tanzania." Dar es Salaam: VETA.

World Bank. 2003a. *Lifelong Learning in a Global Knowledge Economy: Challenges for Developing Countries.* Washington, DC: World Bank.

———. 2003b. "Overall Trends in Secondary Curriculum Reforms in OECD and Balkan Countries." Human Development Network Working Paper, World Bank, Washington, DC.

———. 2005a. *Expanding Opportunities and Building Competences for Young People: A New Agenda for Secondary Education.* Washington, DC: World Bank.

———. 2005b. *Global Economic Prospects 2006. Economic Implications of Migration and Remittances.* Washington, DC: World Bank.

———. 2007. *Cultivating Knowledge and Skills to Grow African Agriculture.* Washington, DC: World Bank.

———. 2009. *Accelerating Catching-Up: Tertiary Education for Growth in Sub-Saharan Africa.* Washington, DC: World Bank.

7

Policy Evaluation: Assessing Innovation Systems and Programs

As is the case with all public policies, science, technology, and innovation programs and policies need proper evaluation. Evidence-based policy making has become increasingly important in recent years, with new techniques being developed to assess individual programs and policies and new methods being adopted to undertake cross-country comparisons of innovation systems. Evaluation has to take place throughout the policy process. Rigorous evaluation is not always easy to achieve, especially for developing countries, because it is resource and data intensive. The benefits, however, can be considerable and can enable public funds to be put to their best use.

It is very difficult to measure the impacts and benefits of innovation policies. Many policies are related, and innovation systems are complex and evolve continuously. As a result, it is often hard to measure outcomes, but such evidence is crucial to establishing whether government policy successfully tackles market failures and provides a positive net stimulus to innovation. Effective evaluation programs should therefore seek to combine a range of evaluation methods, including national and international, quantitative and qualitative, and micro- and macrolevel.

This chapter reviews a range of techniques used to gain information on innovation by firms and economies as a whole and to evaluate programs and policies. It first reviews several widely used international benchmarking indexes that attempt to quantify innovation or innovation capabilities of economies as a whole. The chapter then turns to look at innovation surveys as a means of

---

This chapter was prepared by Désirée Van Welsum, with the contribution of Derek Chen for the section on macro-benchmarking methods.

obtaining information on what level and types of innovation firms actually engage in. It then proceeds with program evaluation, with particular emphasis on the evaluation of public research and development (R&D), and examines the use of field experiments. Last, the chapter reviews both national and regional policy evaluation. The importance of such evaluations is twofold: not only do they help maximize the economic and social efficiency of public spending, but they also provide information on which programs and policies work and under what circumstances that information can be used to learn from experience and inform future policy decisions. In particular, the section on field experiments discusses quantitative methods that have been developed for policy evaluations in emerging and less-developed economies. The particular focus of these techniques is to help understand which policy initiatives are especially effective and might be introduced on a larger scale.

Benchmarking Innovation at the Country Level

In this section, we review some of the more commonly used measures of country-level innovation. By determining the type of data or indicators that are being used by the innovation indexes, we can place the indexes into two broad groups: the first group of indexes is based exclusively on hard or objective data, and the second group of indexes uses a combination of both objective and subjective or opinion-based data. The scarcity of objective data tends to lead to different indexes' gravitating toward the same few available "hard data" indicators. It is unfortunate that all of these hard data indicators are relatively more R&D oriented.

The benefit of using data from opinion surveys is that it provides some indication of performance in an area for which hard data do not exist. One very valid concern with using such data, however, is that the ratings across individuals can never be truly consistent. For example, one person's "very good" may be another's "average," and this inconsistency may not be statistically rectified by increasing the survey sample size. Nevertheless, simple comparisons of country rankings in selected objective- and subjective-based data innovation indexes reveal that the rankings are, to a large extent, correlated. This correlation indicates that both types of indexes provide more or less similar pictures of the level of innovation in countries.

The World Bank Knowledge Economy Index

The World Bank produces one of the longest-running country-level innovation indexes available for a large number of countries. Its innovation index is one of several indexes generated by the Knowledge Assessment Methodology (KAM) (see http://www.worldbank.org/kam). The other KAM indexes include the Education Index, the Information and Communication Technologies (ICT) Index, the Economic Incentive and Institutional Regime Index, and the

Knowledge Economy Index (KEI), which is a simple average of the four preceding indexes. The KAM was launched in 2000 and has been updated at least once every year since. The most recent version of the KAM, KAM 2008, provides data and indexes for 140 countries. The KAM's innovation index is based on countries' performance over three indicators: (a) the number of patents granted by the U.S. Patent and Trademark Office (USPTO); (b) the number of scientific and technical journal articles; and (c) the amount of royalty payments and receipts. These indicators are relatively more R&D-oriented, but this orientation is due to the difficulty of quantifying innovation, as well as to problems in data availability in this domain. Being based on these three indicators allows the innovation index to be more consistently measured across a larger number of countries. In addition, the inclusion of royalty payments widens the definition of innovation to include technological adoption, thereby creating a broader index.

The UNCTAD Innovation Capability Index
In its World Investment Report 2005, the United Nations Conference on Trade and Development (UNCTAD) introduced the UNCTAD Innovation Capability Index (UNICI), which provides a measure of national innovation capabilities. The UNICI itself, which is based completely on hard data, comprises two equally weighted subindexes: the Technological Activity Index, which measures innovative activity, and the Human Capital Index, which measures skills availability for that innovative activity. The UNICI is available for 117 countries for the years 1995 and 2001. The UNICI Technological Activity Index was constructed using three innovation indicators: researchers in R&D, the number of USPTO patents granted, and the number of scientific and technical journal articles; the UNICI Human Capital Index employs the adult literacy rate and the gross secondary and tertiary enrollment rates. Note that the World Bank's KAM education index is also based on the same three education variables.

The UNDP Technology Achievement Index
The Technology Achievement Index (TAI) was developed by the United Nations Development Programme (UNDP) in 2001 to measure a country's ability to create and diffuse technology and build a human skills base, reflecting national capacity to participate in the technological innovations of the network age. The index focused on outcomes and achievements rather than on effort or inputs and is based on four dimensions of technological capacity: creation of technology, diffusion of recent innovations, diffusion of old innovations, and human skills. Each dimension was measured by two objective data indicators. The TAI, published in the *Human Development Report 2001* (UNDP 2001), covered 72 countries but has since been discontinued.

The Arco Technology Index

The Arco Technology Index (ATI) was developed by Daniele Archibugi and Alberto Coco (see Archibugi and Coco 2004) to measure the technology capacities of developed and developing countries. The ATI was built on the basis of the UNDP TAI, but the advantage of ATI is its larger country coverage and the capacity for comparisons across time. ATI 2004 covers 162 countries for two time periods: 1987–90 and 1997–2000. The ATI is composed of eight hard data indicators organized into three domains: technology creation, technological infrastructures, and level of human skills.

RAND Science and Technology Capacity Index

The Science and Technology Capacity Index (STCI) was produced by RAND in 2001 to measure a country's capacity to absorb and use science and technology knowledge. The STCI was updated by Wagner, Horlings, and Dutta (2004) and covers 76 countries. The STCI is constructed from eight hard data indicators that are divided into three domains:

- *Enabling factors,* measuring the environment conductive to the absorption, production and diffusion of knowledge
- *Resources,* contributing directly to science and technology (S&T) activities
- *Embedded knowledge,* measuring the output of S&T knowledge.

The European Innovation Scoreboard Summary Innovation Index

The Summary Innovation Index (SII) has been published annually with the European Innovation Scoreboard (EIS) by the European Commission since 2000 as a comparative assessment of the innovation performance of the member states of the European Union (EU) and other selected countries. The EIS 2007 includes innovation indicators and trend analyses for 37 countries, covering the EU-27 member states as well as Australia, Canada, Croatia, Iceland, Israel, Japan, Norway, Switzerland, Turkey, and the United States.

The SII is constituted of 25 innovation indicators, which are classified into five dimensions:

- *Innovation drivers,* measuring the structural conditions critical to innovation potential
- *Knowledge creation,* measuring the investments in R&D activities
- *Innovation and entrepreneurship,* measuring the efforts toward innovation at the firm level
- *Applications,* measuring the performance of labor and business activities and their value added in innovative sectors
- *Intellectual property,* measuring the achieved results from innovation activities.

Similar to the Global Innovation Index (see below), the EIS arranges the five dimensions into two groups: inputs and outputs. Innovation inputs cover the first three dimensions, and innovation outputs cover the last two dimensions. Similar to the KAM, the SII uses only objective data.

The WEF Global Competitive Index
As 1 of the 12 components of its Global Competitiveness Index (GCI), the World Economic Forum (WEF) produces an innovation index for 134 economies. The GCI, which is published annually with the Global Competitiveness Report, is an index that measures national competitiveness.[1,2] In contrast to the innovation indexes presented above, which are based completely on hard data, the GCI and the associated innovation index are based on both hard data and opinion survey–based data. The WEF innovation pillar is composed of one hard data variable (utility patents) and six opinion survey–based variables (capacity for innovation, quality of scientific research institutions, company spending on R&D, university-industry research collaboration, government procurement of advanced technology products, and the availability of scientists and engineers).

The World Business and INSEAD Global Innovation Index
The Global Innovation Index (GII), currently available for 130 countries, was developed by INSEAD and World Business in 2007 to show the degree to which nations and regions are responding to the challenge of innovation. The GII is made up of 84 variables divided into eight pillars, which are grouped as five input pillars and three output pillars. The five input pillars include institutions and policies, human capacity, infrastructure, technological sophistication, business markets, and capital. These pillars represent factors that enhance innovative capacity. The three output pillars include knowledge, competitiveness, and wealth. They measure results from successful innovation. The GII uses both objective data drawn from various public and private sources such as the World Bank and the International Telecommunication Union and subjective data drawn from the World Economic Forum's annual Executive Opinion Survey.

Country Rankings across Selected Innovation Indexes
While it is important to note differences in the various indexes in terms of their construction and underlying data, it is more important to get a sense of the similarities and differences in terms of their outcomes. To this end, the country rankings of four innovation indexes that are relatively similar in the time period and number of countries covered are compared.[3] Table 7.1 presents the top 10th percentile, the bottom 10th percentile, and the 50th and 60th percentile country rankings for the KAM Knowledge Economy Index, the KAM Innovation Index, WEF's GCI Innovation Index, and the Global Innovation Index.

Table 7.1 Rankings of Economies for the Knowledge Assessment Methodology, Global Competitiveness, and Global Innovation Indexes, 2008–09

Rank	KAM knowledge economy index 2009	KAM innovation index 2009	GCI innovation index 2008–09	Global innovation index 2008–09
Top 10th percentile				
1	Denmark	Switzerland	United States	United States
2	Sweden	Sweden	Finland	Germany
3	Finland	Finland	Switzerland	Sweden
4	Netherlands	Singapore	Japan	United Kingdom
5	Norway	Denmark	Sweden	Singapore
6	Canada	United States	Israel	Korea, Rep.
7	United Kingdom	Netherlands	Taiwan, China	Switzerland
8	Ireland	Canada	Germany	Denmark
9	United States	Israel	Korea, Rep.	Japan
10	Switzerland	Taiwan, China	Denmark	Netherlands
11	Australia	United Kingdom	Singapore	Canada
12	Germany	Japan	Netherlands	Hong Kong, China
50th and 60th percentile				
50	Ukraine	Armenia	Jordan	Brazil
51	Kuwait	Brazil	Ukraine	Turkey
52	Serbia	Serbia	Italy	Oman
53	Brazil	Trinidad and Tobago	Thailand	Barbados
54	Armenia	Turkey	Lithuania	Greece
55	Trinidad and Tobago	Ukraine	Chile	Jordan
56	Macedonia FYR	Mexico	Vietnam	Poland
57	Argentina	Thailand	Slovak Republic	Azerbaijan
58	Russian Federation	Romania	Senegal	Sri Lanka
59	Turkey	Jordan	Malta	Latvia
60	Jordan	Venezuela, R.B.	Colombia	Mexico
61	Thailand	China	Kazakhstan	Croatia
62	Mauritius	Uruguay	Greece	Philippines
63	South Africa	Panama	Poland	Vietnam
64	Oman	Georgia	Nigeria	Trinidad and Tobago
65	Mexico	Jamaica	Turkey	Mauritius
66	Saudi Arabia	Kuwait	Egypt, Arab Rep.	Panama
67	Georgia	Oman	Jamaica	Russian Federation
68	Panama	Moldova	Romania	Romania
69	Moldova	Guyana	Serbia	Nigeria
70	Kazakhstan	Macedonia, FYR	Kuwait	Kazakhstan
71	Jamaica	Tunisia	Uganda	Jamaica
72	Colombia	Colombia	Panama	Bulgaria
73	Peru	Egypt, Arab Rep.	Guatemala	Colombia
Bottom 10th percentile				
111	Zambia	Madagascar	Mozambique	Burkina Faso
112	Mali	Tanzania	Kyrgyz Republic	Moldova
113	Lesotho	Nicaragua	Bangladesh	Cambodia
114	Benin	Cambodia	Guyana	Paraguay
115	Nigeria	Zambia	Mauritania	Ethiopia

Table 7.1 continued

Rank	KAM knowledge economy index 2009	KAM innovation index 2009	GCI innovation index 2008–09	Global innovation index 2008–09
116	Nepal	Tajikistan	Nepal	Albania
117	Burkina Faso	Guatemala	Nicaragua	Kyrgyz Republic
118	Cameroon	Mali	Bosnia and Herzegovina	Bolivia
119	Mozambique	Burkina Faso	Ecuador	Nepal
120	Cambodia	Mozambique	Albania	Mozambique
121	Bangladesh	Bangladesh	Bolivia	Zimbabwe
122	Ethiopia	Ethiopia	Paraguay	Lesotho

Source: Author compilation.
Notes: KAM = Knowledge Assessment Methodology; GCI = Global Competitiveness Index.

Three observations regarding the rankings are noteworthy. The first striking fact is that many economies have broadly similar rankings across the different indexes. These similarities in rankings imply that the indexes are largely correlated and provide similar rankings, even though they are based on different indicators and even on different types of indicators (objective and subjective). That fact consequently suggests that even though innovation is inherently difficult to measure, relative achievements in innovation can be broadly measured so long as reasonable metrics are used.

The second fact is that the positive correlation across the indexes appears to be weaker the further one moves down the rankings. More specifically, nine economies make it to the top 10th percentile (the top innovators) of three or more indexes. In rankings around the median in the 50th and 60th percentile, 15 economies appear in three or more indexes. In the bottom 10th percentile, six economies appear in three or more indexes. One possible reason for the apparent weakening in correlation may be that the collection and availability of data for less advanced countries are more challenging in terms of accuracy and consistency. Poor-quality data could result in a lack of consistency among the different metrics and could lead to significant differences in country rankings, depending on the metrics used.

Third, there does not appear to be a distinct difference in the rankings between the narrower innovation indexes (KAM Innovation Index and GCI Innovation Index) and the broader ones (KAM KEI and the GII). Not surprisingly, this absence of differences suggests that innovation and a strong broader economy go hand in hand or, more specifically, that the latter is a requirement for the former to take place, a point that is heavily emphasized in other chapters of this volume.

It is also important to keep in mind the appropriate use of these innovation indicators and indexes. Various institutions have proposed these indexes and indicators as an easy way to measure the amount of innovation taking place in an economy. However, it should be clear that these indicators are far from ideal, in particular because they may be less relevant for economies that have not attained certain levels of economic development. For example, using fiscal incentives to encourage higher patent counts may be an unsound economic development strategy if the new knowledge associated with the patent is not usable by domestic industries because of their lack of technical knowhow. In general, generating new knowledge may not be the best use of scarce resources in economies that are just embarking on the industrialization process. Instead, a focus on assimilating and adapting existing technical knowledge from abroad to enhance domestic industries would generally be a more appropriate path. In light of the above, policy makers should exercise caution when using these metrics as policy or target variables.

Microlevel Innovation Surveys

Innovation surveys collect information about innovation inputs and outputs in firms. They go beyond S&T statistics such as R&D surveys or patent data by also collecting information on nontechnological innovation and on factors that support or impede innovation efforts. Innovation surveys thus collect data on various types of innovations, the reasons for innovating (or not), the impacts of innovation, collaboration and linkages among firms or public research organizations, and flows of knowledge.

The benchmark definitions of innovation in most surveys reflect those of the *Oslo Manual*, published by the Organisation for Economic Co-operation and Development (OECD), which provides a harmonized framework for innovation surveys to ensure their comparability and quality. While the first edition covered mainly technological product and process innovations, the manual has evolved over time to take new forms of innovation into account.

The current version of the OECD's *Oslo Manual* (2005a) defines *innovation* as the implementation of a new or significantly improved product (good or service) or process; a new marketing method; or a new organizational method in business practices, workplace organization, or external relations. It distinguishes four types of innovation: product, process, marketing, and organizational. Furthermore, the concept of *new* can mean new to the firm, new to the market, or new to the world. Finally, the manual considers the role of linkages and collaboration in innovation: that is, whether innovations are developed mainly by the firm itself, together with others, or mainly by others.

In Europe, the first Community Innovation Survey (CIS) was carried out in 1993, and the sixth in 2008. The survey is now conducted in all EU countries. Similar innovation surveys based on the *Oslo Manual* are conducted in many other countries, including Australia, Canada, the Republic of Korea, New Zealand, Norway, the Russian Federation, South Africa, Switzerland, and Turkey, as well as most Latin American countries. However, the United States does not so far have an official innovation survey based on the *Oslo Manual* framework.[4]

An annex was added to the current edition of the *Oslo Manual* to cover innovation surveys in developing countries. It takes into account the fact that most innovation in developing countries involves dissemination mechanisms and incremental change. Aspects discussed include differences in the structure and functioning of firms and markets, market failures and barriers to innovation, weaknesses in macroeconomic and institutional framework conditions, and the relative weakness of statistical systems.

Innovation Surveys in Developed Countries

Many of the most widely used international statistics on innovation activities do not provide direct evidence on the extent to which innovations are actually introduced. R&D-related indicators are an input measure into the overall innovation process, and patents and scientific publications are intermediate outputs. The formal, internationally agreed definition of innovation, that of the *Oslo Manual,* is output based and reflects the successful commercial introduction of new products and processes.

Direct output indicators of firm-level innovation performance are typically obtained through innovation surveys. The CIS, for example, provides indicators of direct relevance to European policy makers (see box 7.1 for examples) and increases the evidence available for evaluating the need for, and the performance of, innovation policies. The latest available survey, CIS4, covers EU member states and candidate countries as well as Norway and Iceland.

Box 7.1 Examples of Indicators from Innovation Surveys

The Community Innovation Survey provides a very rich data set. Examples of indicators that can be constructed include the share of firms that are

- involved in innovation;
- involved in process, product, marketing, or organizational innovation;
- introducing a good or service new to the firm, new to the market, new to the world;
- performing R&D;
- applying for a patent;
- receiving public funding;
- engaging in innovation cooperation (distinguishing among cooperation with universities, higher education or government research institutes, or foreign partners);
- reporting on important impacts of innovation (improved products, increased range of products, entering new markets);
- reporting impacts of organizational innovation (improved products, reduced response time, reduced costs, improved employee satisfaction).

Additional data cover expenditure indicators such as innovation expenditure and the share of turnover from different types of innovation.

Source: Author compilation.

The survey uses a harmonized questionnaire and survey method based on the *Oslo Manual* and agreed with Eurostat, the European Statistical Office. To maintain the secrecy of enterprise-level information, the microlevel database is confidential and can be accessed only by Eurostat staff. The most recent survey, launched in 2008, is gathering data for the reference years 2006–08 and will collect additional information about "nontechnical" aspects of innovation, such as management techniques, organizational change, design, and marketing issues.

The CIS4 found that around 40 percent of all EU firms undertook some kind of process or product innovation (European Commission 2007). Innovative companies were larger on average than noninnovative companies, accounting for around two-thirds of total employment. Three-quarters of the innovative companies invested in advanced machinery and equipment, and around half in R&D and training. R&D expenditure accounted for only a little over half their total expenditure on innovation. CIS4 data also indicate that around two-fifths of European firms introduced organizational or marketing innovations, a proportion similar to the introduction of other forms of innovation. However, many firms that introduced nontechnological innovations did not introduce product or process innovations.

Business services companies, particularly in computing and related activities and in financial intermediation, comprised a large share of innovative companies. The share of innovative firms in manufacturing differed little from that of the economy as a whole. When companies were questioned about the factors hampering innovative activities, they indicated that the four most important were the costs of innovation, uncertainty about demand, a lack of qualified personnel, and a lack of potential partners, in that order.

In addition to the microlevel indicators from the CIS surveys, related indicators on framework conditions and factors that enable innovation and its diffusion are also very important. Box 7.2 provides some examples.

Innovation Surveys for Developing Countries

The annex of the latest *Oslo Manual* (OECD 2005a) offers guidelines for the implementation of innovation surveys in developing countries. Many developing countries conducted innovation surveys on the basis of the preceding edition. These tended to require adaptations of the proposed methodologies to capture the particular characteristics of innovation processes in countries with economic and social structures different from those of the more developed OECD countries. The first effort to compile these particulars and guide the design of cross-country comparable innovation surveys was made by RICYT (Ibero-American Network on Science and Technology Indicators, or Red Iberoamericana de Indicadores de Ciencia y Tecnología) and resulted in the publication of the *Bogotá Manual*, which was later used in most innovation surveys conducted in Latin America and other regions. The importance

Box 7.2 Additional Sources of Innovation-Related Indicators

Innovation-related indicators include human resources, skills, and knowledge, as these are crucial enablers of innovation, absorptive capacity, and spillovers. They also include "knowledge infrastructure" indicators (such as public sector funding of research, universities, science and technology personnel, number of researchers, and the like). Others include business enterprise expenditure on research and development (R&D) and gross domestic expenditure on R&D as a percentage of gross domestic product (GDP), gross domestic expenditure on R&D financed from abroad, broadband penetration, education and training, participation in lifelong learning, employment in high-technology industries, employment in occupations requiring human resources in science and technology, scientific articles, science and engineering degrees, patents, patents with foreign co-inventors, trademarks, and venture capital, among other indicators. International organizations provide rich data sets with this type of information, in particular the OECD's *Science, Technology and Industry Scoreboard* (OECD 2007d), the OECD's *Science, Technology and Industry Outlook* (OECD 2008d) and the underlying databases, and the European Commission's *European Innovation Scoreboard* (various editions available at http://www.proinno-europe.eu/).

The Information Technology and Innovation Foundation (ITIF 2009) provides another source of cross-country innovation-related indicators. It uses 16 indicators to benchmark countries' competitiveness, including their innovative capacity, based on various innovation-related indicators (1–4) as well as on more general indicators of the macroeconomic framework conditions for competitiveness (5–6):

1. *Human capital.* Higher educational attainment in the population aged 25–34 and the number of science and technology researchers per 1,000 employed
2. *Innovation capacity.* Corporate investment in R&D, government investment in R&D, and share of the world's scientific and technical publications
3. *Entrepreneurship.* Venture capital investment and new firms
4. *Information technology infrastructure.* E-government, broadband telecommunications, and corporate investment in IT
5. *Economic policy.* Effective marginal corporate tax rates and ease of doing business
6. *Economic performance.* Trade balance, foreign direct investment inflows, real GDP per working-age adult, and productivity.

Intangible assets surveys (such as the Intangible Asset Monitor, the Skandia navigator, and the International Accounting Standards Board) also provide useful complementary data, including indicators of human capital, intellectual capital, organizational capital, and relational capital.

Additional firm-level indicators can also be found in private sources, such as the reports and data sets from business consultants such as the Boston Consulting Group, McKinsey & Company, and Booz Allen Hamilton, which also report indicators companies use to measure innovation (outputs). These include revenue growth from new products, sales from new products, profit increase from new products, customer satisfaction with new products, and the return on investment for new products.

Source: Author compilation.

and impact of this standard-setting work were the inspiration for the *Oslo Manual* annex, the preparation of which was coordinated by the UNESCO Institute for Statistics (UIS). Many of the annex's recommendations are based on the experience of countries that have already conducted innovation surveys, most of which are among the medium- and higher-income countries of the developing world where innovation has already become a policy issue. The knowledge gained by these countries should help other developing countries acquire their own experience without having to build exclusively on innovation measurement exercises carried out in developed countries.

The Measurement of Innovation in Developing Countries

The *Oslo Manual* stresses the need for the measurement of innovation in developing countries to be comparable to results obtained in developed countries to enable benchmarking and construction of a coherent international system of innovation indicators. At the same time, the innovation surveys need to take account of the characteristics of innovation in developing countries. In addition to the difficulties of applying existing definitions, challenges can arise in measuring incremental changes that may not result in "new or significantly improved" products or processes or in defining the scope of innovations, since concepts such as "new to the market" may be interpreted differently in environments with less developed infrastructures.

A main reason for conducting innovation surveys in developing countries is to inform public policy making and the design of business strategies. The main focus is on the generation, diffusion, appropriation, and use of new knowledge in businesses. Measurement exercises should focus on the innovation process, rather than on its outputs, and emphasize how capabilities, efforts, and results are dealt with. The efforts of firms and organizations (innovation activities) and their capabilities are equally or even more important than the results (innovations). Factors hindering or facilitating innovation are considered key indicators. "Potentially innovative firms" may be of particular interest in developing countries. They are a subset of "innovation-active firms," that is, those "that have had innovation activities during the period under review, including those with ongoing and abandoned activities." A key element of innovation policies in developing countries is to help potentially innovative firms overcome obstacles that prevent them from being innovative and to convert their efforts into innovations.

Different measurement priorities in developing countries (why, what, and how to measure) guide the design of innovation surveys. Developing countries use surveys to obtain information on the innovation strategies present in the innovation system and to understand how they contribute to strengthening the competitiveness of particular enterprises and to enhancing economic and social development more generally. This effort requires linking the analysis of micro-, meso-, and macroeconomic innovation data to issues such as the technological content of exports, strengths and weaknesses of particular industries or innovation systems, the absorptive capacity of innovation systems, networks, links between education and employment, and indicators of the effectiveness of different public instruments for supporting and promoting innovation.

The *Oslo Manual* finds the concept of innovation capabilities useful for classifying firms and industrial sectors in developing countries. It argues that the most significant innovation capability is the knowledge accumulated by the firm, which is embedded mainly in human resources but also in procedures, routines, and other characteristics of the firm. Because such intangible assets are notoriously difficult to measure, particular attention should be given to the parts of the surveys that directly connect knowledge development and diffusion with innovation capabilities, such as human resources, linkages, and the diffusion and use of ICT. In addition, more complex issues, such as the types of decision-making support systems put in place by the firm's direction and management and the firm's actual potential for knowledge absorption, should also be examined. For an accurate measurement of firms' innovation efforts, it is also essential to collect information about the intensity of innovation activities Therefore, details should be obtained about the innovation activities undertaken by the firm and, where possible, data on expenditure by innovation activity.

Organizational change is extremely important in developing countries, where the absorption of new technologies, mostly incorporated in machinery and other equipment, can require significant organizational change. Questions on the implementation of organizational innovations should therefore be supplemented with questions on human resources and training and on the incorporation of ICT. This process can help provide an indication of an enterprise's innovative capabilities.

Adapting Existing Innovation Surveys

Innovation surveys should be adapted for developing countries in three main areas: ICT, linkages, and innovation activities. Surveys should specifically address ICT. If specific surveys on ICT in businesses are not available, however, innovation surveys should inquire about available infrastructure, the purpose and use of ICT (separating front- and back-office activities), the existence of internal ICT management and development capabilities, and ICT expenditure and its relation to organizational innovation.

For a firm's external linkages, a proxy measure of complexity can be developed by crossing the "type" and "objective" of the linkages. This process establishes a matrix of *linkage agents* (universities, technical and vocational training institutions, technological centers, test labs, suppliers, clients, head office, enterprises belonging to the same group, other firms, consultants, R&D firms, public S&T agencies) and *types of linkages* (open information sources, acquisition of knowledge and technology, and innovation cooperation, supplemented by complementary activities, particularly access to new sources of financing and to commercial information).

For measurement of innovation activities, "hardware purchase" and "software purchase" should be covered separately as should "industrial design" and "engineering activities," "lease or rental of machinery, equipment and other capital goods," "in-house software system development," and "reverse engineering." Data should also be collected on the composition (by qualification, type of occupation, and gender) and management of human resources. For the latter, information on actions taken by firms with regard to training, including the resources involved, is important. Data can be collected to obtain information on the innovative capabilities of enterprises, not only on training activities linked to innovation but also on general training in areas such as management and administrative training, ICT, industrial security, and quality control.

Methodological Issues

The design and planning of innovation surveys in developing countries need to take account of the relative weakness of statistical systems. Because linkages between surveys and data sets tend to be weak or nonexistent, it is difficult to use information from other surveys in the design of the exercise and in the analysis of its results. The weakness, or sometimes lack, of official business registers, which are normally used as sample sets, is another example of this type of problem. It is important to involve national statistics offices in innovation surveys. If the statistical system lacks appropriate data about firm performance, some basic variables (for example, questions on sales or on turnover) can be included in the innovation survey to enable analysis of the relation between actions taken by firms for innovation and market performance (competitiveness).

Personal interviews (instead of mail or phone surveys) by adequately trained staff (for instance, undergraduate or graduate students) are recommended, since they have a positive impact on the response rate and on the quality of the results obtained. Interviews conducted by qualified staff can also provide the respondent with help in completing the questionnaire and increase response rates, particularly in countries where postal services may not be reliable.

The questionnaire can be designed with separate sections so that different persons in the firm can reply to different sections. Guidance for the respondent should be included with the questionnaire. It may be necessary to clarify certain concepts and provide a definition of terms used, and the wording needs to be adapted to the knowledge and experience of an "average" respondent. In certain cases, questionnaires may need to be formulated in more than one language.

It is generally recommended that innovation surveys be conducted every two years, but in developing countries every three or four years may be more appropriate. If possible, they should be timed to coincide with the major international innovation surveys, such as Europe's CIS, to obtain comparable data for similar time periods. It is also good to update a minimum set of variables every year (such as the main quantitative ones) if resources permit. A less costly strategy is to attach a significantly reduced questionnaire to an existing business survey. In some cases, simplified questionnaires can be designed to cover small firms to encourage their participation in innovation surveys.

The results of the innovation surveys should be published and distributed widely to encourage businesses to participate in future rounds and to increase awareness and use by researchers and policy makers. Diffusion mechanisms need to be included in the budget early in the exercise. Finally, an adequate legislative basis for collecting innovation statistics can help ensure the success of such exercises.

Program Evaluation

Evaluation of government action for innovation can be focused primarily on specific programs. After discussing methodological trends and issues, we will detail countries' experience in evaluating tax incentives for business R&D and public R&D support in various forms, including the support of R&D institutes.

Trends and Issues

Evidence-based policy making and the effective evaluation of public policy have become increasingly important in recent years, especially for science, technology, and innovation. These areas are becoming widely recognized as key drivers of economic growth and competitiveness. They also help reach socioeconomic objectives. It is important to evaluate not only whether the policy was implemented as planned but also whether it had the expected impact.

Program and policy evaluation is important, because it is necessary to make optimal use of public funds by maximizing the desired outcomes and ensuring that scarce resources are efficiently allocated. It is also important to gather information on which programs and policies work, and under what circumstances, to learn not only from success but also from failure. Insights into the determinants and magnitude of successful program and policy outcomes can be used to inform future policy decisions.

However, it is difficult to measure the impacts and benefits of government policy. Because many programs, policies, and policy areas are related, and because innovation systems are complex and evolve continuously, outcomes are not easy to measure because they may differ in the short, medium, and long term and because it is a challenge to establish causality. Effective evaluation programs therefore seek to combine a range of evaluations: national and international, quantitative and qualitative, micro- and macrolevel (box 7.3). Moreover, as innovation programs and policies vary greatly across countries and agencies in their content and objectives, evaluation methods and approaches often need to be adapted to individual circumstances.

In a program evaluation, it is important to bear in mind the immediate problem or market failure it aims to address, as well as the wider context, such as the program's contribution to overall innovation goals and the

---

Box 7.3 Effective Policy and Program Evaluation Challenges for Designing Evaluation Schemes

The difficulty of designing an effective evaluation can be illustrated by looking at the evaluation of public programs that support research, as these are an important aspect of innovation policies. The difficulties include selection bias in the attribution of funding, the attribution of research results that may draw on previous work to varying degrees, potential knowledge spillovers and absorptive capacity, the "additionality" of public funding, and potential "crowding out" effects.

A further challenge is to identify and measure not only the economic but also the wider socioeconomic benefits and to determine the time scale and time lag of the evaluation as the outcomes identified may depend on whether short-, medium- or long-term effects are examined. Isolating the effects of particular policies can also be difficult, given the increasing interaction between different policies and programs across a wide range of areas not limited to innovation policy.

Deciding on the evaluation criteria and benchmarks and who should carry out the evaluation is also not straightforward. For example, for a peer group evaluation, innovation users (business, government) and international representation are ideal, but those who also competed for the funding should not be included.

In evaluations of larger programs or projects, it becomes difficult to identify impacts, especially when the beneficiaries of the research are not those that perform it. In addition, one or a small number of successful projects can skew the distribution of the impact results of a portfolio of projects. In evaluating programs, the so-called project fallacy is another common problem, as the sponsored organization, which carries out the publicly funded research as part of a larger program of work, may be inclined to overattribute effects or deliverables to the funded part of the research to please the sponsor (OECD 2006).

Quantitative versus Qualitative Assessment

Quantitative assessment does not tell the whole story, and it is difficult to disentangle relations and correlations. Furthermore, no fully specified dynamic general equilibrium models of innovation exist. Econometric models tend to assume a linear process that does not take into account the complexity of innovation. Qualitative assessment is important but is difficult, as innovation performance depends on the characteristics of the country's economy, innovation systems, and institutions and because cultural factors also play an important role (corporate culture, entrepreneurship culture, "general" culture). The two approaches are complementary, and an effective policy evaluation should combine elements of both.

> **Box 7.3 continued**
>
> **National versus International Assessment**
>
> National assessments can combine a detailed analysis of microdata with survey results. Overall assessment of national innovation policies and systems (such as the evaluations carried out in the OECD's national innovation policy reviews) can complement more detailed studies.
>
> International assessment may include international benchmarking; for example, the European Commission's Innovation Scoreboard and the World Bank's Knowledge Economy Index, Knowledge Index, and scorecards as well as the ITIF innovation and competitiveness benchmarking study (ITIF 2009). Cross-country empirical studies, such as the OECD's Going for Growth studies and Innovation Microdata Project, can complement the international benchmarking, as can qualitative surveys, such as that of the World Economic Forum (WEF 2008).
>
> The efforts of international organizations greatly facilitate and enhance the scope for international policy learning. The OECD and UNESCO have a long-standing commitment, dating from the 1960s—alongside the European Union, UNCTAD, and the World Bank—to provide a platform for international learning. The recent success of emerging economies is likely to have been aided by international policy learning.
>
> *Source:* Author compilation.

possibility of unintended effects, interactions, or trade-offs with other programs. Furthermore, in impact assessment, it is important to understand why and how these occurred in order to gauge whether the program would lead to similar outcomes elsewhere. However, in the absence of strong institutional support and encouragement, underinvestment in program evaluation is likely, especially when the desired outcomes of policy interventions are not narrowly or precisely defined and when the impacts concern broader sectoral or economy-wide outcomes or take relatively longer to materialize (Ravallion 2009). Nonetheless, support for rigorous evaluation is increasing, as illustrated by the evaluation of government support mechanisms, publicly funded research, and use of field experiments.

Fiscal Incentives for R&D

Many governments continue to offer fiscal support for private sector R&D through grants or R&D tax incentives. These are longstanding and widely used policy instruments for stimulating innovation. Evaluating the outcomes of government-supported projects is difficult because it may be necessary to take account of their wider social benefits and to know what the situation would have been in the absence of public support. The latter is a particularly important hurdle. Evaluations of government support are also complicated by the fact that it may take time, even many years, for the benefits to appear, but judgments regarding the use of the public money cannot always wait that long. Finally, wider socioeconomic benefits and other potential spillovers that are hard to identify and measure are possible. Thus, many challenges must be overcome to achieve effective ex post evaluations of public programs that support research.

Selection bias is another issue. One way to overcome it might be to compare firms that have received funding with similar firms that have not. However, if the decision on funding evaluates the quality of proposals correctly, these would also have been the most likely to succeed in the absence of funding. Therefore, this approach does not necessarily get around the selection bias problem. Identifying factors that determine the probability of selection but not the probability of a successful outcome would make ex post evaluations less biased.

Another difficulty when evaluating the impact of public support for R&D is to identify and take account of potential knowledge spillovers. These may include both economic and socioeconomic benefits, especially when the recipient of public support produces innovations that are used by economic actors not included in the support program. In the absence of data with which to test this hypothesis, the impact of public support programs may be underestimated. In developing countries, these spillover mechanisms may be relatively less important.

A central issue in the evaluation of public support is the "additionality" of public funding, that is, the extent to which public support leads to a higher overall level of R&D expenditure than would otherwise have occurred. A so-called crowding-in effect may also occur if public support enables firms to carry out projects they would otherwise have been unable to finance. At the same time, a "crowding-out" effect may occur if firms that receive public support reduce the amount of funding they would have invested themselves; in which case, public support does not bring about additional R&D.

Tax incentives for R&D may be less likely to result in increased crowding-out effects than direct subsidies, because they operate by reducing the marginal cost of R&D rather than as a potential substitute for funding raised elsewhere, for example, on capital markets. Nonetheless, their impact on real resources may be relatively small, at least in the short term, as they may also help increase the prices of inputs in fixed supply, such as the wages of skilled researchers. They may also distort private sector project decisions if the design of the tax credit gives firms an incentive to undertake projects with a particular payback period.

Empirical Studies on the Impact of Fiscal Incentives. The literature on the impact of government subsidies, tax incentives, and public research programs has been reviewed in many studies. These suggest little consensus on the effectiveness of such instruments. All studies, however, emphasize the sensitivity of the conclusions to the control variables included in empirical assessments and the level of aggregation of the data set used. For example, evidence of crowding-out effects is more common in firm-level studies than in studies at higher levels of aggregation. It may also be the case that complementarities between publicly financed and privately financed R&D are due to the effect of the former on the input prices of the resources used by the latter. While Guellec and van Pottelsberghe (2000) find evidence of a positive net overall effect from public funding on the growth of privately financed R&D, some forms of funding are found to have a positive effect while others have a negative effect. Finally, some studies have found that the effects of government funding vary with firm size, although results again differ.

Jaumotte and Pain (2005a) have studied cross-country differences in business sector R&D and patenting and shown the importance of initial conditions on the effects of subsidies on innovative activities. They find a small positive effect on R&D from higher direct subsidies, especially when the share of corporate profit is small. In this case, the availability of public funding can help alleviate potential financial constraints. However, at other times, higher subsidies reduce innovative activity. The authors also find that more generous tax relief for R&D has a positive impact on the amounts of both R&D and patenting, with the impact often greater than that of additional direct funding. However, these results are sensitive to the exact specification of the regressions. Using sectoral data from the CIS, they find a significant positive correlation between public funding and the shares of innovator firms and of turnover accounted for by new products.

Jaumotte and Pain (2005b) suggest that tax incentives may be effective, at least in some circumstances, but they fail to show that the social gains from such programs outweigh the associated compliance and administrative costs, although the wider spillover effects of higher R&D on productivity growth raise the likelihood that they do. They also note a higher probability of research duplication when the support takes the form of tax relief rather than grants. Furthermore, new and small firms may be at a relative disadvantage if tax incentives are the only type of support, since they may have relatively little taxable income.

Even if tax relief for R&D is effective, other issues still need to be considered. As for direct grants and subsidies, a complete evaluation would also need to take into account the budgetary costs for the public sector. These need to be balanced by offsetting changes in other fiscal instruments (for any given overall budget balance target), which will also have economic effects. Even if fiscal instruments are effective, the wider question is whether the gains from supporting innovation are greater than the potential gains from supporting other activities or the (deadweight) costs of raising the necessary revenues (Jaumotte and Pain 2005a).

Finally, little is known about whether fiscal incentives for R&D have additional effects arising from their impact on the international location decisions of research-intensive multinational firms. If tax relief affects location decisions, countries that do not offer it may be at a disadvantage (Poot and others 2003). The extent to which the benefits from cross-border knowledge spillovers require local research capabilities also matters; these considerations may imply a stronger argument for tax relief in smaller countries.

Public R&D

The increased emphasis on evidence-based policy making means a greater need to understand and measure the impact of public sector R&D, notably to ascertain whether public spending on R&D is efficient and whether it contributes to the achievement of social and economic objectives. Public R&D is also increasingly used to address global challenges, such as climate change and the environment. However, it remains difficult to determine and measure the various benefits of R&D investment for society. Furthermore, because the benefits of public R&D can take some time to materialize, especially for basic rather than applied, research, it is difficult to determine the appropriate time for measuring the impact of public R&D and for identifying and quantifying its socioeconomic benefits.

The impacts of public R&D investment have been assessed using econometric analysis and case studies. However, the techniques used and the underlying assumptions determine, in part, the results. Particular challenges include establishing causality, capturing spillovers (international, sectoral, interdisciplinary), the unknown and varying time lags between the investment and the outcome, identifying the main actors and appropriate indicators of outcomes, and the evaluation of results. To some extent, these difficulties reflect the public good nature of public R&D investment and public knowledge more broadly: that is, the fact that it is not depleted when shared and it is difficult to exclude others from its use.

To date, few microeconomic studies address the impacts of public R&D on private sector productivity, and their results are not very conclusive. However, studies of the impact on private sector R&D have demonstrated strong returns to private investment and strong spillover effects that generate substantial economic benefits. Jaumotte and Pain (2005a) find evidence suggesting that research in the nonbusiness sector is an important component of innovation, both directly, as reflected in patenting, and indirectly, through its wider effects on private sector research. Even though an expansion in public sector research can help push up the wage costs of business sector researchers, this effect is more than offset by the positive impact on their efficiency.

The extent of collaboration between business and public research organizations, as proxied by the share of nonbusiness R&D expenditure financed by industry, has increased over time in almost all OECD countries. The work of Jaumotte and Pain also suggests that higher funding shares by the business sector provide an additional stimulus to private sector innovation, in addition to the direct effects from higher R&D spending in the nonbusiness sector. Data from the CIS also show that collaboration between the public and the private sector increases the share of turnover from new products. These aggregate findings need to be complemented by more detailed analyses of specific programs and different forms of research collaboration to gain a closer understanding of some of the mechanisms at work.

Recently developed indicators provide a means of assessing not only the economic but also the social impacts of public investment in R&D. They link government budget appropriations or outlays for R&D, which classify public budget figures according to socioeconomic objectives, to other data sources. They can help show the contribution of public money to achieving national socioeconomic objectives (OECD 2008d). The next step in assessing the impact of public R&D will be to link data on public R&D budgets by socioeconomic objectives to other data sources, such as scientific publications and patents. Definitions and practices will need to be better harmonized before the contribution of public R&D to socioeconomic objectives can be more fully understood.

Public support for R&D can be channeled in a variety of ways. OECD (2006) distinguishes four levels of evaluation of publicly funded research: (a) institutes and groups, including research departments, teams, laboratories; (b) institutions and operators, including public research organizations and research councils; (c) programs and procedures; and (d) research and innovation systems. Box 7.4 identifies a set of emerging cross-cutting issues.

Research Institutes and Groups. Evaluation tends to be carried out in accordance with one of two common models: the one-off model (such as the Max Planck *Gesellschaft's* approach to creating new groups and using committees of peers) and the periodic recurrent model (such as France's INSERM, which periodically reviews bottom-up proposals from research groups). However, there is an increasing shift toward the latter (OECD 2006), with evaluation evolving in two directions: embedding evaluation within overall strategic exercises (for example, the bottom-up strategic plans of the Spanish national research agency CSIC are reviewed by thematic panels) and taking a more transversal approach to the allocation of core grants (like the German Helmholtz Association's competitive process based on program-oriented funding with interdisciplinary programs evaluated by review panels).

National or subnational governments' evaluations of university research and research institutes have also changed to improve the allocation of core grants at the national level (for example, the United Kingdom's Research Assessment Exercise, a disciplinary peer review exercise, which has inspired similar models in Hong Kong, China, and in New Zealand) and to search for critical mass and excellence, with public funding increasingly concentrated in a few institutes or centers (such as the U.S. National Science Foundation

> **Box 7.4 Emerging Cross-Cutting Issues in the Evaluation of Publicly Funded Research**
>
> Changes in evaluation practices have been driven by four major trends: tighter public governance, competition for research funding, increased focus on the interfaces between fields of research and the economy and society, and increased political acceptance and integration of evaluation outcomes owing to better evaluation methods and tools. The analysis of the impact of these changes on evaluation of publicly funded research points to five issues:
>
> - Clarification of the differences and interactions among indicators, benchmarking, and evaluation is needed.
> - The increasing tendency toward internationalized peer review may lead to a rapid normalization of criteria for funding and evaluating research, at the risk of losing specific aspects of local settings.
> - The object of evaluation needs to be situated in its proper context (scope, timing) to avoid project fallacy problems (see box 7.3).
> - The impact of evaluation depends on the context in which it is implemented and on whether it is a one-time event or an institutionalized part of a regular policy process. In addition, context matters, in particular whether the interests of key stakeholders are aligned with evaluation goals. Operating below the political level can be useful for getting results accepted, timing and matching the decision cycle are important, and the evaluation has to be relevant and robust to be credible.
> - The success of an evaluation can be measured by its effects, including the intended and unintended consequences of the public intervention as well as that of the evaluation.
>
> *Source:* OECD 2006.

Engineering Research Centers, the U.K. Research Councils, Canada's Networks of Centers of Excellence, Australia's Cooperative Research Centers, Sweden's competence centers, and the Dutch Top Technology Institutes).

Research Councils and Public Research Organizations. Research councils and public research organizations can be differentiated according to their functions in the research system and the type of research they carry out. National research councils mainly *fund* research, while public research organizations *perform* research. Hybrids both fund and carry out research. Some focus on basic research, while others are industry oriented. These institutions' assessments of the impact of public R&D have tended to be relatively successful at quantifying impacts. The methodologies used in impact assessments include surveys, input-output analysis, a combination of top-down (contribution of funding to productivity growth) and bottom-up (return to funding through main benefits' transmission channels) approaches, and simulations on computable general equilibrium models.

Research Programs. Research programs are one of the main instruments used by developed countries to implement research and innovation policies. They may fund basic or more applied research in general or in a specific sectoral context, with or without a commercial objective. Two of the most important research programs in terms of resources are the European Union's Framework Programme and the U.S. Advanced Technology Program (ATP). The nature and scope of the research carried out under these two programs are very different.

The EU Research and Technological Development (RTD) Framework Programme (FP) is the main multi-annual R&D funding program in Europe. The FP7—that is, the seventh four-year framework program—is more ambitious than the previous programs as it brings all research-related EU initiatives together under a common umbrella. It has a budget of €53.2 billion and runs from 2007 to 2013. It funds both basic and applied research and aims at enhancing the research capacities and results of all stakeholders: private companies, individual researchers, universities, public research institutions, and foreign actors. Impact assessment tends to rely on econometric modeling. The FP7 also uses an ex ante or prospective calculation of the impacts of expenditure generated by a general equilibrium model using impact scenarios drawn up by the European Commission.

In the United States, the ATP, which started in 1990, aims to accelerate the development of innovative technologies for broad national benefit through partnerships with the private sector. It provides cost-shared funding to industry to speed up the development and dissemination of challenging, high-risk emerging technologies that can yield promising commercial possibilities and widespread benefits. It was designed specifically to help U.S. firms translate inventions in universities or national and corporate laboratories into new products, processes, and services able to compete in rapidly changing world markets. Between 1990 and September 2004, the ATP held 44 funding competitions and provided US$2.2 billion in grants, complementing the US$2.1 billion provided by industry. Impact assessment of these awards is carried out by the Economic Assessment Office (EAO), which tracks the progress of funded projects for several years after ATP funding ends and identifies the benefits, both direct and indirect, that ATP award recipients deliver. Direct benefits are achieved when technology development and commercialization are accelerated, leading to private returns and market spillovers. Indirect benefits are considered to include publications, conference presentations, patents, and other means of disseminating knowledge.

The EAO uses a variety of methods to measure the investments of the ATP, including surveys, detailed case studies, cost-benefit analysis, statistical analysis, comparison of ex post benefits with ex ante expected benefits, the tracking of knowledge created and disseminated through patents, and informed judgments. Because the evaluation of emerging technologies is a relatively new field, existing tools often have to be modified or new ones developed. The ATP also relies on the Business Reporting System (BRS), a data collection tool for tracking the progress of its portfolio of projects and individual participants, from the ex ante baseline to the end of the project and beyond. Progress is assessed against business plans, projected economic goals, and other ATP criteria.

Research Systems. Two recent trends in the evaluation of research systems are the application of detailed evaluation tools to the research or innovation system as a whole to answer particular policy questions, and country reviews of national innovation systems and policies. Examples are the review of the Finnish innovation support system by Georghiou and others (2003), the evaluation of Japan's First and Second Basic Plans, the indicators developed for the U.K. government's 10-year investment framework for science and technology, and the U.S. Government Performance and Results Act and Program Assessment Rating Tool (OECD 2006).

Field Experiments

Field experiments and pilot studies of policies and programs are undertaken to evaluate how they actually work and how their effects might differ from expectations. This assessment is essential for ensuring that only the most effective microprograms are scaled up to national or international levels.

In econometric studies of policy interventions, a particular problem is to know what would have happened to the "treated" group (that is, the group subject to the intervention in question) in the absence of the intervention. A credible impact evaluation has to address this issue. The work of Duflo and her collaborators (Duflo 2004, 2006; Banerjee and Duflo 2008) has recently popularized randomized evaluations as a possible way to address this problem.

Randomized evaluations are intended to help overcome various types of selection bias when measuring the impact of a program or policy intervention by randomly allocating individuals to a "treatment" group of individuals who benefit from the program and a "comparison" group of individuals who do not. For the method to be effective, the random selection of both groups is essential. The outcomes are then compared across the treated and the comparison groups. This approach can be used to test not only the overall effectiveness of a particular program but also the effectiveness of different parts of the program, as some parts may be especially effective while others are not. Duflo (2004) argues that randomized evaluations can be used in many different contexts, provided that the programs have clearly defined objectives and are targeted to individuals or local communities. Programs that affect all individuals or communities as a whole are not suitable, because it is not possible to define a random group that is not subject to the program.

Field experiments and randomized evaluations by Duflo and others have already provided some important insights. Experimental evidence has confirmed that individuals respond to incentives and will try to pervert them if they can do so at little cost. Experiments have also highlighted features that are important in the design of incentive schemes, as opposed to their effect. For example, people are more responsive to an immediate reward, even if it is small, than to a longer-term reward. More research is needed to examine the role and impact of delayed rewards.

The extent to which people learn from each other is another important issue and one with clear implications for innovation and technology diffusion. In developing countries, the impact of learning on technology adoption has been examined in the context of agriculture: in this case, an experiment identifies how social learning affects the development of a technology within a group of farmers and follows its subsequent adoption by them and the members of their network (this treatment group is selected because it faces common unobserved shocks). Such experiments can be designed to examine specific questions or mechanisms and the conditions in which they might work.

Field experiments can be used to test the predictions of theories, and randomized program evaluations can be used to test the effectiveness of more complex policies, including the combination of a variety of policy levers that have not necessarily been tested or even implemented. Ideally, the results of field experiments and the underlying theories would also inform the design of "combination" policies, so that the two approaches are both policy relevant and complementary. Field experiments need theory to derive specific testable implications and to give a general idea of the interesting questions. Field experiments also make it possible to test empirical predictions. Scaling up by generalizing from experiments is the next step to consider. Well-designed program evaluations are, in effect, international public goods. They provide information to other countries about what might work and what might not. As a result, they are very important for international agencies that seek to introduce related programs in different countries.

Several reasons explain why the results of a well-executed experiment may not always be generalizable (Duflo 2004; Banerjee and Duflo 2008). First, the experiment may affect the treated or the comparison sample, for example, if the provision of inputs temporarily increases morale among beneficiaries, thereby improving performance (the so-called Hawthorne effect). While this factor would bias randomized evaluations, it would also bias other types of evaluation, including econometric techniques such as fixed-effect or difference-in-differences estimates. The two groups may also be temporarily affected by their participation in the experiment (the "John Henry effect"). However, such effects are less likely to be present for large-scale evaluations if the time span is long enough or if the program is large.

Finally, it is never possible to replicate a project identically: circumstances vary, and ideas need to be adapted to local contexts.

Duflo's work also points to a number of important recommendations for the design and implementation of evaluations. These include reducing the number of evaluations; conducting credible evaluations in key areas, combined with randomized evaluation in other areas as opportunities occur; and establishing dedicated evaluation units in (international) organizations. She also argues that it is as important to publish negative results as it is to publish positive results and calls for institutions to ensure that negative results are also systematically disseminated, as is already the case for medical trials.

Innovation Policy Reviews

The importance of policy evaluation is twofold. First, it is important to learn from experience which policies and programs work and which do not and under what circumstances. These lessons are especially important as circumstances may change rapidly, and, as new forms of innovation emerge, innovation policies need to reflect these developments. Second, policy evaluation is essential as a guide for public spending on R&D and for resource allocation more generally. Many types of evaluation are possible, the quality varies widely across projects and countries, and feedback of the results into policy making is not always sufficient. A lack of transparency is also a relatively common hindrance.

The National Level

The evaluations of national policies are largely based on the experience of the OECD. They began some 50 years ago in the 1960s, primarily with science policy, and have shifted gradually toward review of broader innovation systems and policies.

The OECD's Innovation Policy Reviews. The OECD's reviews of national innovation policies provide a comprehensive assessment of the innovation system of individual countries.[5] They focus on the role of government and provide recommendations on improving policies that affect innovation performance. Each review identifies good practices that might be useful in other countries. An evaluation of a country's innovation policies and systems is prepared and then peer reviewed by an OECD committee of government officials and national experts in the field of innovation policy.

The reviews undertaken to date provide some important insights into the efficacy of innovation policies and the use that can be made of them. They look not only at government policies for stimulating innovation directly but also at broader factors, such as the overall governance of the innovation system and the extent to which the changing nature of innovation affects the linkages between innovation and economic performance and, hence, policy priorities. The reviews, in conjunction with associated work by the OECD and other international organizations, also draw attention to some important issues for developing economies. These include the strategies that can be adopted to build up innovation capabilities, policies to move up the value chain, and the importance of history and path dependence on former industrial experience or economic regime. Box 7.5 presents some of the key initial findings of the OECD reviews on market and governance arrangements for innovation and policy instruments and the policy mix.

Reviews of the overall innovation system are an important complement to studies of individual programs and policies. Not only do they provide a broader perspective on the activities of governments, but also they make it possible to assess the overall coherence of the policies adopted to support innovation. These may include policies that affect innovation indirectly, such as competition policy and the openness of the economy to international trade, investment, and migration. Such economy-wide factors are an important part of the innovation system.

Strategies to Build Innovation Capabilities. Countries may develop innovative capabilities as part of their catch-up strategy, with a range of positive effects. Developing domestic human resources and other forms of scientific capabilities increases a country's attractiveness as a location for foreign investment and enhances its absorptive capacity, raising the extent to which domestic companies and institutions can take advantage of spillovers and technology transfer from inward foreign investment. At the same time, this process enables the country to diversify its activities and reduce its dependence on any one activity or sector. It will also enable it to link more closely with the activities of globalized economies and tap into new markets.

The innovation systems of developing and emerging economies share certain weaknesses, including a lack of skilled human resources, inadequate innovation capabilities in business firms, and poor coordination among industry, universities, and public research organizations. These weaknesses need to be addressed in innovation policies, the implementation of which will require good policy governance. The OECD's innovation policy reviews of Chile, China, Korea, Mexico, and South Africa and provide examples of effective policy and governance reforms aimed at developing new areas of comparative advantage. Korea's experience, for example, illustrates the importance of significant stocks of science and technology capabilities for implementing imitation strategies, for moving up the value chain, and for speeding up the catching-up process (OECD 2009). China illustrates the benefits of large-scale investment in science and technology infrastructure, including human resources for science and technology (OECD 2008b).

Box 7.5 National Innovation Strategies: Lessons from OECD Country Reviews

Market and Governance Arrangements for Innovation

Improving framework conditions: Lack of competition acts as a barrier to innovation in many countries, but there is too little awareness of the role of competition policy in fostering innovation.

Policy coordination and participatory governance: These are important for ensuring effective policy coordination and effective participation by stakeholders.

Leadership: Involvement of the highest level of government is needed for securing policy attention and commitment.

Commitment: It is important to ensure that public funding for innovation is not "crowded out" by short-term demands.

Stability and predictability of institutions and policy delivery: While innovations in the policy framework are sometimes necessary, frequent changes tend to be counterproductive.

Evidence-based policy making: It is important to make effective use of reviews and evaluations. However, policy learning is easily disrupted and often difficult to institutionalize.

Steering and funding of public research organizations (PROs): The role of PROs needs to be redefined and their connection to the business sector improved to enhance their contribution to the overall performance of the innovation system.

Policy Mixes and Instruments

Striking a balance in policy instruments: Balance is important for stimulating business innovation, as policies are often introduced along several dimensions. Some are top-down, especially when the need for a change in direction is clear, but others are bottom-up. Some aim at improving economy-wide capabilities, such as policies to reduce financial barriers to investment in innovation, while others have specific policy objectives, such as tax credits for R&D.

Building capabilities: This can be done, for example, by reducing financial barriers to investment in innovation.

Direct and indirect support measures: Ideally, these two types of tax incentives should be applied in a complementary way to make the best use of their respective advantages, but this is not always the case.

Bottom-up and top-down approaches: These approaches should be complementary. Bottom-up approaches should be used for standard types of innovation projects and for gathering information and inducing self-organization in new areas, for example, by competitive calls. Top-down approaches should be used for changes in policy directions.

Different types and combinations of support: This support can include individual project-based support, ad hoc support, consortium-based, and longer-term support. Consortia are also useful for triggering behavioral change, such as cooperation between different types of actors.

Competition for funding: The shift toward competitive funding has provided powerful incentives for PROs and universities, while safeguarding a degree of stability and maintaining capabilities.

Source: Adapted from Guinet and Keenan 2008.

Long-term reform is not easy, although the potential gains may be large if it is successfully carried out. Becoming locked into wrong technologies or infrastructure is a potential danger, as is the danger that special interest groups will capture the project. For that reason, the governance of innovation projects matters.

Moving Up the Value Chain and Diversification. Another area of interest for developing countries is the use of innovation policy to move up the value chain. As globalization intensifies competitive pressures, many countries respond by trying to diversify their economies and move up the value chain. For example, Mexico and Hungary have an important manufacturing base, driven by inward foreign investment, which largely functions as an export platform (to the United States and the European Union, respectively). Hungary produces and exports many medium- and high-technology goods, in spite of the relatively low R&D intensity of the country's firms. This fact indicates that national innovation policy should aim for better integration of the foreign-owned sector into the national innovation system, including universities and public research organizations, and for improved absorptive capacities of domestic small and medium enterprises.

The economies of some countries are highly specialized, which may present a risk in the long term. The economies of Chile, Mexico, and Norway, for example, are largely based on natural resources, and Luxembourg is dominated by its financial sector. Diversification may also be desirable for countries with a small domestic market or a remote geographical location, such as New Zealand. National innovation policies can help meet such challenges. Norway has seized opportunities for knowledge-intensive activities in and around the oil and gas sector (OECD 2008c). Chile and New Zealand are adopting measures to aid a shift toward more innovation-based growth strategies (OECD 2007a, 2007b). Developing human resources is crucial to any strategy for innovation-based growth. For example, Chile, where the lack of skilled human resources constitutes a significant bottleneck, is adopting measures to raise educational standards to international levels, among others. In New Zealand, more emphasis needs to be put on improving framework conditions and stimulating market-led innovation throughout the economy, including by stimulating entrepreneurship and developing management, marketing, and distribution skills.

History and Path Dependence. History and path dependence are a significant issue for developing countries. Existing institutions have norms and routines that are reflected in their day-to-day operations. While such features provide stability and can thus be a positive factor, they may also result in inertia and prevent institutional reform. All countries face such risks, irrespective of their degree of development or the state of their innovation policies and institutions. If the nature of innovation and the dominant technologies change, as they do at present, this fact needs to be reflected in the innovation policies and strategies pursued.[6]

Reported Evaluation Practices. In Norway and Switzerland, for example, evaluation is common and attempts to follow internationally accepted best practice. Programs are often legally required to undergo ex ante and ex post evaluation, as well as ongoing monitoring during implementation. In Switzerland, domestic and foreign experts are involved in the evaluation process and often contribute to the development of evaluation methodologies. In Norway, evaluation is actively promoted by key agencies such as Innovation Norway and the Research Council of Norway.

The importance of evaluation has been recognized in many other countries, such as Chile and Korea, and efforts are underway to catch up with practices elsewhere. Korea has recently introduced a large-scale evaluation system, involving a combination of program evaluations and meta-evaluations (that is, those with strict performance targets). Interim evaluation results have been used to modify the resources made available to particular programs. Other countries have as yet made relatively little effective use of evaluations of innovation policies and programs.

It is very important for the results and findings of evaluation exercises to be used in subsequent evidence-based policy making. However, the differences in the extent to which policy making is evidence based are significant. Effective evidence-based policy making is not easy to implement, as it requires resources and substantial expertise, as well as clearly defined objectives. Outcomes also need to be measureable, which again may require investment in new resources and institutions.

How Can Developing Countries Use These Reviews? The OECD country reviews of innovation policies cover a variety of countries at different stages of development and innovation performance. As the reviews evaluate different policies and programs, their implementation, and the governance of the innovation system and also identify best practices and make recommendations, countries can learn about what works and what does not and under what circumstances. These results can inform the design and implementation of innovation policy in developing countries if they are adapted to local characteristics. However, policy measures that are effective in one country may be ineffective or inappropriate in another, depending, for example, on institutional factors, industry specialization, and size. A country's innovation performance depends not only on its performance in each element of the national innovation system but also on how the elements interact. OECD analysis suggests that no single combination of elements is successful: what matters is the cohesiveness of the system for innovation performance and how well the country performs in each of the main dimensions.[7]

Recommendations from the OECD reviews of particular importance for developing countries include improving framework conditions for innovation (that is, fostering competitive and open product, labor, and capital markets); implementing and enforcing intellectual property legislation; ensuring a supply of appropriately skilled people by improving access to higher education (including vocational training); improving incentives for firms to invest in training; building capacity in small and medium enterprises; and promoting and supporting entrepreneurship.

UNCTAD also carries out science, technology and innovation policy reviews specifically designed to help developing countries identify and adjust their policies and institutions to support technological transformation, capacity-building, and enterprise innovation. At the time of writing, reviews had been completed for Colombia, Iran, and Jamaica.

Regional Level
New microlevel work coordinated by the OECD has looked at the regional dimension of innovation, including firm location and linkages (OECD 2008a). Linkages among geographic areas and between firms result from the flow and transfer of intellectual assets and knowledge spillovers, which often require proximity. Evidence points to significant variations in the inventive performance of regions, with a high concentration in certain regions of continental Europe, North America, and Japan. The development of inventive activities in countries tends to take place in a small number of regions, and highly inventive regions usually cluster together. This spatial dependence has increased over time. Moreover, the inventive performance of regions is directly influenced by the availability of human capital and R&D expenditure. Cross-country differences point to the importance of national innovation systems and of linkages within firms across regions, as the most inventive regions have relatively more multiregional companies among their innovative firms.

Governments are increasingly aware that the regional dimension of innovation matters for strategies that use innovation to promote growth. The OECD is carrying out work on regional innovation to help policy makers from different backgrounds at both national and regional levels. Objectives include strengthening the evidence base for policy making, improving the use of resources in different regional contexts, ensuring coherence between innovation and other policy objectives, and assessing the impact of policies at the regional and national level. Current work includes an ongoing series of reviews of innovation in regions from national and regional perspectives (Italy, Mexico, and United Kingdom) and an analysis of innovation indicators using the OECD Regional Database's innovation data set. Regional policy initiatives are evaluated in OECD (2007c).

The origin of national and EU programs to support clusters and regional specialization can be found in regional, S&T, and industrial policies. Several programs that originated in S&T policy specifically support large-scale collaborative R&D projects to stimulate the most promising technology sectors in regions in which key institutions, researchers, and firms are concentrated. However, the evaluation of these approaches is often inadequate, especially since not all programs are evaluated and tools to measure impacts are often lacking. As a result, it is difficult to assess whether these programs are appropriate, realistic, and flexible enough to achieve their goals. The stated goals of cluster and regional specialization programs are often vague or broad (OECD 2007c), complicating the identification of appropriate participants, duration, targets, budgets, and funding. Cluster policies may also lack the private sector engagement on which their long-term effectiveness depends.

Overall, there are three main issues for policy and program design, based on practices across OECD countries: the degree to which the programs are appropriate, realistic, and flexible enough to achieve their goals; policy coherence within and across different levels of government; and the importance of private sector engagement to the ultimate outcomes.

Conclusions

It is clear from the above discussion that evaluation practices of innovation systems and programs are yet embryonic in emerging and developing countries. However, it is also clear that methods, surveys, and reviews adapted to their needs are being increasingly elaborated and implemented. More and more countries are using them and providing pioneering examples that can inspire the whole community. Several stand out as particularly important:

- *The development of "macro" benchmarking methods and indicators.* These take due account of emerging and developing countries' particular situations, do not measure innovation performance or capabilities exclusively with R&D-related indicators, and allow an accurate appraisal of improvements over years.
- *The implementation of innovation surveys.* These surveys capture evolutions of fundamental importance for emerging and developing economies, such as the diffusion of new and basic technologies and improving productivity, welfare, or the environment. The use of the newest tracking methods, such as geographic information systems, should be strongly encouraged.
- *The systematic evaluation of policy programs.* These take into consideration intangible developments, such as network and competence building, that always precede visible technical or economic achievements. They also make use of innovative approaches to examining program implementation and impact, such as field experiments and tests.
- *The use of national policy assessments.* These assessments use standard approaches such as international peer reviews that involve foreign expertise,

including experts from the developed world and the donor community, to stimulate mutual learning processes.
- *New types of indicators.* Finally, it is of the utmost importance to develop new types of indicators that go beyond the usual quantitative measurement of economic growth to integrate more qualitative dimensions. Significant attempts in this direction are summarized in box 7.6.

Box 7.6 Beyond GDP: Alternative Measures and Indicators of Economic and Social Progress

Gross domestic product is widely used by economists and the public at large to gauge the health and welfare of a nation. However, "if ever there was a controversial icon from the statistics world, GDP is it. It measures income, but not equality, it measures growth, but not destruction, and it ignores values like social cohesion and the environment" (OECD 2005a). Challenges to the use of GDP as a standard measure of comparison between countries reached a new high after the recent global economic crisis and the rise in consciousness over climate change (for example, GDP treats loss of ecosystem services as a benefit instead of a cost). Some "alternative" measures that attempt to include the social dimension already exist, although GDP is often used as a basis. These include, among others, the UN Human Development Index (HDI) and the Bhutan Gross National Happiness Index (GNH).

The most widely used alternative measure is the HDI, which is a composite index that combines normalized measures for three dimensions: (a) life expectancy at birth, as an index of population health and longevity; (b) knowledge and education, as measured by the adult literacy rate (with two-thirds weighting) and the combined primary, secondary, and tertiary gross enrollment ratio (with one-third weighting); and (c) standard of living, as measured by the natural logarithm of GDP per capita at purchasing power parity.

For its part, the GNH is an attempt to define quality of life in more holistic and psychological terms than GDP, and is used by Bhutan in its development strategy. The concept of GNH is based on the premise that the true development of human society takes place when material and spiritual development occur side by side to complement and reinforce each other. The four pillars of GNH are the promotion of sustainable development, preservation and promotion of cultural values, conservation of the natural environment, and establishment of good governance.

Reflecting the general dissatisfaction with GDP as a measure, some initiatives have also proposed a revision of the measure itself. In February 2008, for example, the French president, Nicolas Sarkozy, asked a commission of esteemed economists and statisticians to look more closely at GDP, identify its limits as an indicator of economic performance and social progress, consider what additional information may be required to produce more relevant indicator(s), and assess the feasibility of alternative measurement tools. The underpinnings of this initiative were clear: "What we measure affects what we do. If we have the wrong metrics, we will strive for the wrong things. In the quest to increase GDP, we may end up with a society in which most citizens have become worse off" (Stiglitz 2009). The commission has recently

continued

> **Box 7.6 continued**
>
> submitted its report, which gives a series of recommendations for improvements to the existing measure of GDP, as well as the possible construction of new indexes that will better measure social well-being and sustainability of growth. Some recommendations include looking at income and consumption rather than at production, as these are more likely to reflect material living standards; emphasizing the household perspective, as household income has often been quite different—and much lower—than GDP growth; giving more prominence to the distribution of wealth in measurement of economic progress; broadening measurement to nonmarket activities; and increasing measurement of sustainability.
>
> *Source:* Author, based on Stiglitz 2009; Stiglitz, Sen, and Fitoussi 2009.

At a time when the world community is experiencing major challenges in coping with a deep economic slowdown and a mounting environmental crisis, it is more important than ever to develop methods that allow a better allocation of resources. Demonstrating by rigorous methods what works and what does not work in the field of innovation is paramount, since innovation is the basic factor of economic growth and more generally for adaptation to social challenges.

Notes

1. The WEF defines competitiveness as the collection of factors, policies, and institutions that determines the level of productivity of a country and that, therefore, determines the level of prosperity that can be attained by an economy.

2. The 113 variables included in the GCI are grouped into 12 pillars, each of which reflects one aspect of competitiveness. The 12 pillars are institutions, infrastructure, macroeconomic stability, health and primary education, higher education and training, goods market efficiency, labor market efficiency, financial market sophistication, technological readiness, market size, business sophistication, and innovation.

3. To enable a proper comparison across the indexes, the indexes had to be adjusted so that each index ranks a common set of countries. In other words, for this exercise, any country that was omitted from any one index was dropped from the remaining three. This process resulted in a set of 122 countries common to all four indexes.

4. The United States does not carry out the equivalent of a "community innovation survey." Instead, it conducted a pilot survey in 1994, a survey on innovation in the information technology sector in 2001, and, more recently, a business R&D and innovation survey in 2008.

5. See http://www.oecd.org/sti/innovation/reviews.

6. See also the work carried out under the OECD Innovation Strategy: http://www.oecd.org/innovation/strategy.

7. To assess national innovation performance within the context of national innovation systems, the OECD recommends analysis of the following 10 areas: demand; human resources; finance; physical inputs; access to science, technology, and business best practice; ability and propensity of firms to innovate; effectiveness of market processes; networks, collaboration, and clusters; institutions and infrastructure; and business environment (OECD 2005b).

References and Other Resources

Archibugi, Daniele, and Alberto Coco. 2004. "A New Indicator of Technological Capabilities for Developed and Developing Countries." CEIS Research Paper 15 (44) from Tor Vergata University, Centre for Economic and International Studies, Rome.

Banerjee, A. V., and E. Duflo. 2008. "The Experimental Approach to Development Economics." NBER Working Paper 14467, National Bureau of Economic Research, Cambridge, MA.

Duflo, E. 2004. "Scaling up and Evaluation." Paper prepared for the Annual World Bank Conference on Development Economics, "Accelerating Development," May 21–23, 2003, Bangalore, India.

———. 2006. "Field Experiments in Development Economics." Paper prepared for the World Congress of the Econometric Society, January, Boston, MA.

European Commission. 2007. *European Innovation Scoreboard 2007: Interactive Benchmarking.* http://www.proinno-europe.eu/.

Georghiou, L., K. Smith, O. Toivanen, and P. Yla Antttila. 2003. *Evaluation of the Finnish Innovation Support System.* Helsinki: Ministry of Trade and Industry.

Guellec, D., and B. van Pottelsberghe. 2000. "The Impact of Public R&D Expenditure on Business R&D." STI Working Paper 2000/4, OECD, Paris.

Guinet, J., and M. Keenan. 2008. "National Innovation Strategies: Lessons from OECD Country-Specific Work." Presentation for the OECD–World Bank conference, "Innovation and Growth," Paris, November 18–19, 2008. http://www.oecd.org/dataoecd/60/17/41709674.pdf.

ITIF (Information Technology and Innovation Foundation). 2009. *Benchmarking EU and US Innovation and Competitiveness.* Washington DC: ITIF.

Jaumotte, F., and N. Pain. 2005a. "Innovation Policies and Innovation in the Business Sector." Economics Department Working Paper 459, OECD, Paris.

———. 2005b. "An Overview of Public Policies to Support Innovation." Economics Department Working Paper 456, OECD, Paris.

OECD (Organisation for Economic Co-operation and Development). 2002. *Proposed Standard Practice on Research and Experimental Development: Frascati Manual.* Paris: OECD.

———. 2005a. *Guidelines for Collecting and Interpreting Innovation Data: Oslo Manual.* Paris: OECD.

———. 2005b. *Innovation Policy and Performance: A Cross-Country Comparison.* Paris: OECD.

———. 2006. *Science, Technology and Industry Outlook 2006.* Paris: OECD.

———. 2007a. *OECD Reviews of Innovation Policy: Chile.* Paris: OECD.

———. 2007b. *OECD Reviews of Innovation Policy: New Zealand.* Paris: OECD.

———. 2007c. *OECD Reviews of Regional Innovation: Competitive Regional Clusters: National Policy Approaches.* Paris: OECD.

———. 2007d. *Science, Technology and Industry Scoreboard.* Paris: OECD.

———. 2008a. *Creating Value from Intellectual Assets: Synthesis Report.* Background report for the meeting of the OECD Council at Ministerial Level. Paris: OECD.

———. 2008b. *OECD Review of Innovation Policy: China.* Paris: OECD.

———. 2008c. *OECD Review of Innovation Policy: Norway.* Paris: OECD.

———. 2008d. *Science, Technology and Industry Outlook 2008.* Paris: OECD.

———. 2009. *OECD Review of Innovation Policy: Korea.* Paris: OECD.

Poot, T., P. den Hertog, T. Grosfeld, and E. Brouwer. 2003. "Evaluation of a Major Dutch Tax Credit Scheme (WBSO) Aimed at Promoting R&D." Presentation at the FTEVAL conference, "The Evaluation of Government Funded R&D," Vienna, Austria, May 15–16.

Ravallion, M. 2009. "Evaluation in the Practice of Development." *World Bank Research Observer* 24 (1): 29–53.

Stiglitz, Joseph E. 2009. "Towards a Better Measure of Well-Being." Available on FT.com (Sept. 13, 2009).

Stiglitz, Joseph E., Amartya Sen, and Jean-Paul Fitoussi. 2009. Report by the Commission on the Measurement of Economic Performance and Social Progress. http://www.stiglitz-sen-fitoussi.fr/documents/rapport_anglais.pdf.

UNCTAD (UN Conference on Trade and Development). 2005. *World Investment Report 2005*. Geneva: UNCTAD.

UNDP (United Nations Development Programme). 2001. *Human Development Report 2001: Making New Technologies Work for Human Development*. New York: UNDP Report Office.

Wagner, Caroline S., Edwin Horlings, and Arindam Dutta. 2004. "Can Science and Technology Capacity Be Measured?" Unpublished manuscript. http://www.google.com/search?q=Can+Science+and+Technology+Capacity+Be+Measured&ie=utf-8&oe=utf-8&aq=t&rls=org.mozilla:en-US:official&client=firefox-a.

WEF (World Economic Forum). 2008. *Global Competitiveness Report 2008–2009*. Geneva: WEF. http://www.weforum.org/pdf/GCR08/GCR08.pdf.

Part III
Policy Implementation

8

Policy Implementation: The Art and Craft of Innovation Policy Making

Putting in place on a large scale the foundations of innovation policy as described in former chapters is a daunting task for developing countries. They do not have the necessary resources and the educated cadres, and more generally, the institutional situation is not fit for it. Therefore to help them cope with such challenges, we will discuss two crucial points: how to develop a pragmatic innovation agenda, and how to build an institutional framework for change.

Adapting Best Practices to the Local Context: The Pragmatic Innovation Agenda

Organizational and technological innovation involves doing new things in an existing context. Even in economies with poor institutions, such as Belarus and the Islamic Republic of Iran, where the institutional and investment climate is very difficult, surprisingly dynamic, innovative, and export-driven start-ups and spin-offs are present. Their success depends crucially on the specific local context, because the instruments that can facilitate innovation (shared vision, incubation, and angel and early-stage venture capital networks, among others) work differently in the Islamic Republic of Iran from the way they work in Argentina or Ukraine, for example, which are roughly comparable middle-income economies. This critical dependence on local specifics is one characteristic of pragmatic innovation agendas. Another is the open-ended nature of the relevant

---

This chapter was prepared by Yevgeny Kutznetsov.

policies and instruments. Blueprints for innovation are useful only to the extent that they can be adapted to changing circumstances. This is how China—a paragon of pragmatic innovation—introduced incremental and gradual changes that were ultimately strategic and radical into its innovation system.

The recognition that local institutional contexts are not merely important but critical requires reconsidering the familiar reliance on "best practice" and its adaptation. If the context is crucial, a successful practice in one country signifies, at best, a promising approach in another: "best practices" no longer exist, only "promising practices." If best practice is highly contextual, no institutional recipes exist; therefore, finding a best practice requires experimenting and taking risks. A process that emphasizes a pragmatic search for solutions is called *self-discovery*—the process of trial and error through which an enterprise or entrepreneur determines what markets it can (or can become able to) serve (Hausmann and Rodrik 2002; Hausmann, Rodrik, and Sabel 2007).

Self-discovery applies not only to enterprises and private sector entrepreneurs. Just as a private entrepreneur has to discover a cost structure that will allow him or her to enter a new market, the public sector needs to seek new institutional configurations to support private self-discovery. The public sector also needs to take calculated risks, which may fail, and be accountable for the results. Self-discovery of new practices by the productive sector (with the private sector entrepreneur at the center) and self-discovery of an appropriate institutional framework to support it (with public sector entrepreneurs at the center) are two sides of the same coin.

The first section of this chapter looks at self-discovery for innovation. The second focuses on the transformation of the institutional context.

Agents and Processes of Self-Discovery

Private entrepreneurs and productive enterprises are at the center of the self-discovery process. Their risk taking and experimentation are supported by an innovation system: a network of organizations, rules, and procedures that affect how a country acquires, creates, disseminates, and uses knowledge. Key organizations participating in the private sector's self-discovery process are universities, public and private research centers, and policy think tanks. For the innovation system to be effective, the private sector must require knowledge, and effective links between research and development (R&D) and industry are vital for transforming knowledge into wealth. Therefore, self-discovery is a collective process that takes place through networking. Interactions among the different organizations, firms, and individuals are critically important. Ireland offers a good example of the main aspects of the self-discovery process. Its recent financial and economic crisis does not make its exemplary path of the past decades any less relevant.

As is well known, Ireland demonstrated that one of the poorest members of the European Union, highly dependent on agriculture and low-end manufacturing, could successfully turn its economy into a provider of high-technology services. Ireland's transformation is attributable to sustained and well-targeted investment in education and to a policy framework favorable to foreign direct investment (FDI), notably in the information and communication technology (ICT) sector. At 20 percent of gross domestic product (GDP), Ireland's net inflows of FDI are one of the world's highest, second only to Sweden. The country has become one of the most dynamic knowledge-based economies in Europe and is the second-largest exporter of software. With an average rate of GDP growth of 8.9 percent over the period 1995–2002, the "Irish miracle" is not attributable solely to the government's investment in education and its efforts to attract FDI. Substantial European Union (EU) assistance also helped Ireland attract investments relevant to a knowledge economy. Today, it is the headquarters of many European technology giants, and Dublin has taken advantage of its well-developed network infrastructure to become the hub for European telephone call centers. Ireland has thus come a long way from its traditional low-end manufacturing economy. Yet, while it was extremely successful in attracting major multinationals, their links to the Irish economy remained limited. To become a full-fledged knowledge economy, Ireland had to strengthen indigenous innovation. In response to this challenge, Ireland increased investments in education and innovation and made a major commitment with its National Linkage Promotion program (see box 8.1). After an initially slow start, multinationals increased local purchases significantly. This program illustrates a self-discovery process stimulated by appropriate procurement measures.

Box 8.1 Private and Public Sector Entrepreneurs Come Together: An Irish Experience

In the wake of a highly successful foreign direct investment (FDI) program, Ireland faced the challenge of how to deepen FDI involvement and how to leverage the technology then being used to develop an indigenous technological capability. In response, the Industrial Development Authority took a calculated risk by bringing together a group of multinational companies and potential suppliers through a systematic search process that came to be known as the National Linkage Promotion Program (1987–92). The key problem in developing potential suppliers is that one is "doomed to choose" (Hausmann and Rodrik 2006): that is, one must choose among potential suppliers simply because developing large numbers of them is wasteful. This process involves risk that needs to be shared by the government and the private sector. Three main groups were involved in the program:

- *Government:* Government provided the political imperative and charged various state agencies with supporting the program and cooperating. Budget lines were established, and the Department of Industry took a close interest in the program's operation and effectiveness. Input at this level was essential for

continued

> **Box 8.1 continued**
>
> maintaining political visibility and support for the program. A total of eight agencies contributed staff and assistance, in part to help small and medium enterprises (SMEs) navigate the bureaucracy when seeking the best and most appropriate assistance. Staff members from each agency had to shed familiar bureaucratic routines and behave entrepreneurially to fast-track the many applications for assistance and to fine-tune the services being offered to meet the specific needs of both the customers and their suppliers.
>
> - *Industry, primarily MNCs (through FDI):* The principal sector targeted was electronics, since it was the largest and most dynamic and had the greatest propensity to source locally. Industry cooperation was sought, and the MNCs (multinational corporations), through the Federation of Electronic Industries, contributed to program costs in the first two years. Companies were lobbied at high levels by senior agency executives and government ministers. Incoming companies were introduced to the Linkage Promotion Program's executives so that local sourcing opportunities could be discussed and developed. MNCs were also asked to provide technical assistance, in association with state technical agencies.
>
> - *SMEs:* A rigorous assessment procedure was used to select participating companies. It included an analysis of existing or potential capabilities against perceived supply opportunities, a detailed examination of financial management, and an assessment of existing management and of the firms' potential.
>
>   An essential part of the program was the development by linkage executives of close relationships with key MNCs. Because of the number of agencies involved in the program, a well-balanced and multifaceted team of experts in management, business development, technical issues, accounting, and banking was the key to success. This array of skills allowed the team to carry out the initial assessment and selection of suppliers (in close cooperation with the MNCs) and also to carry out early-stage development workshops with the SMEs.
>
> - *Outcomes:* Over the five years of the program, locally sourced materials in electronics increased from 9 percent to 19 percent of MNC purchases. While the total population of MNCs in Ireland was about 900, approximately 200 proved to be effective participants in the program, both through purchases and through their support. The core group of 83 supplier companies participating in the program dramatically outperformed other similar companies on average. This outcome was partly a function of the selection process, partly a function of intensive support, but largely due to interaction with demanding customers who forced them into a competitive mode. Over the period, these companies achieved average growth in sales of 83 percent, average productivity improvement of 36 percent, and average employment growth of 33 percent.
>
> *Source:* Author.

However, over the past three years, attractive wages in China, India, and Eastern Europe have weakened Ireland's competitive advantage, and many global companies have scaled back or canceled their plans for Irish operations. Ireland has had to fight hard to reclaim its status as a major destination for outsourcing and has done so by leveraging the brainpower, productivity, and flexibility of its workforce, in short, by achieving its transformation to a fully fledged knowledge-based economy.

The success of this strategy is already apparent, and a number of large multinational companies have returned or relocated or plan to relocate to Ireland in the near future. Companies such as Dell—which employs about 4,000 people in Ireland but which also began outsourcing to India and elsewhere—have not always found the quality they hoped for. As a result, countries like Ireland, which, in parallel to strong marketing campaigns, have strengthened their knowledge base through concentrated investments in R&D and education, have seen large multinationals return, and, more important, have turned out products and services higher on the value chain. Today, investment is going into higher-level jobs in pharmaceuticals, biotechnology, and digital media. In contrast, countries like Poland, not long ago an attractive location for foreign investment, are beginning to lose their share of FDI owing to weak marketing capacities and their failure to "sell" their sources of competitive advantage.

The example of Ireland highlights three main issues of pragmatic innovation as self-discovery: the first-mover problem, the critical mass problem, and the restructuring problem.

The "First-Mover" Problem. Change invariably starts with first movers (firms and other actors), those who are the first to recognize and capture new opportunities, such as Dell in the case of Ireland. In countries with weak institutions and low knowledge endowments (low-income economies), a central problem is to find first movers able to demonstrate what can be achieved in spite of obstacles and a sometimes hostile institutional environment.

The Critical Mass Problem. Scaling up and learning from the experience of first movers and pilot projects require creating critical mass by building constituencies for reform and change. This effort involves raising awareness among key groups of what is at stake and making a strong case for the need for reform. In addition, a coherent governance structure must be institutionalized to ensure coordination among the various private and public agents. Top-down vision and leadership, implementation, and follow-up are elements of success. All of these were necessary to achieve the serious investments in R&D and education in the Irish example.

Two analytical constructs drawn from management science have proved particularly useful in aggregating and scaling up first movers: clusters and supply chains, also known as value-added chains. Innovation clusters (see chapter 10) are groups of firms, research centers, and universities that conduct knowledge-intensive activities and cooperate to achieve economies of scale and scope. A value-added chain (see chapter 9) is the full range of activities required to bring a product or service from conception and design, through the different phases of production (involving a combination of physical transformation and the input of various producer services), marketing, and delivery to final consumers. It is usually defined for particular products (automobiles, electronics, garments, pharmaceuticals), and it typically crosses industries. Each stage of production is much more closely linked with upstream and downstream industries on the value chain than with other producers in the same industry. The two concepts share the view that economic activity is not coordinated solely by means of signals generated by an impersonal marketplace but that such activity also involves direct coordination through face-to-face communication.

The critical mass problem is important for countries with intermediate knowledge endowments and institutional capabilities. It is particularly acute in large middle-income countries such as Argentina, Brazil, and Mexico in Latin America and in Eastern European economies.

The Restructuring Problem. Restructuring requires the identification of new innovation domains (innovation clusters and value chains). As many countries have discovered, successful innovation clusters (such as the forestry cluster in Finland or the garment and furniture clusters in Italy) or value chains (such as the electronics supply chain in Ireland) do not guarantee success. Even highly innovative clusters can decline as new and more successful competitors emerge. Here, this restructuring is considered in the light of a search for new innovation domains. Understood in this way, the problem is faced almost entirely by economies with advanced innovation capabilities.

Diversity of National Innovation Agendas

Because the self-discovery process considered here is closely related to specific institutional circumstances, the strength and sophistication of public and private institutions are one variable that pragmatic agendas need to take into account. Institutional endowments, however, are hard to measure, and very imperfect proxies are used. Another variable is a country's knowledge endowments, comprising education, innovation, and information technology (IT) (measured, roughly, by the knowledge economy index; see chapter 7).

In what follows, these two variables are used to arrive at a taxonomy of pragmatic innovation agendas. In the short run, a pragmatic policy agenda considers, on one hand, the country's level of technology and, on the other, the conditions of private sector development, and, based on those seeks a functional fit between a country's knowledge and its institutional endowments. For instance, Argentina and the Russian Federation have a paradoxical combination of weak institutions and relatively high knowledge endowments. To achieve a functional fit between their knowledge and institutional endowments, they may need to adopt somewhat untraditional institutions. In the long run, the pragmatic agenda becomes a self-reinforcing virtuous cycle of simultaneous enhancement of institutional and knowledge capabilities.

The Republic of Korea's transformation from a system of crony capitalism at the beginning of the 1960s (and abysmally low endowments on both counts) into an emerging innovation leader is an example of such a virtuous cycle.

The following discussion distinguishes three levels of technology development and three levels of institutional development. The evolution of technology (table 8.1, vertical axis) is reflected in the familiar distinction (de Ferranti and others 2003) between technology adoption (appropriate for low knowledge endowments), technology adaptation (for intermediate knowledge endowments), and technology creation (for the high knowledge endowments). Little technological capability exists at the technology adoption stage (Central America except Costa Rica, for example, and Sub-Saharan Africa).

Table 8.1 Diversity of Pragmatic Innovation Agendas

Level of innovation and human capital capabilities	Strong investment climate and institutions — Long term	Tolerable and improving investment climate and institutions — Medium term	Poor investment climate and institutions — Short term, survival
High Technology creation	Innovation leaders agenda: • Development of proprietary technology through promotion of innovation clusters • Examples: Finland; Ireland; Israel; Republic of Korea; Portugal; Singapore; Spain; and Taiwan, China	Critical mass agenda: • Increase of value added of natural resources wealth and technology commercialization • Example: Russian Federation	• Leveraging pockets of dynamism • Examples: Argentina (1990s), Belarus, the Islamic Republic of Iran, Russian Federation (1990s)
Medium Technology creation and technology adaptation	—	Critical mass agenda: • Development of innovation clusters and high value-added supply chains • Examples: middle-income economies of Latin America (Argentina, Brazil, Chile, Mexico), Asia (Malaysia, Philippines, Thailand), Eastern Europe, and South Africa	—
Low Technology adoption	Creation of knowledge endowments: • Making investments in higher education and technology adoption • Examples: oil-rich Gulf countries	"Exports as a springboard" agenda: • Development of nontraditional exports as entry point for institutional and technology development • Examples: Bolivia, Central America (except Costa Rica), rural regions in India and China, Kazakhstan, Republic of Korea in the 1960s, Mauritius, Mexico in the 1970s, Pakistan, Paraguay, Vietnam	Institutional context agenda: • Creation of basic institutional infrastructure through a diversity of entry points • Creation of demonstration effect to show that innovation does matter, in particular in health, education, agriculture, and crafts • Examples: most of Sub-Saharan Africa and most Central Asian states

Source: Author.

In contrast, at the technology creation stage, a critical mass of national science and technology (S&T) capabilities is relevant or potentially relevant for business (the Asian Tigers and Russia, for example). Between these extremes, the technology adaptation stage is characterized by a critical mass of qualified engineers and technical staff (advanced Latin American and post-Socialist economies: Argentina, Brazil, and Mexico; Eastern European countries that recently joined the European Union; and Ukraine).

In terms of institutional endowments, a distinction can be made between strong and weak links within the innovation system, using the share of business R&D in total R&D as a proxy. In an efficient innovation system, the business sector takes the lead in R&D financing and execution (Asian high performers, for example), while in a dysfunctional innovation system R&D is performed by the public sector (such as in post-Socialist and post–import-substitution economies). Countries with extremely weak institutions, in particular public sector institutions (table 8.1, col. 4), are a special case. Here, the binding constraint is a difficult and often unpredictable investment climate (as in Belarus and the Islamic Republic of Iran, or in Argentina and Russia in the 1990s), which supersedes any other considerations.

The planning horizon of business sector actors is a good proxy for the quality of institutions. Poor institutions are correlated with a short-term planning horizon and survival. Strong links in the innovation system are correlated with a long-term decision-making horizon; concerted action—interorganizational links—rarely pays off in the short run. The medium-term planning horizon is correlated with a system that lies between the two extremes. Table 8.1 illustrates seven broad policy agendas that help show the diversity of circumstances under which countries construct their self-discovery of innovation agendas. The table simply aims to give a sense of the variety of possible approaches. The first three policy situations focus on moving exceptions, or potential first movers (which exist but are isolated), into the mainstream. The remaining situations all focus on the first-mover problem: facilitating the emergence of exceptions—pockets of excellence and dynamism in a hostile environment—to provide an example to follow, scale up, and diffuse. The specific nature of such first movers differs. When both knowledge endowments and institutions are rudimentary, the first movers are institutions of excellence in education and public service delivery. When the institutional environment is better, first movers are export-oriented firms, like those that initiated the radical transformation underway in China and Vietnam. Finally, the peculiar situation that combines a long-term planning horizon with low knowledge endowments (oil-rich Arab economies) calls for a first-mover agenda in higher education, innovation, and IT.

Critical Mass Agenda: Developing Innovation Clusters and Value Chains. The critical mass agenda applies to countries that have technical capabilities (engineering and applied research) and export-driven manufacturing and natural resources, often as subsidiaries of multinationals. They include most of the Eastern European post-Socialist economies (such as Poland or Hungary), large Latin American economies (Argentina, Brazil, Chile, and Mexico), emerging Asian Tigers (Malaysia and Thailand), and advanced regions of China and India. In these countries, human capital costs are relatively high. They are often squeezed between the lower-cost technology adoption countries and the more advanced economies of the Organisation for Economic Co-operation and Development (OECD). The sense that they are at a turning point is strongest in higher middle-income countries such as Chile, Hungary, Mexico, and Poland. They have isolated pockets of dynamism and innovation, illustrated by the phenomenon of developing country multinationals, firms from developing countries that expand abroad on the basis of their innovation capabilities (Techint, a steelmaker in Argentina; CEMEX, a pioneer in just-in-time cement production in Mexico; and Infosys, the information processing paragon in India). While these pockets of innovation signal the country's potential, they remain exceptions. This situation makes the issue of links (value chains and clusters) a central focus of the policy agenda.

This policy diagnosis is not new. Virtually all these countries recognize the fragmentation of their innovation systems and their failure to develop links on the basis of innovative "first movers" as a central problem. They have tried many approaches and adopted most best practices. These practices are rich and diverse: a supplier development program to promote value chains (exemplified by the Irish program) and a variety of innovation sites (technology incubators, business development centers, innovation zones, and so on, as noted in chapter 10). A long process of policy learning and experimentation has revealed the same amount of heterogeneity and internal diversity in the performance of innovation programs and policies as in the performance of firms. A few are successful, but most do not effectively address the central problem of the fragmentation of actors in the innovation process. It is quite easy to develop sites and much more difficult to articulate innovation networks. Technology incubators—which rent office space to technology start-ups and provide business development services—are widespread, but very few succeed in developing vibrant early-stage networks that help techno-entrepreneurs develop the managerial, technical, and financial skills they need to grow their fledging start-ups.

As a result, policy makers in these countries suffer from "recommendation fatigue": they have seen and tried almost everything available, generally with disappointing results. They realize that copying best practices does not work and that they need to adapt "promising practices" more creatively in local institutional contexts. They are discovering that they have to embark on the self-discovery process.

Given the accumulated stock of programs and policies, a policy priority should be to recombine industrial capital, human capital, and policy assets.

The notion of recombination proceeds from an observation that a wealth of industrial assets, talent, and public programs is already present (Gu and Steinmuller 1996; Stark 1996). The priority is to make sense of what exists by recombining the viable assets into sensible programs rather than to invest in new assets and programs.

This objective implies drawing upon the variety of existing small and medium enterprises (SMEs), R&D, labor retraining, and innovation programs to make them work together toward a common goal with clear performance benchmarks. Ireland's National Linkage Promotion Program (box 8.1) is an example of such a framework program: it is not just another business development program but a program that draws upon and taps into existing programs. It encourages links among programs and facilitates changes in them by making the main actors the beneficiaries and providing clear feedback loops to detect and correct errors. If, for instance, a large share of the potential SME suppliers chosen to participate in the program fails to become actual suppliers, something is wrong both with the framework program and with the support programs on which it depends. At that point, all relevant stakeholders must come together to deal with such problems, a practice not generally adopted by SME and innovation programs. The issue of framework programs will be addressed in more detail in the section on creating a conducive framework.

The aspect of the agenda that addresses moving up the value chain and making the transition from global sourcing to proprietary knowledge presents two quite different cases. Most of the countries concerned have very dysfunctional innovation systems and a business sector that performs little R&D. They need to recombine their technological capabilities and capitalize on them to create wealth. Increased public R&D spending is valid only if it translates into business R&D. This is rarely the case. In Thailand and Malaysia, however, as in Finland and Ireland in the 1980s, innovation is fairly efficient, but knowledge creation is weak. In such a case, an increase in the public R&D budget may be advisable.

Often the agenda for reaching critical mass is purely institutional. India boasts emerging innovation clusters and vibrant equity finance; yet seed and early-stage financing for technology start-ups is in its infancy. Therefore, the government put in place an ambitious fund-of-funds program to encourage private venture funds to consider smaller projects that are more risky and involve high transaction costs. Brazil, Chile, Mexico, and Russia are also experimenting with such programs to promote techno-entrepreneurship. The general principles behind these programs are similar, but each is structured pragmatically to reflect local circumstances.

Critical Mass Agenda: Leveraging Natural Resources and S&T Endowments. While all the characteristics of the critical mass agenda described above apply to Russia, its self-discovery process, involving the construction of innovation clusters and value chains along with the supporting institutions, is something of a special case. Russia has unusually high endowments of both natural and S&T resources. Although the latter have deteriorated significantly in the past 18 years, the federal government is attempting to enhance and restructure them. This combination is not easy to manage, because it calls for a "double transformation." On the one hand, the country needs to commercialize its S&T capabilities in products and services valued by the market (development of clusters), and on the other hand, it needs to develop value chains to move toward greater processing capacity and more value added from natural resources. Russia is not unique in this respect. The export structure of Australia, Canada, New Zealand, and Norway is dominated by primary resources.[1] Yet each of these countries diversified backward into capital goods and higher education to become world leaders in mining and oil management, capital goods production, and higher education for the primary resource sector. Finland, with its world-class forestry cluster and a cluster of firms around Nokia, is a model of success in this respect and Russia's neighbor. Inspiring as they are, though, these countries are of little immediate relevance for Russia's self-discovery.

What makes Russia different is the combination of high endowments and weak institutions. The legendary success of Tekes in Finland and similar examples of focused action are irrelevant to policy action in Russia today. Rent seeking is so pervasive and so creative that coordination devices that have proved helpful elsewhere—such as interministerial innovation councils—easily degenerate into cartels or into forums where each agency defends its turf rather than developing joint agendas for action.

At a subnational level (Novosibirsk, St. Petersburg, and Tomsk), innovative start-ups and promising innovation programs and initiatives to support them do exist (see box 8.2). Yet these firms and programs remain exceptions. The federal government, which recognizes that isolation and lack of knowledge sharing by local institutional experiments are a major problem, has instituted a grant scheme to encourage drawing lessons from these initiatives and sharing promising practices at the local level.

Critical Mass Agenda: Leveraging Pockets of Dynamism. When the gap between fairly strong knowledge endowments and unusually fragile and unpredictable institutions is large, the challenge is to leverage pockets of dynamism. Countries in this situation are countries in decay, characterized (at least until recently) by the flight of both human and financial capital. In the 1990s, Argentina, Armenia, and Russia were examples of institutional instability. Today, this is the case in Belarus and the Islamic Republic of Iran.

As for innovation performance, the picture is not uniformly bleak. Some highly successful innovation-based companies are first movers. For instance, EPAM in Belarus is an information-processing firm that now boasts more than 3,000 employees, with offices in Hungary, Russia, and the United States.

> **Box 8.2 Turning Scientists into Entrepreneurs: Moscow University's Science Park**
>
> Moscow University's Science Park was established in 1991, as a joint venture of Moscow State University, the Russian Ministry of Science, and the private sector. More than 30 companies in software development, laser technology, and biotechnology currently work in the park. These firms benefit in several ways and are also shielded from interference by the state because inspectors harassing the firms had to deal first with the park's administration.
>
> They also benefit from a clustering effect through their access to the university's human capital and R&D. Synergy among tenants has also been beneficial: communication between seemingly unrelated tenants has produced at least two new commercial ideas. The park also provides access to modern telecommunications, including a satellite teleprompter, and office infrastructure.
>
> Business development services appear to be less important to park tenants. Those services are mainly available through private service providers, a practice consistent with international best practice. The park does not provide financing to tenants.
>
> The park evolved in stages. Initially, Russian start-ups moved in. These firms generated interest among foreign investors, with whom they formed joint ventures. This foreign direct investment helped the park expand. Global companies came as both shareholders in the park and cosponsors of its expansion. The park's third office facility, for example, is being constructed jointly with Samsung, and Intel plans to cosponsor a contest for the best commercial idea.
>
> The park's success is due to several factors, but especially strong leadership and incremental growth. Rather than beginning with a single grand project, the park's leaders started small and established credibility. Only then were they able to attract brand-name tenants and investors.
>
> *Source:* Author.

It aspires to become the next Infosys and has the growth dynamics to do so in the long run. How does the firm not simply survive but grow rapidly in such an unstable institutional environment? The main reason is that it was "born global," created by a Belarussian emigrant living in New York who appreciated the creativity and problem-solving skills of Belarussian engineers and software developers. "Born global" innovation-based growth was pioneered by Israel, among others, to overcome the constraints of an inhospitable and often hostile investment climate and firms' lack of marketing skills. Under this strategy, only R&D and production are carried out in the problematic country, while marketing and access to finance take place overseas. The same strategy is now applied, or rather is being rediscovered, by nascent firms in other countries with difficult environments.

The issue for these countries is to leverage existing pockets of dynamism through science and technology parks, technology incubators, and other bridge institutions to help entrepreneurial individuals articulate their vision. These countries' policy and institutional experimentation may be quite intense and diverse, but it is often parochial and isolated from relevant experience elsewhere in the world. Few know, for example, that the Islamic Republic of Iran boasts an early-stage venture capital program to support techno-entrepreneurship. It is a tiny but reasonably structured and quite commendable initiative; yet the relevant officials seem to be unaware of similar initiatives in almost all middle-income countries, including Armenia, India, and Russia. Thus, although self-discovery is occurring, it is strikingly isolated. Not only is the wheel being reinvented all the time (to a certain extent, that is the essence of self-discovery), but also there is little awareness that others too are reinventing the wheel. Openness to promising worldwide practice is a priority for these countries. By implication, multinational organizations can help improve the institutional environment in these countries by more actively incorporating them in South-South networks for sharing relevant and promising practices.

Emerging Innovation Leaders Agenda. Emerging innovation leaders is an agenda for countries as diverse as Ireland, Italy, the Republic of Korea, and Spain. As examples of successful and recent catch-up, these countries need to strengthen their R&D by investing more in fundamental and applied research and advanced human capital. Since an extensive literature addresses the restructuring of innovation systems in OECD economies, this subject is not addressed in detail here. Suffice it to say, the restructuring of and search for new innovation domains can be quite a daunting policy challenge. As semi-industrialized economies such as Brazil and China advance their innovation agendas, they become increasingly formidable competitors of established and sophisticated innovation clusters in OECD economies. The focus of this agenda is on restructuring and searching for new innovation domains to ensure higher value added and raise the population's standard of living.

Institutional Context Agenda: Nurturing Actors to Become Levers for Change. To spur growth and innovation, the countries of the "bottom billion" (Collier 2007)—that is, the poorest billion people living in mostly landlocked countries with very weak institutions, including most of Central Asia and Sub-Saharan Africa—need government intervention most. Yet these are precisely the countries in which institutions are the weakest, a context in which any intervention is likely to fail. An "infernal trap," a low-level equilibrium that blocks both technological and institutional learning, is often the consequence.

A central problem of these countries is a pervasive body of entrenched interests. To deal with this problem requires actors sufficiently well acquainted with the institutional context yet not dependent on those interests. Such actors can become levers for beginning the transformation of this difficult institutional environment in order to escape the low-level trap. Successful diaspora members can often serve as such levers (see box 8.3). On the one hand, as natives of the country, they have a good understanding of the institutional reality. On the other, they do not depend on rents from natural resources or a government position and are consequently not dependent on entrenched interests.

Box 8.3 Diaspora Member Creates First-Mover Institution in Tertiary Education

After living in the United States for nearly 20 years, Patrick Awuah moved back to his native Ghana to start a new university to educate Africa's next generation of leaders. Awuah had left Ghana in the mid-1980s, when the country was under military rule. He graduated from Swarthmore College with an engineering degree in 1990. Soon after, he joined Microsoft, moved to Seattle, and became a millionaire before he was 30. Having achieved economic well-being, a solid reputation, and a fulfilling family life, he decided to relocate to Ghana. When asked about his motivation for returning to Ghana, he mentioned the birth of his son: "Having a son caused me to reevaluate all my priorities," he says. "This was something that was eating at me. What kind of world is it that my son is going to grow up in? And how is Africa represented in that world?"

His goal was to establish a university of Ivy League quality in his home country and train the next generation of African leaders, with a focus on ethical entrepreneurship and integrity. Awuah used his U.S. contacts and his professional knowledge to develop and assess his business plan. He found a team of University of California, Berkeley, MBA students and management consultants to conduct a feasibility study. He and his family invested more than half a million dollars in the Ashesi project and another US$4 million more through private, U.S.-based networks, including former colleagues at Microsoft, private corporations, and foundations.

Ashesi is a private university in a leafy residential suburb of Ghana's capital city, Accra. Its campus and facilities present a stark contrast to Ghana's five public universities, where enrollment has soared to 65,000 since 1990 and where overcrowded lecture halls, substandard student residences, rising tuition fees, and poor staff salaries have led to angry protests and frequent strikes. However, tuition at public universities is also much cheaper than the US$4,500 in fees that Ashesi charges.

Ashesi has small classes, well-trained and well-paid staff, and international partnerships with top-tier universities such as New York University and with the Council on International Educational Exchange. About 80 percent of the university's students are from Ghana. The rest are from other nations in Africa. About half the students receive financial aid. In 2005, four years after enrolling its first crop of freshmen, Ashesi issued its first diplomas to a graduating class of 20 students. Ashesi offers two four-year degrees, in computer science and in business administration, both of which also emphasize a broad foundation in liberal arts. As one student described the experience,

> You're like raw gold. The school is like a furnace. The heat from all the courses, from the professors, from the projects that you undertake—you come out as a refined substance, you come out glittering. You dream beyond your world.

Source: Author.

Entry points in this situation need to be both diverse and modest: *diverse* to make up for the likelihood of many failures and *modest* to minimize the costs of failure. In this context, interventions such as export processing zones, microfinance initiatives, distance learning–based training initiatives, and the like are akin to a venture capital portfolio in which most initiatives are expected to fail. Yet development returns from those that succeed compensate for the many that fail. The low-level trap of stalled institutional learning is discussed in more detail in the later section on institutional framework.

Exports as a Springboard Agenda. Countries and regions with stable enough institutions may use exports as a springboard agenda, as they transform low unit labor costs into marketable products and services. This is a well-known strategy, owing to the highly visible successes of the East Asian Tigers and Japan. Exports and export growth are a natural benchmark for open-ended pragmatic measures and policies, and export growth provides a clear and unambiguous feedback loop between innovation policies and outcomes. Countries as diverse as Armenia, Bangladesh, and Vietnam are successfully pursuing this innovation agenda.

Export processing zones (EPZs), which provide a more stable environment than that in the rest of the country, are one policy instrument of this agenda. However, they are often ineffective and are rightly criticized for distorting incentives and inviting fraud and corruption. Yet EPZs can be designed in many ways. The traditional design is a territorial enclave whose implicit objective is to minimize interactions with an unpredictable, unstable, and corrupt domestic economy. Second-generation EPZs have been successfully piloted in African countries (Madagascar, Mauritius) and offer an incentive regime for all exporters in the country to expand the market-friendly framework to the entire economy. They include a substantially reduced tax and regulatory burden and light, nondistorting assistance. Such an incentive regime also produces a constituency for reform, consisting of first movers and others who benefit from enhanced private sector dynamism. This constituency is likely to push for further reform, including the reform of enterprises outside the EPZ.

Knowledge Endowments Agenda: Creating First-Mover Institutions. Oil-rich economies in the Persian Gulf have strong institutions (in the sense that they have a long-term strategic planning horizon for decision making), yet very modest knowledge endowments. Hence, their agenda is to leverage oil revenues to create internationally competitive higher education and R&D organizations. The first priority for Gulf countries is to build a few organizations of excellence, if only to reduce their dependence on imports of human capital (from India and other economies).

Self-discovery is very intense in these countries. New organizations are created with lavish funding, and intense knowledge transfer is underway, as some of these new organizations are almost entirely staffed by foreign experts. However, there is too much adaptation, even replication, of best practices from elsewhere rather than true self-discovery in the sense of experimenting with novel, yet existing institutional features that reflect the local culture. The innovation agendas of these countries provide a curious mirror image of countries in self-imposed isolation such as the Islamic Republic of Iran. Whereas the latter are strikingly original in their reinvention of the institutional wheel, the oil-rich countries shun or downplay institutional experimentation. They are prepared to pay whatever price is necessary for the best experts and best global practice, assuming, at times naively, that best practice will remain just that in any context.

Table 8.2 offers Saudi Arabia suggestions of possible paths for building indigenous innovation capabilities. Saudi Arabia can buy technologies from abroad, improve domestic ones, develop joint ventures with foreign partners, or develop its indigenous R&D. Specific actions can help these different options, which are not mutually exclusive, take concrete form.

Structuring the Self-Discovery Process: The Subnational Dimension

The heterogeneity of both private and public sectors has two crucial policy implications. The spatial differentiation of economic activity, typically linked to industrial specialization, means that a focus on national indicators and institutions can obscure critical transformations occurring at a subnational level. Likewise, the state, in developing as well as in developed countries, is not a unified whole. Rather, it consists of multiple, differently organized units with varying political and economic resources, jurisdictions,

Table 8.2 Possible Innovation Paths for Saudi Arabia

Strategic option	Possible policy action
Improve: improvement of existing products by adding new features and value-added services	Put in place a multiskilled and multi-industry support group to help Saudi industrials make minor innovations that generate big rewards.
Research: support for Saudi research and breakthrough innovation through regional and national funding as well as private research conducted by Saudi industrials	Develop an innovation scheme to promote public-private partnerships and industry-university collaboration, focusing on funding of seed stage for potential niche research topics.
Venture: Saudi industrial venturing by sourcing entrepreneurship ideas providing incubation, and building innovative prototypes that could become successful products on the international markets	Support entrepreneurship through national awards, and support projects at the seed stage with appropriate grants.
Buy: purchase of corporate external venturing through capital investment (as done by the Gulf Venture Capital association within the Gulf countries and through mergers and acquisitions)	Link with global value chains through foreign direct investment, and encourage application-oriented research (e.g., the recent agreement between KACST and IBM).

Source: World Bank Institute 2008, adapted from Chebbo 2008.
Note: IBM = International Business Machines; KACST = King Abdul Aziz City for Science and Technology.

and interests. As a result, economic and institutional change begins in certain locations or domains and advances through partial and incremental (microlevel) reforms that only aggregate into larger-scale transformations over time.

Successful innovation performers are said to be sociocultural islands: homogeneous and, by implication, relatively small national economies (such as Denmark; Finland; Ireland; Israel; the Republic of Korea; Norway; Sweden; and Taiwan, China) are said to be important predictors of growth and performance (Aubert and Chen 2008). The insight here is that similar "islands," such as Bangalore, also emerge in a highly diverse economy and that the reform process is largely a matter of building bridges from them to the rest of economy. Only by disaggregating innovation policies and their interactions with (parts of) the equally differentiated public and private sectors is it possible to see whether and eventually how these policies can rebuild institutions for economic development. This is the first of the two implications mentioned above. Incremental microreforms capable of reshaping institutional frameworks and triggering growth and reforms are the subject of the section on institutional frameworks.

A flexible, decentralized policy process that takes the diversity of circumstances into account is needed. National innovation policies often fail because they are too crude and general to be relevant to economic actors with widely different interests and capabilities. In contrast, a decentralized innovation system envisions new and varied roles for federal, state, and local authorities and for civil society.

The following recommendations are directed toward a clearly decentralized context in which the subnational level has sufficient autonomy and the federal government is active:

- *Subnational level.* Piloting of new innovation initiatives (entry points) at the state and regional level that (a) grow out of discussion and debate within civil society; (b) draw upon active participation of the private sector in financing, conception, and operation; and (c) build in mechanisms for evaluation and improvement in light of the state's own experience, experience elsewhere in the country, and experience abroad.

- *Federal level.* Continuous monitoring, evaluation, and technical assistance to state-level initiatives that (a) consolidate existing programs; (b) make demand by the states and private sector clients the driver of federal programs; (c) ensure flexible federal budgeting, capable of adapting to demand and building upon experience, and mix federal funds with those of other sources; (d) build in mechanisms for evaluation, for identification of best practices at the state level and for collecting, evaluating, and disseminating international experience; (e) incorporate feedback from the states to improve federal instruments continuously; and (f) attract top-caliber talent to promote and facilitate the planning process, to administer the operating program, and to guide experimentation and evaluation.

- *Civil society.* National process of vision building and construction of shared agenda for change that puts into practice the new decentralized incentive framework to promote innovation using altogether three approaches: matching grants, benchmarking of the business and innovation environment, and competition for federal funds.

Matching Grants. Under this principle, the central government agrees to match every dollar, up to a certain limit, that subnational governments dedicate to innovation and economic development projects, decided in collaboration with private actors, on the condition that those actors match the subnational contribution as well. The idea is simply that if the regional government and economic actors are willing to put their own money at risk in financing the projects they define together, the national government can assume that their choices are well considered and should back the project as well. The advantage of this method is to impose some discipline on project selection with little or no increase in red tape. The regional economic actors, public and private, have an incentive to sort through their priorities—and identify potential problems—and the national government acts only to ratify their provisional decisions as they emerge.

Benchmarking of the Business and Innovation Environment. This approach also aims to discipline project selection while holding bureaucracy and the politics of clientelism generally in check. But it does this not by ratifying actors' decisions but by providing information on economic performance that causes them to reflect on their possibilities in new ways. The provision of this crucial information can take place through the creation of a so-called league table of regional economic performance that covers several topics:

- *Business registration*—costs (for all areas, including time, formal and informal types of payments and contributions, including bribes), procedures required, delays
- *Business licensing*—numbers and types of licenses required, cost, time, and payments required
- *Acquisition of business premises*—procedure, costs, constraints, and delays
- *Business inspections*—types (and agency responsible), costs, number, and process followed.

Competition for National Funds. The main feature of this approach is competition among subnational entities for national funds for innovation and economic development. The entities would receive the funds based on the quality of their proposals, so that excellent proposals would have more funding than less compelling programs. A national public-private innovation council would grade the quality of the proposals. Proposals would share their strategies so that even losers would gain knowledge. Administration of such a competition may require the ability to make impartial project selection that may not be readily available. In this case, this would not be an immediate option but a possibility to consider for the future.

Promising and best practices that have emerged recently tend to be a combination of these three approaches. For instance, sectoral funds for innovation receive government funding for research on the basis of matching-fund contributions from private sector and subnational government and, of course, on the basis of the quality of the proposals. Funding for science proposals is administered as a contest, in which the main criterion for winning is demonstration of interorganizational links, such as university-industry connections. Such contests can be quite elaborate. An initiative to establish innovation and technology zones in Russia, for example, started with a contest between subnational entities that took into account all three criteria (matching contributions, prior performance, and quality of the proposal). The selection process was difficult, as only four proposals were chosen out of dozens submitted; yet as long as decisions are transparent and credible, such procedures stimulate local creativity while working to meet national innovation objectives.

How to Create a Conducive Institutional Framework: The Virtuous Cycle

Creating an institutional framework conducive to innovation is not, in general, something that can be made through a clear blueprint rigidly prepared in advance and then closely followed. It is more a search process that begins with microreforms—well designed and conducted—that lead progressively to virtuous circles.

Reshaping an Institutional Framework as a Search Process

Conventional economic development focuses on endowments: with an appropriate endowment (good institutions, good investment climate, cultural dispositions, property and trade laws, rule of law), economies grow. Those that lack such endowments do not grow. But the surprising frequency of spontaneous growth episodes in "poorly" endowed economies, the sharp disparities in regional development within national economies subject to the same general rules, and the periodic successes of economies that change their institutional endowments by growing (China) rather than by fixing endowments to grow, all strongly suggest fundamental flaws in this all-or-nothing view of endowments.

This section develops an alternative view, according to which the institutional framework is necessarily changed through the implementation of innovation programs and policies. The challenge is to monitor this institutional change on a microlevel and scale it up.

In the case of countries at the advanced technological frontier (such as Finland or the United States), reaching that frontier improves an industry's prospects if subsequent development builds on that frontier. However, a general result of what is loosely called the information revolution—the widespread diffusion of powerful computers and telecommunications networks—is the increasingly unpredictable direction of technological development. The easier it becomes to explore technological frontiers and to survey results across these frontiers, the greater the chances of multiple, competing solutions to any given problem, each better on some dimensions than the others, but none dominant on all. Hence, one good solution cannot be expected to lead, by a natural progression, to another. In other words, the more knowable the world as a whole becomes, the less confident one can be about the kind of knowledge that will prove useful in engaging its parts.

By the same token, the more development depends on applying knowledge from domains traditionally unrelated to an industry's core activities, the less meaningful the very idea of a technological frontier—*it is everywhere and nowhere*—and the less confident one can be that leadership today guarantees leadership tomorrow. In these circumstances, it may well be more important to be able to search effectively across domains than to dominate the generation of ideas and technologies in any one of them. The decline of the centralized corporate research laboratory (in which stable project groups could pursue a line of research for a decade or more) and the rise of the ad hoc research consortium (which brings together expertise from previously separate domains) is one widely noted result of this transformation. This transition is one example of a changing institutional framework for innovation in advanced settings.

At the other end of the spectrum are countries with highly dysfunctional institutions and low knowledge endowments. Most are in Sub-Saharan Africa, where almost nothing works effectively. Interventions and policies tend to fail because of interlocking institutional traps: pervasive problems with security (strife and civil wars), high costs of access to ports, and other binding constraints. Here again, finding solutions requires cutting across several domains and thinking outside the box, experimentally, and innovatively. Even then, as Collier (2007) notes, most policies are likely to fail, simply because the institutional environment is so difficult and the constraints are so numerous and interlocking. Few interventions succeed; the institution of higher education in Ghana described in box 8.3 is one that did. In a dysfunctional environment, what is needed is a venture capital perspective on institutional formation: a search for ideas in different domains, innovation and experimentation, and an understanding that most projects will fail yet the few that succeed will provide a development payoff that counterbalances the failures. As Collier (2007) remarks, governments and development businesses are extremely risk averse, failure is discouraged and perceived as a mistake, and learning by experimenting is alien to the development bureaucracy culture. Yet the required search and experimentation processes needed in advanced settings (near the technological frontier) and in "the bottom billion" are strikingly similar.

In such risk-averse contexts, change comes often from members of diasporas who are not linked to established domestic institutions but have experienced innovation and newness in the course of building a new life in the countries to which they have immigrated. Therefore, members of diasporas can be very effective change agents, as illustrated by Chilean and Taiwanese examples (see box 8.4).

A pragmatic agenda for change often implies focusing on bottom-up entry points (the immediate policy agenda), scaling them up to ensure coordination

---

Box 8.4 Members of the Diaspora Trigger Changes in Innovation Systems

In 1997, Ramón L. García, a Chilean applied geneticist and biotechnology entrepreneur with a PhD from the University of Iowa, contacted Fundación Chile, a public-private entity charged with technology transfer in the area of renewable resources. García is the chief executive officer of InterLink Biotechnologies, a company based in Princeton, New Jersey, which he cofounded in 1991. After jointly reviewing their portfolios of initiatives, Fundación Chile and Interlink founded a new, co-owned company to undertake long-term R&D projects. These projects are focused on the transfer to Chile of technologies important for the continuing competitiveness of its rapidly growing agribusiness sector. Without García's combination of deep knowledge of Chile, advanced U.S. education, exposure to U.S. managerial practice, and experience as an entrepreneur, the new company would have been inconceivable.

The fact that skilled expatriates can create enormous benefits for their countries of origin has gained attention in recent years, owing to the conspicuous contributions that the large, highly skilled, manifestly prosperous and well-organized Chinese and Indian diasporas have made to their home countries. García's collaboration with Fundación Chile, however, suggests that diasporas do not need to be large to produce an impact: 10 similar initiatives could transform entire sectors of the economy in relatively small countries like Chile. Moreover, García's collaboration with Fundación Chile suggests that even small, informal diaspora networks linking small home countries with their talent abroad have some important institutional resources and may prove capable of developing more.

As of January 2008, García had created three biotechnology firms with Fundación Chile. ChileGlobal, a network of about 100 high achievers of Chilean origin, was established in 2005 to institutionalize contributions that similar efforts can make to the Chilean innovation system. The story does not end here, but rather begins. ChileGlobal recently organized a workshop to promote mentoring between innovation start-ups in Chile and Chilean high achievers abroad. As a sign of recognition of both ChileGlobal and the Chilean diaspora, key participants of the workshop were received by Alejandro Foxley, then foreign minister and vice president. Somewhat unexpectedly, the intricacies of establishing an early-stage venture capital industry became a focus of the discussion. Foxley requested members of

continued

> **Box 8.4 continued**
>
> ChileGlobal participating in the meeting to lead an informal working group that involved public agencies active in this area. The working group is examining issues that need to be addressed in Chile's institutional environment: the focal point is a low-key and highly focused reform effort.
>
> In this endeavor, the Chileans can study (but not copy) the well-known experience of Taiwan, China, with creating an early-stage venture capital (VC) industry. When the Taiwanese government decided to promote a VC industry in the beginning of the 1980s, it had neither the capacity nor a blueprint for doing it. Many were opposed to the idea because the concept of venture capital was foreign to traditional Taiwanese practice, in which family members closely controlled all of a business's financial affairs. Entrenched interests wishing to maintain the status quo were strong. Through intense interaction with the Taiwanese diaspora in Silicon Valley new institutions such as the Seed Fund (with an initial allocation of NT$800 million, later complemented by an additional NT$1.6 billion) provided matching capital contributions to private VC funds.
>
> Two American-style venture funds—H&Q Asia Pacific and Walden International Investment Group—were created in the mid-1980s. They were managed by U.S.-educated overseas Chinese who received invitations to relocate to Taiwan, China. Once the first venture funds proved successful, domestic IT firms created their own VC funds. Once those started to pay off, even the conservative family groups began investing in VC funds and the IT businesses.
>
> A search network consisting initially of dynamic and forward-looking members of the Taiwanese government and leading overseas Chinese engineers in Silicon Valley was central to the emergence of a modern VC industry in a country dominated by conservative and risk-averse business groups.[a] This network did not have a blueprint; yet it had a role model (Silicon Valley) and a clear idea of what to do next. By defining each step along the road, the network became wider and eventually incorporated skeptics and opponents.
>
> The extension of diaspora entrepreneurs' projects from cofounding joint firms in home countries to cocreating the institutional infrastructure that allows these firms to flourish is a natural progression. The initial objectives of Ramón García and his Taiwanese peers were both modest and specific: to advance their professional interests by setting up technology firms in their home countries. Yet as the constraints of the home country institutional environment became apparent to them, they worked to advance institutional reform to remedy some of the constraints. The successful growth of knowledge-based firms and the creation of an appropriate institutional environment became two sides of the same coin. Innovation entrepreneurship blossomed into institutional and policy entrepreneurship.
>
> *Source:* Author.
> a. A search network is defined as a network for identifying successive constraints and then the people or institutions that help mitigate, at least in part, the difficulties associated with these constraints.

and focused action (the medium-term policy agenda), and then moving on to major reforms (the longer-run policy agenda). The art and craft of policy making are to sequence the various horizons of a policy agenda to achieve a virtuous circle of growth and reforms. A pragmatic agenda is needed to get around the institutional rigidities faced by many developing economies and to create momentum for change by fostering stakeholder awareness, gaining a consensus on tackling some key obstacles at the national level, and strengthening demand for institutional change. It is then possible to move ahead with concrete, manageable, bottom-up approaches that can serve as demonstration projects to advance the larger agenda. The process starts with microreforms. A framework program can then be used as a vehicle to scale up microreforms to a critical mass. Finally, it may become possible to reshape even national institutions.

Using Microreforms as an Entry Point

As the first section of this chapter argues, the heterogeneity of private and public sectors in developing economies is crucially important yet often overlooked. Also discussed is the considerable spatial differentiation of economic activity. Economic and institutional change therefore begins in certain locations or domains and advances through partial and incremental (microlevel) reforms that are aggregated to become larger-scale transformations only over time. A small example from India illustrates how a microlevel reform can facilitate the matching of collaborators and how reform can diffuse.

In the early 1990s, Indian products were generally suspect because they were considered to be of low quality. Quality problems in software were an important obstacle to collaboration between local suppliers and customers in world markets. In software, the problem was not specific to India. Anticipating this problem, an Indian engineer from the Software Engineering Institute (SEI) at Carnegie-Mellon University traveled to Bangalore to speak to software firms about the institute's recently introduced Capability Maturity Model (CMM) for software engineering process improvement. The core of the model was a process of periodic peer review of development "pieces" to ensure, by ongoing clarification of specifications, that the rate of error detection was higher than the rate of "error injection." Many firms immediately picked up the idea and sponsored conferences and consultations on the topic. By the end of the decade, virtually all large Indian software companies had adopted the CMM. Today, India is widely recognized for its high-quality software development processes; it has more SEI-CMM Level V (the top level) certified companies than any other country (Saxenian and Sabel 2008).

The development of a globally competitive software services and technology industry in Bangalore involved a multiplicity of similar microlevel reforms, both within the cluster and outside it. Such changes occur incrementally, without any guarantee that they will continue. But as the Taiwanese example illustrates, when changes endure, they have the potential to alter the institutional fabric of the economy.

Since microreforms may not continue and may not necessarily be scaled up, they often escape the notice of policy makers. Yet such entry points are ubiquitous, particularly in countries with a difficult institutional environment. The "born global" start-ups in Belarus and the Islamic Republic of Iran discussed above are examples of such microreform. Born global firms create a search network that adapts global best practice to a local and often hostile environment and involves government along the way. The Belarussian government, for instance, is creating an ambitious IT park as a platform for such firms in the IT area to grow. The Islamic Republic of Iran has more than 100 science park and technology incubators, all providing a micro-environment for start-up firms. The number of such establishments suggests an intention to promote and scale up nascent microreforms.

Such parks and incubators have very heterogeneous performance and quality, and they may remain enclaves in an otherwise unfriendly institutional environment. But, as the example of the Moscow University Science Park in box 8.2 indicates, the best-performing are more accurately described as *exclaves*—extensions of the world economy. Such parks can become a demonstration case for others to follow, but, of course, this does not happen either automatically or necessarily.

Providing an Environment for Microreforms to Flourish: Framework Programs
How can diverse but fragile microlevel reforms be scaled up to the level of clusters and value chains? Economists call this the "mezzo level," between change at the microlevel and solid reform at the national level. Framework programs provide an environment for microreforms to continue and scale up (World Bank 2001). The Irish National Linkage Promotion Program and the Taiwanese program to create a venture capital industry discussed above are examples of framework programs. Unlike typical government programs or initiatives, framework programs have two distinct features.

First, they start from existing institutions and programs. By linking better-performing segments of private and public sectors, they alleviate institutional constraints and allow the advocates of change to institutionalize their agendas. Both the Taiwanese reform and the Irish linkage efforts were initially viewed with skepticism. Yet they drew on existing organizations and programs and created sustained dynamics (in backward links with SMEs and venture capital funds), which eventually won over the skeptics. What started as a microreform went on to create national change.

Second, by searching for outside-the-box solutions to familiar problems, the institutional framework itself is reshaped. There appeared to be no institutional space for a venture capital industry in Taiwan, China, in the 1980s, so tight was the grip of established large actors (large firms and banks). The institutional framework for venture capital emerged on the organizational periphery as several venture funds. The institutional framework for a venture capital industry and the venture capital industry itself emerged simultaneously, in a dynamic virtuous circle.

Development of China illustrates how framework programs can lead to deep institutional reform. By 1980, China had developed a massive but largely incoherent R&D system. The reform program initiated in 1985 consisted of two framework programs to encourage microlevel reforms and experimentation. On the one hand, "technology markets" were established to align R&D institutes with industry needs. On the other, operational subsidies from the government were gradually reduced. Various forms of autonomy were introduced in R&D institutes (in terms of personnel, research projects, and acceptance and use of contractual fees). The technology markets, which were central to the initial programs, have largely failed. Both buyers and sellers had difficulty engaging in market transactions. Buyers were not able to absorb the transferred technology, and sellers of technology could not earn enough to secure their R&D institutes because the market was too small.

In response, in 1987 policy reform began to promote the merger of R&D institutes with existing enterprises or enterprise groups. This effort was again largely a failure. Huge gaps between the disparate parties, owing to differences in work culture and administrative affiliations, were hard to overcome. Yet budget constraints arising from the drastically reduced subsidies to R&D institutes (the second prong of 1985 reform) opened a policy space for a variety of spin-offs. First, individual scientists and engineers created spin-offs from their parent R&D institutes. These ventures were later followed by organizational spin-offs. In 1988, the Torch Program was launched to encourage spin-off enterprises, called NTEs (new technology enterprises), from existing R&D institutes and universities. NTEs became an institutional vehicle for bringing together the most dynamic segments of the R&D establishment: R&D institutes, universities, S&T staff, and local governments. Local governments invested in new and high-tech industry zones as support institutions for NTEs. Scientists and engineers, often with the support of their parent institutions, developed commercial applications of their inventions and expertise.

The Chinese strategy simultaneously freed up a policy space for dynamic new elements to emerge (from this perspective, the draconian reduction of subsidies was the key, as it created a motivation to search and experiment) and took explicit measures to encourage diverse pilots and organizational spin-offs. This strategy worked well because it was almost ideally suited to leveraging the tremendous heterogeneity of the Chinese economy and innovation system.

Gradually reducing subsidies to existing players freed up policy space and motivated dynamic segments of the system to search for new solutions and approaches. Explicit measures to promote spin-offs created and institutionalized search networks—networks of diverse individuals and organizations looking for new solutions. In the case of NTEs, search networks brought together federal government officials (who monitored the results of the experiment), industry, R&D institutes, and local governments, which contributed critical resources such as high-tech industry zones but also reaped rewards of high growth. The result was "double transformation": high growth due to self-discovery and diffusion to new segments of the economy and reform of the established institutional structures. A double transformation generates a diversity of gradual step-by-step reforms that can lead to extraordinary changes.

Another example of double transformation, taken from the other end of the heterogeneity spectrum, is the transformation of rural industry and the role of the Spark program (see box 8.5). The cascade of institutional changes begins in the 1970s with an agricultural reform that recognized peasants' control over the plots they currently worked and permitted them to sell, at market prices and for their own account, any surplus above target levels. The result was a sustained increase in agricultural productivity and a rise in rural incomes. In the 1980s, another wave of reform allowed for investing the proceeds of agricultural improvement in town and village enterprises (TVEs): manufacturing firms, owned by municipalities or co-owned by them and private parties, that produced for both domestic and export markets. Again, proceeds in excess of tax obligations to higher authorities were retained by the enterprise and available to its stakeholders. The TVEs continued to expand through the mid-1990s, competing with state-owned firms and adding to modest pressure exerted by the central state for their reform. TVEs unleashed creativity at the lower end of the heterogeneity scale in China's rural industry. Measures to promote search networks to bring together dynamic segments from diverse areas were important (box 8.5).

Further up the heterogeneity scale, these changes were accompanied and accelerated by partial reforms of the financial system and the opening of export-processing enclaves to foreign firms and joint ventures (another example of a framework program to promote microreforms). At the high end of the productivity spectrum, reform of the innovation system through recombination, described above, resulted in a dramatic change.

The outcome of these framework programs to promote self-discovery and experimentation is a profusion of new institutions that create incentives for investment and efficiency-enhancing behavior in domain after domain, without ever creating traditional institutional preconditions such as stable property rights, the rule of law, and the like. China is privatizing state firms very haltingly, has only recently recognized private corporate property as a distinct legal category, and makes little pretense of an independent judiciary.

Creating Frameworks for Change: Strategic Incrementalism

Can other countries learn from China's experience in harnessing its heterogeneity and creating new institutions, promoting growth, and undertaking reforms?

Box 8.5 A Framework Program to Promote Experimentation in a Rural Setting: The Spark Program

With the emergence of a rapidly growing and dynamic rural nonstate enterprise sector in the early 1980s and with the Chinese government's determination to make more productive use of science and technology developed in China, the Ministry of Science and Technology initiated the nationwide Spark program in 1986. Its overall objective was to transfer technological and managerial knowledge from more advanced sectors to rural enterprises to support continued growth and development in the nonstate rural enterprise sector, mostly town and village enterprises (TVEs), and to help increase output and employment. The program has since spread to virtually every province in the country and has helped develop 66,700 projects and many more individual enterprises within them. Some 20 million people have found employment in rural areas. Possibly the greatest impact has been the increase in annual per capita income of the rural population in the areas where the program has been active. In a TVE in Jingyang County in Shaanxi, for example, per capita income increased almost threefold in five years.

Under Spark, training courses were conducted, and modern training centers were established with up-to-date computer equipment, video production facilities, and language and scientific laboratories. The TVE sector demanded training for rural enterprises, and the Spark training program responded with appropriate teaching methodologies, such as instructional packages and materials, curriculum, and audio and video productions. A computerized technical information system was also set up, with thousands of technical databases for rural enterprises. These networked systems provide technical, economic, marketing, and sales channel information to TVEs. Broadcast-quality videos of Spark science and technology programs were also developed for TVEs and farmers. To evaluate Spark projects in a systematic way, the project offered technical evaluation training to staff in national, provincial, and local program offices and equipped them with analytical techniques and sources of information to allow them to offer quality help to rural enterprises. Another major objective of the program was the diffusion of technical and managerial knowledge from successful projects to nonproject beneficiaries.

The most dynamic segments of China's rural industries are drawn to the program because it increases their productivity and helps them expand. Spark's most successful projects have become pillar industries in their respective "Spark-intensive areas" and have led to vertical and horizontal integration of related industries either in their own localities or in other provinces. The program provides a way to diffuse and scale up local success stories. It found a way to leverage the tremendous heterogeneity of China's rural economy. Spark not only amplifies its better-performing segments but also connects them by assembling packages of managerial, marketing, and technical services.

The Spark Program has also become a focal point for leveraging the best and the most relevant outcomes of China's massive but not particularly efficient agricultural research system, so that the system provides incentives for staff to carry out research programs that serve the needs of rural clients.

Source: World Bank 1998; Huang and others 2004.

Most countries have many examples of microreforms of their institutional framework that emerge in unexpected settings or proceed in unanticipated ways. Various means of scaling up microreforms into changes in the national institutional framework for innovation can be proposed:

- Institutionalizing search networks of leading actors, including continuous monitoring of the progress of reform and benchmarking to determine what is feasible
- Systematically evaluating programs and projects
- Designing and implementing a portfolio of strategic pilots that probe the economic potential of projects and establishing benchmarks for action
- Initiating an innovation foresight process.

Scaling up microreforms is strategic incrementalism (see figure 8.1): change proceeds gradually, step by step, but its long-term outcome is dramatic. Each of the components mentioned in the list above is considered in turn.

Institutionalization of Networks of "Champions"

Change is driven by individuals (champions) who are willing to risk their reputation on the results of reform. The first priority for this group is to conceptualize, in a series of focused discussions on the "next steps," the nature of the reform they are collectively promoting. The second priority is to include important decision makers from key national decision-making bodies in these deliberations.

Figure 8.1 Elements of Strategic Incrementalism

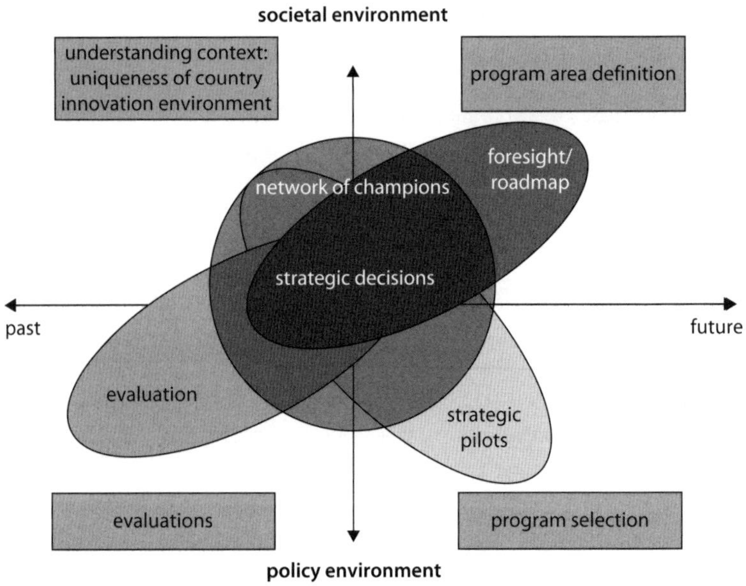

Source: Author.

Systematic Evaluation of Programs

Evaluation is a management tool that links the impact of programs to budget allocation decisions. Key issues in the design of a national evaluation program include cost-effectiveness criteria. It is important for the evaluation criteria to be transparent, for objectives to be clear, and for the application of the criteria to be measurable. The cost effectiveness of the process must also be factored in in the design of an evaluation process. Incorporating widespread use of ICT could be a step toward greater cost effectiveness.

The monitoring and evaluation processes should be separate to avoid potential conflicts of interest in implementing a national procedure for program evaluation. While monitoring and ex post reviews should be carried out by a neutral third party, ex ante evaluation can be carried out within the program itself to facilitate linking the program's key financial decisions to evaluation. International projects should be evaluated and monitored in the same manner as national projects, bearing in mind national benefits, objectives, and demands.

Optimally, 3–5 percent of the program budget should be allocated to evaluation. While evaluation is ideally a management tool, as mentioned above, this amount for evaluation may initially prove difficult. Programs found to be inefficient through evaluations will resist regular review procedures. This resistance raises the problem of managing entrenched interests. Making the feedback public will help resolve the problem. The mere fact of public access to impartial evaluation results will provide a strong disciplinary element and pressure actors to change established procedures and improve performance.

Portfolio of Strategic Pilots

Strategic pilots should examine new organizational models, test their feasibility, and in this way harness the unique features of the innovation system. These pilots are strategic because they introduce the features of a reformed innovation system: accountability for results, built-in incentives for collaboration, and structures of governance. All those are important for the continuous redesign of the pilots.

Innovation Foresight Process

The foresight process attempts to identify potential future opportunities for the economy or society arising from innovative science and technology and considers how future technology can address society's key challenges. Early efforts to use this type of foresight approach were carried out in the United Kingdom in the 1990s but have now been widely adopted throughout the EU and elsewhere. They have proven particularly useful in defining longer-term needs and helping develop the creative links from which innovations emerge. The process includes several elements:

- A steering group comprising leaders from the three main constituent communities—government, academia and business

- A secretariat to identify the main participants (usually through some variant of a co-nomination exercise), to initiate and shape discussions (writing initial position papers, arranging and orchestrating working groups) and to draw together, in conjunction with the working groups, individual contributions into an integrated summary
- An organized program of semi-autonomous working groups, by topic lines (which reflect a mixture of key needs and strategic technologies) to undertake analysis, evaluate evidence, reach conclusions regarding the timeframe, and produce a summary report on their evidence, findings, and prognosis
- An integrative effort, usually conducted by chairpersons of the working groups and the secretariat, to integrate the groups' efforts and develop a conclusion, which usually suggests lines of action and priorities for resource use over the shorter term.

In parallel with the written papers, the collaborative process leads to cohesion and broad, although not necessarily universal, ownership of the strategic lines and priorities for future action. It provides government, academia, and business with a point of reference for their future efforts. In some cases, for example, the exercise has addressed the regional level rather than the national level. These exercises are repeated, and the analyses and conclusions are updated using the same procedures—sometimes reduced in scale and scope, for example, in cycles of three to five years—to ensure that they remain relevant and take into account intervening scientific progress and changes in the needs of society.

Promotion and dissemination efforts are then initiated to ensure widespread awareness of the findings and conclusions of the reports. This process adds to the shared vision of goals and reduces information asymmetries across target audiences. It also enables the findings to be incorporated into public policy and budgetary cycles and into strategic decision making in the enterprise sector. Academic bodies have also used the reports to determine allocations and priorities for selective efforts in research and teaching.

With the foresight reports as a guide, the steering group may use its prestige and influence with the concerned executive agencies to direct resources toward programs that are recognized in the findings of the reports. Monitoring and evaluation would follow, along the cyclical lines described earlier.

One of the consequences of the foresight process is the articulation of poorly structured issues of concern to everyone. For instance, the first foresight process in the United Kingdom unexpectedly identified the widespread ramifications of an aging population in its conclusions. In India, a nationwide foresight process might start by focusing on the country's thematic challenges, such as access to clean water or road congestion.

Let us now come back to the agents of change—search networks of champions. These networks are consolidated through the deliberative evaluation of projects and programs, the next-step discussion that takes into account the lessons emerging from the implementation of projects and relevant international best practices. Nokia, a leading multinational, can serve to illustrate this principle. Many of its labs (called *lablets*, a term borrowed from Intel, which pioneered them) are co-located at major research universities. Their success is judged by the impact they produce in attracting talented young graduate students. But that cannot be the only criterion for evaluation: it is possible that talent is being attracted to the selection of topics, rather than to the opportunity for conducting potentially interesting research for the private sector in general and for Nokia in particular. Indeed, if such applied research is conducted, a discussion of how it could be relevant for Nokia and how to attract relevant graduates to Nokia needs to take place. What is usually expected from a formal evaluation is not a set of figures but dialogue: a mini-innovation foresight on its own, combining both an appraisal of individual talents and interests, and an assessment of the relevance for the firm.

Each strategic pilot (like the innovation foresight process itself) should be regularly evaluated. Uncoordinated and isolated programs can be drawn into the process, and these programs can be coordinated into an overall strategy. Ideally, a body should exist to pool information and draw lessons from specific pilots and projects from different domains of innovation.

Summary of Policy Principles

Analysis of policy making and policy implementation is only now entering the literature on innovation. The following is a summary of the main principles underpinning the present analysis and recommendations:

- *Rely on better-performing segments of existing institutions to leverage reform and change.* Institutions in developing countries may be dysfunctional, but they are not uniformly so. Within a given ministry, some segments or individuals perform better than others. These can be leveraged to transform a difficult institutional environment.

- *Use search networks to link better-performing segments of the economy.* Search networks are networks of individuals and institutions that solve complex problems by finding individuals who already are working on the solution to (part of) the problems. Strategic pilots make it possible to institutionalize emerging search networks to bring together champions from private and public sectors and (possibly) the country's talent abroad. Search networks encourage change and reform by linking together the better-performing segments of national economies.

- *Pursue the goal of "double transformation."* Double transformation involves the creation of an appropriate context for reform. Reforms that start from the better-performing and more entrepreneurial segments of the economy are more likely to succeed. The demonstration effect makes the diffusion of reform to other segments of the economy easier. It also neutralizes the resistance of vested interests. Growth is more likely to provide space for self-reinvention at least among some segments of the entrenched interests and define their position in a new reform scenario.

- *Impose top-down measures to free up policy space.* Programs, policies, and projects cannot be multiplied ad infinitum. The introduction of new pilots means cutting down on existing programs, not only to provide budget space to trigger piloting and experimentation but also, and more important, to provide the correct incentives for players to perform. Underperforming projects are scaled down, and released resources are reallocated to test new approaches.

- *Follow the bootstrapping approach, at once humble and ambitious.* This approach involves a bold vision and strategic change in the long run through a gradual process of implementation of incremental bottom-up changes in which a favorable balance of risks and returns encourages initial steps at many entry points. In this process, each move increases the chances of initiating a virtuous cycle of institutional reforms and private sector development. Policy makers considering bootstrapping need to be prepared for the emergence of unexpected coalitions for reform.

The prevailing view of reform starts with the design of a blueprint for change, a blueprint with a known outcome. In the "strategic incrementalism" approach advocated in this chapter, the institutional outcomes are open-ended, and attempts to create a blueprint are viewed as outdated central planning. To detect problems and errors, policy makers should constantly monitor and benchmark the process of reform and restructuring.

Note

1. In Australia, New Zealand, and Norway, the share of the processing industry in overall exports ranges from 22 to 28 percent. The average for OECD economies is 82 percent.

References and Other Resources

Aubert, Jean-Eric, and Derek H. C. Chen. 2008. "The Island Factor as a Growth Booster: A Mental Advantage Econometrically Revealed." *Journal of Intellectual Capital* 9 (2): 178–205.

Chebbo, Maher. 2008. "Corporate Innovation: The Engine for Economic Growth in Knowledge Economy." Presentation at the First Annual Conference of the Arabian Knowledge Economy Association, Jeddah, Saudi Arabia, January 12–13.

Collier, Paul. 2007. *The Bottom Billion: Why the Poorest Countries Are Failing and What Can Be Done about It.* New York: Oxford University Press.

de Ferranti, David, Guillermo E. Perry, Indermit Gill, J. Luis Guasch, William E. Maloney, Carolina Sánchez-Páramo, and Norbert Schady. 2003. *Closing the Gap in Education and Technology.* Washington, DC: World Bank.

Gu, Shulin. 1996. "The Emergence of New Technology Enterprises in China: A Study of Endogenous Capability Building via Restructuring." *Journal of Development Studies* 32 (4): 475–505.

———. 2006. "Policy Process and Recombination Learning: China in the 1980s and 1990s." Draft prepared for the Asia Innovation Forum. May 18.

Gu, Shulin, and W. Edward Steinmueller. 1996. "National Innovation Systems and the Innovative Recombination of Technological Capability in Economic Transition in China: Getting Access to the Information Revolution." UNU/INTECH (United Nations Institute for New Technologies) Discussion Paper 2002–3, Maastricht, the Netherlands.

Hausmann, Ricardo, and Dani Rodrik. 2002. "Economic Development as Self-Discovery." NBER Working paper 8952, National Bureau of Economic Research, Cambridge, MA.

———. 2006. "Doomed to Choose: Industrial Policy as Predicament." Draft for presentation at the Blue Sky Seminar, Harvard University Center for International Development, Cambridge, MA, September 9.

Hausmann, Ricardo, Dani Rodrik, and Charles Sabel. 2007. "Reconfiguring Industrial Policy: A Framework with an Application to South Africa." Center for International Development Working Paper 168, Harvard University, Cambridge, MA.

Huang, Can, Celeste Amorim, Mark Spinoglio, Borges Gouveia, and Augusto Medina. 2004. "Organization, Program, and Structure: An Analysis of the Chinese Innovation Policy Framework." Economics Working Paper 17, Department of Economics, Universidade de Aveiro, Aveiro, Portugal.

Rodrik, Dani. 2007. *One Economics, Many Recipes.* Princeton, NJ: Princeton University Press.

Saxenian, Anna Lee. 2002. *Regional Advantage.* Cambridge, MA: Harvard University Press.

———. 2006. *The New Argonauts: Regional Advantage in a Global Economy.* Cambridge, MA: Harvard University Press.

Saxenian, Anna Lee, and Charles Sabel. 2008. "Venture Capital in the 'Periphery': The New Argonauts, Global Search and Local Institution Building." *Economic Geography* 84 (4): 379–94.

Stark, David. 1996. "Recombinant Property in East European Capitalism." *American Journal of Sociology* 101(4): 993–1027.

World Bank. 1998. "China: Rural Industrial Technology (Spark) Project, Implementation Completion Report." Report 18126, World Bank, Washington, DC.

———. 2001. *Think Globally, Act Locally: Decentralized Incentive Framework for Mexico's Private Sector Development.* Research Report. Washington, DC: World Bank.

World Bank Institute. 2008. "Establishing a Knowledge Economy in Saudi Arabia." Draft policy note, World Bank Institute, Washington, DC.

9

Promoting Competitive and Innovative Industries

In the Lao People's Democratic Republic, the Jhai Coffee Farmer Cooperative, a Fair Trade certified cooperative promoted by the government in the southern part of the country, produces specialty coffee that is exported mainly for wealthy customers in France and Japan. Grown on the country's Bolaven Plateau, which offers consistent rainfall, cool temperatures, and rich volcanic soil at an elevation of 1,300 meters, the coffee beans are handpicked and washed, with washing machines imported from Germany and adapted for local use. In Croatia, tourism represents nearly 20 percent of gross domestic product (GDP) and is one of the fastest-growing sectors. With its beautiful islands and shoreline, cultural attractions, excellent facilities for travelers, and a highly visible marketing campaign supported by the government, the country attracts more than 10 million foreign tourists a year. In Brazil, Embraer has, by forging a successful partnership with the government, found a niche by building small regional jets of a type not produced by either Boeing or Airbus and by selling many planes to middle- and low-income countries, including India and China. The latest version of its executive jet has been very popular among corporate executives and the very wealthy, making Embraer an emerging symbol of Brazil's competitiveness. In Kenya, Safaricom, the country's most popular mobile phone company, has taken advantage of the government's liberalization of the telecommunications sector to sign up more than 11 million customers. In June 2008, it raised over US$800 million in the largest initial public offering to date in Sub-Saharan Africa. Safaricom

---

This chapter was prepared by Ronald Kim.

and other regional companies like Celtel have helped Africa become a global leader in using mobile phones for payments and remittances.

Innovation, a Global Phenomenon

There are countless other examples of innovation in even the most remote and poor places: innovation is a global phenomenon that is not limited to wealthy countries, and it represents tremendous opportunities and challenges. Specific innovations are the foundation of competitive industries, the source of exports, and a substitute for imports. The promotion of competitive industries requires the mastering of all areas involved in the quality, competitiveness, and delivery of products: design, production process, certification and standards, marketing, transport, and the like. These stages can often be customized first for domestic and regional markets and then scaled up to meet the demands and requirements of international markets. Latecomers can use various approaches to catch up, and some countries have done exceptionally well. Singapore, for example, with a land mass of less than 700 square kilometers, has used its innovation and competitiveness to export some US$300 billion in goods and services, more than Russia with a land mass of 16 million square kilometers.

It is important to emphasize several recent and growing trends related to innovation. First, globalization has accelerated and changed dramatically in recent years. Globalization used to mean that "business expanded from developed to emerging economies. Now it flows in both directions and increasingly also from one developing country to another. Business these days is all about competing with everyone from everywhere for everything" (*Economist* 2008). A growing number of companies from developing countries are now among the world's largest. In fact, some of the most recognized firms and brands in developed countries have recently been bought by emerging market companies, such as Inbev's purchase of Budweiser, Mittal's takeover of Arcelor, Lenovo's acquisition of the IBM personal computer line, Tata's and Sichuan Tengzhong Heavy Industrial Machinery's purchase of Land Rover/Jaguar and Hummer, respectively.

Many firms in developing countries have also gone beyond the low-labor-cost model. Examples include several Indian software companies that have become world leaders in business information technology (IT) services; Haier, a Chinese manufacturer that is moving into high-end appliances and electronics; and Desarrolladora Homex, a Mexican home builder that is replicating its business model in countries such as the Arab Republic of Egypt, where it recently signed an agreement to build 50,000 low-cost homes in Cairo. While China has specialized in the low-cost labor model, many companies are now working closely with the government to enter new sectors. Box 9.1 provides details on China's recent entry into wind power.

> **Box 9.1 China, an Emerging Leader in Wind Power**
>
> China is the world's fastest-growing wind energy market, with an average annual growth of 56 percent in the past seven years. The country has now reached fifth place for installed wind energy capacity, with 5.9 gigawatts at the end of 2007. China's wind power sector has developed significantly since the adoption of the National Renewable Energy Law in 2005. The government also enacted a series of policies to facilitate the development of wind power. An important step has been to improve regulation of wind power pricing and to disperse industry worries about excessively low bidding hindering further development.
>
> Given China's substantial coal resources and the relatively low cost of coal-fired generation, reducing the cost of wind power production has been a crucial issue, addressed through the development of large-scale projects and increased local manufacturing of wind turbines. Wind power projects larger than 50 megawatts are approved through the National Development and Reform Commission, based primarily on price and the share of domestic components used (70 percent of the components should be made in China). The provincial power grid company guarantees purchase of all electricity produced by the project. The government also supports wind power through tax incentives and subsidies.
>
> Policy incentives and the government's prioritization have sent a clear signal to the market. Before 2005, only a few small turbine producers existed, and most turbines and components were imported. At the end of 2007, there were 40 Chinese manufacturers, accounting for about 56 percent of equipment installed during the year. The Chinese turbine-manufacturing capacity is expected to more than double over the next five years, and by 2012 the country will be able not only to meet domestic demand but also to become a major exporter of wind turbines. As the stability of the sector attracts greater investment, wind power may be able to compete with coal generation by as early as 2015.
>
> *Sources:* Global Wind Energy Council 2008; Li 2008.

The aim of this chapter is not a comprehensive examination of innovation and the impact of government policies but an illustration of the powerful role of innovation in specific industries. The chapter looks at how innovation is nurtured and sustained through formal and informal interactions and through partnerships between the private and the public sector, with examples that highlight relevant issues. Rwandan coffee, Chinese automobiles, tourism in Costa Rica, and IT services in Vietnam are used to analyze the dynamics of innovation in specific instances and how these dynamics result in competitive industries.

The agricultural sector can offer developing countries important opportunities. By starting from the production and export of commodities, countries can gradually climb up the value chain and develop value-added activities, a rise that requires strong organizational capabilities in all concerned actors. The development of manufacturing export industries can benefit from the involvement of foreign investors and imports of capital goods, as these facilitate the gradual transfer of competencies and technologies. In the services sector, information technology has created considerable opportunities, in software services, business process outsourcing, and call centers, for example. Again, the point is to position the company in the global value chain according to its competencies and infrastructure. Tourism also presents excellent opportunities for countries that can exploit their climate and natural landscape, historical assets, and cultural heritage if they have key prerequisites such as transport and hotel infrastructure, safety, and qualified personnel.

Success in agriculture, manufacturing, or services is a long-term process that requires tremendous learning and discovery. Failures are numerous. On the bright side, the case of Mauritius may offer some insights into the learning process (see box 9.2). By analyzing the role of seven key dimensions of the innovation system in each of these three sectors, this chapter shows how success is achieved:

- *Vision and leadership.* Political system and stability; strategic focus
- *Framework conditions.* Overall economic and institutional regime, taxation and incentives, competition

---

Box 9.2 Mauritius, Reinventing for Survival in the Global Economy

Until recently, sugar cane and textiles dominated the economy of Mauritius. Then Europe began dismantling the sugar preferences that benefited countries like Mauritius with above-market prices. The end of the Multi-Fiber Agreement resulted in the loss of 30,000 jobs in the textile industry. While sugar and textiles are still important industries, the country has had to diversify and deepen its commitment to other sectors:

- *Tourism:* With its world-renowned beaches and upscale resorts, Mauritius attracts many high-end tourists from Europe and Asia.
- *IT services:* A growing number of workers are employed at call centers and in data entry, software and web development, telemarketing, and information processing.
- *Financial services:* The country now hosts 20 banks, and the introduction of Islamic banking has brought investment from oil producers in the Persian Gulf.

To achieve this transformation, the government simplified and cut taxes, reduced regulations, lowered or removed tariffs, enacted new laws to make it easier to hire and fire workers, and invested heavily in upgrading the information and communication technology infrastructure. The result has been a partnership between the government and the private sector to promote the country as an excellent place to do business, especially for investing in Asia and Africa. In fact, Mauritius ranked 28 in the World Bank's 2008 ranking of "ease of doing business." The country is also attracting major Indian and Chinese investors to set up export processing zones and special investment zones.

Sources: Newfarmer, Shaw, and Walkenhorst 2009, 184–85.

- *Education and research.* Human resource capacities, training, institutions of higher education
- *Infrastructure.* Business support and services, finance and venture capital, information and communication technology
- *Industrial system.* Type and mix of companies
- *Intermediaries.* Information brokers and disseminators, research institutions
- *Demand.* New markets, finding a niche, opportunity.

Agriculture

Agriculture has a powerful and pervasive place in nearly every developing country. Although it represents only 4 percent of global GDP, it plays a fundamental role in sustainable development and poverty reduction. Three out of every four poor people in developing countries live in rural areas—2.1 billion live on less than US$2 a day and 880 million on less than US$1 a day—and most depend on agriculture for their livelihood. In agriculture-based countries, agriculture generates on average 29 percent of GDP and employs 65 percent of the labor force. The production, trade, and consumption environment for agriculture and agricultural products is increasingly dynamic and evolving in unpredictable ways. This evolution must be seen in the context of current food security problems, which have led to a surge in prices, plunging more than 100 million people into poverty, and the need to double agricultural output by 2050 to feed a growing world population. If farmers and companies are to cope, compete, and survive, they need to innovate continuously. Consequently, for the foreseeable future the growth strategy for most agriculture-based economies has to focus on improvements.

For many countries, where the development of world-class manufacturing and service industries is unlikely, at least in the short to medium term, agriculture needs to tackle fundamental problems: poor infrastructure, inaccessible markets, poor storage methods, lack of processing facilities, and the relative lack of fertilizer and seeds. Some countries have initiated programs to strengthen basic technology and capacity building to improve agriculture and rural livelihoods. Rwanda, for example, views agriculture as the driver of poverty reduction and economic development, and the focus has been on raising agricultural productivity, alternative energy, water conservation, food processing and storage, public health, and technical and vocational education (World Bank 2007b, 2008a; Watkins and Verma 2008).

Malawi offers insight into how remarkable change can be achieved by addressing basic priorities in agriculture. After a disastrous corn harvest in 2005, almost 5 million of Malawi's 13 million people needed emergency food aid. In what has been called an amazing turnaround, its farmers produced record-breaking corn harvests in 2006 and 2007 (Dugger 2007). Soon the country was selling more corn to the World Food Programme of the United Nations than any other country in southern Africa and exporting hundreds of thousands of tons of corn to Zimbabwe. The key intervention was the government's provision of coupons to 1.3 million farm families, which allowed them to purchase three kilograms of hybrid corn seed and two 50-kilogram bags of fertilizer at a third the market price (Bourne 2009). Malawi's successful use of subsidies is a reminder of the unparalleled importance of agriculture in alleviating poverty and demonstrates how critical public investments are—including those in fertilizer, improved seed, farmer education, credit, and agricultural research—in the basics of a farm economy.

Indeed, Africa provides numerous examples of what happens when the fundamentals of agricultural development "come together." In Uganda, for example, more than 100,000 farmers are growing sunflower seeds, which are being crushed into cooking oil locally, displacing palm oil imported from Asia. Cotton production in Zambia has increased 10-fold in a decade, bringing new income to 120,000 farmers and their families, nearly 1 million people in all. Exports of flowers from Ethiopia are growing so rapidly that they threaten to surpass coffee as the country's leading cash earner. In Kenya, tens of thousands of small farmers who live within an hour of the Nairobi airport grow French beans and other vegetables, which are packaged, bar-coded, and air-shipped to grocers throughout Europe. Exports of vegetables, fruits, and flowers, largely from eastern and southern Africa, now exceed US$2 billion a year, up from virtually nothing a quarter of a century ago (Zachary 2008). In Africa, while agricultural firms can and do export, their numbers are relatively small. Because exporting itself increases learning and raises the quality of the product, exporting firms can achieve rapid productivity growth. When domestic markets are too small to support competition, learning from exporting is a powerful tool (Collier 2007).

It is increasingly acknowledged that agriculture has changed significantly over the past 25 years, with new markets, innovations, and roles for the state, the private sector, and civil society. In the so-called new agriculture, private entrepreneurs, including many smallholders, are linking producers to consumers and are finding new markets for staple food crops and export commodities. This vision of agriculture requires rethinking the roles of producers, the private sector, and the state. Production is carried out both by smallholders, who are often supported by cooperative organizations, and by labor-intensive commercial farming, which sometimes offers a more productive and efficient model. The state's role, through enhanced capacity and new forms of governance, is to correct market failures, regulate competition, and engage strategically in public-private partnerships (World Bank 2007a).

Agricultural Innovation Systems

The new agriculture—with its increasingly complex agricultural markets, networked knowledge, and competitive advantage linked to capacities for knowledge application—emphasizes institutions, coordination, and improved links among the main actors of the innovation system. As a complementary frame of reference, the agricultural innovation system (AIS) approach recognizes that many types of innovation—related to technology, organizations and partnerships, processes, products, and marketing—can take place at any time in different parts of the overall system. Promoting innovation in agriculture requires coordinating support for agricultural research, extension, and education; fostering innovation partnerships and links along and beyond agricultural value chains; and enabling agricultural development. The AIS approach emphasizes technology and knowledge generation and adoption rather than simply strengthening research systems and their outputs. At the same time, it looks at the whole range of actors and factors needed for innovation and growth and assumes that innovation derives from an interactive, dynamic process that increasingly relies on collective action and multiple knowledge sources at diverse scales to leverage the resources of the private sector, civil society, and farmers' associations.

Thus, the structure, quality, and dynamics of the AIS drive the agricultural sector as a whole. For an understanding why some agrarian economies lag while others leapfrog, it is crucial to note that the AIS, coupled with land and climatic characteristics, determines specific innovation and production outcomes. Neither potential profit margins in new markets nor regulations to facilitate entering export markets are necessarily sufficient to transform a subsector into a productive, innovative, and competitive engine of growth. Instead, it is the interplay of all the actors and institutions involved—the characteristics of the industry, transport conditions, policies and the enabling environment—that determines the level of innovation and competitiveness that emerge (see figure 9.1).

The AIS approach implies that innovation can appear at any point along the value chain as the result of the mediated or coordinated interaction of various actors. It does not necessarily depend on any specific government role or action. Nevertheless, because public policies directly influence the national competitiveness of firms and the health of value chains, the innovation system requires a comprehensive set of pro-innovation agriculture, trade, science and technology, finance, and education policies. Well-crafted and coordinated public policies can facilitate, steer, and reinforce innovation by providing incentives and structures for individuals, companies, and institutions.

Agribusiness provides the inputs, expertise, and services needed for farm production and for the markets for farm products. In many developing countries, it fills the vacuum caused by the retreat of the state-run operations, which once provided essential input and marketing services. It also provides employment and entrepreneurial opportunities in rural and urban areas and contributes to the growth of micro- and small enterprises through the establishment of market links. As the key interface between markets and rural

Figure 9.1 Conceptual Diagram of an Agricultural Innovation System

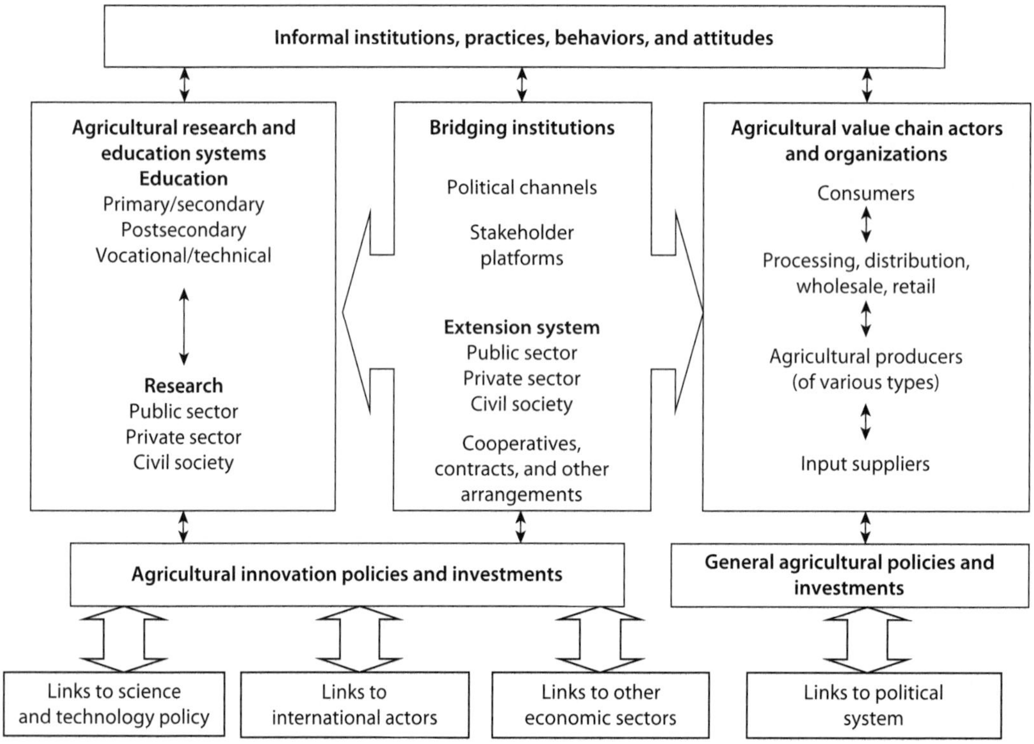

```
┌─────────────────────────────────────────────────────────────────────┐
│          Informal institutions, practices, behaviors, and attitudes │
└─────────────────────────────────────────────────────────────────────┘
       ↕                       ↕                            ↕
┌──────────────────┐  ┌──────────────────┐  ┌──────────────────────┐
│ Agricultural     │  │ Bridging         │  │ Agricultural value   │
│ research and     │  │ institutions     │  │ chain actors         │
│ education systems│  │                  │  │ and organizations    │
│ Education        │  │ Political channels│ │ Consumers            │
│  Primary/secondary│ │                  │  │      ↕               │
│  Postsecondary   │  │ Stakeholder      │  │ Processing,          │
│  Vocational/     │  │ platforms        │  │ distribution,        │
│  technical       │  │                  │  │ wholesale, retail    │
│     ↕            │  │ Extension system │  │      ↕               │
│                  │  │  Public sector   │  │ Agricultural producers│
│ Research         │  │  Private sector  │  │ (of various types)   │
│  Public sector   │  │  Civil society   │  │      ↕               │
│  Private sector  │  │                  │  │                      │
│  Civil society   │  │ Cooperatives,    │  │ Input suppliers      │
│                  │  │ contracts, and   │  │                      │
│                  │  │ other arrangements│ │                      │
└──────────────────┘  └──────────────────┘  └──────────────────────┘
       ↕                                            ↕
┌───────────────────────────────────────┐  ┌──────────────────────┐
│ Agricultural innovation policies      │  │ General agricultural │
│ and investments                       │  │ policies and         │
│                                       │  │ investments          │
└───────────────────────────────────────┘  └──────────────────────┘
   ↕                ↕                ↕                    ↕
┌─────────┐   ┌─────────┐    ┌─────────┐          ┌─────────┐
│ Links to│   │ Links to│    │ Links to│          │ Links to│
│ science │   │international│ │ other  │          │political│
│and tech-│   │ actors  │    │economic │          │ system  │
│nology   │   │         │    │ sectors │          │         │
│ policy  │   │         │    │         │          │         │
└─────────┘   └─────────┘    └─────────┘          └─────────┘
```

Source: Adapted from Spielman and Birner 2008.

households, agribusiness firms link agriculture to industry (OECD 2007) and respond to opportunities growing out of the liberalization of economies and the globalization of trade, thereby assisting a country's agricultural producers in moving up the value chain in various markets (see box 9.3).

High-Value Commodities

Competition and new markets play an instrumental role, especially in global markets where the return on investment and the possibility of moving up the value chain are greater. Global competition also imposes quality standards, leading to major improvements in all parts of the value chain, including cultivation, harvesting, processing, testing and quality assurance, storage and transportation, and marketing. Developing countries can exploit their latecomer status to close the gap with developed countries in particular commodities or subsectors through the application of more knowledge-intensive and market-driven production technologies (table 9.1).

In addition, public sector support for interactions, collective action, and broader public-private partnership initiatives is often critical. Combining links

Box 9.3 Main Messages from *Agribusiness and Innovation Systems in Africa*

The following points summarize the conclusions of an analysis conducted on agricultural innovation systems in Africa on the basis of in-depth case studies implemented in several countries (Larsen, Kim, and Theus 2009).

1. Evolving domestic and regional markets offer new opportunities for agribusiness and farmers, including potential staple food sectors. The rise of new domestic and regional markets offers new opportunities for agribusiness and farmers to sell their products and raise incomes, supplementing or replacing production for export markets when the transaction, investment, and compliance costs are too high for participation. These markets display a remarkable degree of innovation.

2. Innovation in formal markets requires significant adaptation, coordination, and collaboration. The entire value chain is critical. The need to maintain grades and standards within the value chain, not only in export markets but also in evolving domestic and urban markets, drives innovation in agribusiness.

3. Context-specific public sector programs and the prospect of higher profit margins are crucial to integrating smallholder farmers into more innovative formal markets. In a variety of countries, failed initiatives to open new markets by introducing more advanced technology and adapting supply and organizational systems suggest that technology must be appropriate to the specific context and that "push" strategies and initiatives are successful only if markets offer sufficient profit margins for agribusiness.

4. The structure, quality, and dynamics of the innovation system drive agribusiness and the agricultural sector. Neither potential profit margins in new markets nor the regulations for entering export markets are sufficient to transform a subsector into a productive, innovative, and competitive growth engine. Instead, the innovation potential of an industry is determined by the specific interplay of the different actors and the overall environment in which it operates.

5. The state needs to build institutional capacity, align investment priorities with wider economic strategies, and provide more access to finance, particularly in rural areas, to create a functioning enabling environment for agribusiness innovation.

6. To promote innovation, the public sector could further support interactions, collective action, and broader public-private partnership programs. Both formal markets and infrastructure in Africa put a premium on organizational innovation for agribusiness, especially in high-value and cash crop subsectors, after the postliberalization public sector retreats to play a more regulatory and facilitating role. Meanwhile, the private sector takes over the value chain, leaving coordination to the processing industry, in some instances aided by nongovernmental organizations; and producers and processors, cooperatives, and other organizations achieve critical mass and economies of scale.

Source: Larsen, Kim, and Theus 2009.

Table 9.1 Leading Exporters of High-Value Commodities in Developing Countries

Commodity	Countries
Coffee	Brazil, Colombia, Indonesia, Mexico, Vietnam
Farmed fish	China, India, Indonesia, Philippines, Vietnam
Flowers	Colombia, Ecuador, Kenya, Mexico
Tea	China, India, Indonesia, Kenya, Sri Lanka,
Wine	Argentina, Chile, South Africa

Source: Author.

within and beyond the value chain often contributes to an innovation strategy that focuses on strengthening interactions between key public, private, and civil society actors. Box 9.4 on Malaysia's palm oil industry demonstrates this strong collaboration.

Many places around the world offer the appropriate climate and physical landscape for coffee production, and, in fact, coffee is grown in 60 countries on nearly every continent. In 2007, total worldwide coffee production was 7 billion kilograms. Coffee is the most popular beverage worldwide with over 500 billion cups consumed each year ("Top 100 Espresso" 2008). It is also the world's second-most valuable commodity exported by developing countries, after oil. Global earnings are estimated at US$60 billion annually, and over 25 million people worldwide are employed in the subsector. In Brazil alone, 5 million are involved in the cultivation and harvesting of more than 3 billion coffee plants and in milling, processing, and exporting.

While the coffee industry focused for many years on increasing production, there is now greater emphasis in many countries on improving quality—through Fair Trade practices, for example—as a way of entering new markets and benefiting from higher profit margins. Accordingly, a greater percentage of harvested coffee beans is being designated "premium" and receives higher prices from companies like Starbucks.[1] Specialty coffee sales are increasing by 20 percent a year and account for nearly 10 percent of the US$20 billion U.S. coffee market.

The Example of Rwanda

Many developing country governments support their coffee industry in both targeted and general ways, and coffee often represents a large percentage of their foreign exchange. For example, it is currently Rwanda's most important export crop, accounting for more than a third of its GDP and 75 percent of its export income. Its lack of minerals and other natural resources, its landlocked state, the current low level of industrialization, and the weak purchasing power of the population largely explain why agriculture and livestock will remain the key to faster economic growth and sustained development. With this in mind, the government has given priority to rural development and the

> **Box 9.4 Malaysia's Palm Oil Industry**
>
> Malaysia is the largest producer and exporter of palm oil in the world, accounting for about half of global production. Oil palm cultivation originated in western Africa, where climatic conditions are ideally suited for this crop and where cooking oil was first extracted from oil palms. Even though palm trees are not native to the country, Malaysia was able nonetheless to become a global leader in the industry, largely due to export promotion policies adopted by the government, the active role of the private sector and research and development (R&D) institutions, and the exemplary coordination of different actors.
>
> Export-oriented industrialization began in 1968 with the enactment of the Investment Incentives Act. The exemption of processed palm oil from export duties after 1976 encouraged firms to switch from crude to processed palm oil. Under the Industrial Master Plan of 1986, exporters received substantial corporate tax exemptions, with the most successful able to avoid paying taxes altogether. Furthermore, export-oriented firms benefited from preferential credit schemes. Generous financial incentives were provided to facilitate R&D in manufacturing. A tax allowance of 50 percent on qualifying R&D expenditures was offered over a period of 10 years, and tax exemptions were provided for employee training. In addition, firms enjoyed access to R&D carried out in the Palm Oil Research Industry of Malaysia, the Malaysian Agricultural Research and Development Institute, and universities. Government grants have been extensively used by university academics to undertake R&D on palm oil products with joint support from firms. Sustained financing made it possible to introduce new products, such as biodiesel or specialty fats, while the scope of R&D efforts expanded to include productive recycling of waste, environmentally friendly manufacturing, and higher value in existing products.
>
> The relationship between the government and the private sector has been characterized by strong collaboration, with the latter able to shape a number of industry directives, including the contingency strategies to regulate supply in response to prices. The private sector and industry associations have also been instrumental in lobbying the government to coordinate overseas promotion efforts as well as in institutionalizing sustainable practices.
>
> *Sources:* Adapted from Chandra 2006; Malaysian Palm Oil Association, http://www.mpoa.org.my/.

agricultural sector. It has put in place policies and institutions to drive agribusiness in profitable subsectors.

In 1998, the Rwandan government recognized that the viability of smallholder coffee production depended on making the industry a producer of premium quality, fully washed Arabica beans. On the initiative of the Rwandan diaspora and with support of the private sector and major donors such as the U.S. Agency for International Development (USAID), the European Union, the World Bank, and the International Fund for Agricultural Development, it created a detailed medium-term plan to transform the industry. The government liberalized the coffee sector, facilitated the formation of cooperatives, and emphasized quality standards. The medium-term plan included the construction of private and cooperatively owned coffee-washing stations, replacement of older coffee trees, improvement in production techniques, and easier access to finance and to a supply of water and energy. Coffee-washing stations and specialty coffee had been known in the country since the 1950s. However, in the absence of supportive policy, the marketing and promotion of Rwandan coffee languished until quality coffee was judged to be in the national interest and became a focus of policy and strategy. In early 2002, there were two washing stations; now there are more than 100.

After the sale of coffee was liberalized in 1999, exporters were free to transact business without export taxes or undue government involvement. The government's Office of Rwandan Industrial Crops-Coffee has elaborated a national coffee policy, has established quality standards and classification systems, and issues certificates of origin and quality. It also helps importers and roasters establish contacts with Rwandan exporters and facilitates eventual transactions.

Since 2000, university and research institute partnerships, with funding provided by USAID, have introduced many innovations in the production of high-end, specialty coffee with a focus on quality. Rwandan coffee growers' associations have gained access to new markets and introduced quality standards for sales to these markets. Innovations have included (a) quality-enhancing technologies and innovation in production, field management, and transportation; (b) improvements in processing, including sorting of coffee cherries, floating and grading before depulping; fermenting and cleaning to remove mucilage and dried areas; (c) regular sampling and testing according to area, date, and cooperative; and (d) organization and management to strengthen the cooperatives' finance, marketing, and extension programs (Rukazambuga 2008). The government has also initiated or implemented some major interventions:

- *Rural financing and impact.* Microcredit banking services and small businesses are springing up near cooperatives, and primary school enrollments have increased as heads of households are better able to meet school fees. These resources are particularly important as many Rwandan coffee farmers have been widowed or orphaned due to the earlier civil war and the AIDS pandemic.

- *Infrastructure support.* Improving coffee quality through innovation requires infrastructure beyond the capacity of cooperatives. Grants were used to build the first facilities (coffee-washing stations, dry-processing units, cupping laboratories); however, later facilities were built using cooperatives' own funds from income or with bank loans. These facilities ensure quality control from production to export.

- *Capacity building of farmers.* Awareness building and training at the community and cooperative level on innovations for specialty coffee occur at all stages of the value chain and emphasize good practices used in other coffee-producing countries in areas such as harvesting, processing, cupping, and marketing.

- *Partnerships, links, and intermediary organizations.* The success of innovations applied to specialty coffee can be attributed to the combined efforts of many partners: universities such as the National University of Rwanda, Michigan State University, and Texas A&M; the district authorities; the National Agricultural Research Institution; nongovernmental organizations (NGOs); donors such as USAID; buyers (such as Community Coffee of USA and Union Roasters); cooperative members; and government agencies. A partnership between the former Maraba District authorities and the National University of Rwanda convened all stakeholders to focus on the coffee project. The willingness and flexibility of donors to support quality-enhancing activities and to offer advice on market access and partnership with buyers led to links between cooperatives and the international specialty coffee market.

Manufacturing

Manufacturing is a very wide-ranging sector that encompasses the production of such diverse products as automobiles, jet engines, household appliances, clothes, jewelry, and paper products. A crucial source of economic growth and jobs for many countries, manufacturing is a key driver of exports for low-, middle-, and high-income countries alike, including the United States, where manufacturing makes up more than two-thirds of exports and has contributed more to growth than any other sector of the country's economy during the past 20 years. Manufacturing makes up about 18 percent of GDP worldwide. In China, where industry plays an instrumental role, the number is 33 percent. In Germany, exports of a wide range of manufactured goods are still the engine of the German economy. Box 9.5 illustrates the example of how gold jewelry production has developed into a vibrant industry in Turkey.

The garment industry is also an interesting example. In some developing countries, garments account for a large proportion of total exports. Global garment exports are valued at more than US$310 billion a year, or about 3 percent of total world merchandise trade. A distinctive feature of the industry is the number of countries that are highly dependent on garment exports. In 2004, clothing provided more than 40 percent of total merchandise exports for Bangladesh, Cambodia, El Salvador, Lesotho, Mauritius, and Sri Lanka. Reliance on the garment industry for both jobs and export revenues makes these

> **Box 9.5 Gold Jewelry in Turkey**
>
> Turkey has recently emerged as the third-largest gold jewelry manufacturing center and second-biggest exporter in the world. Turkey's production has been growing steadily over the past 16 years, after the government liberalized gold imports, established a modern gold exchange and a gold refinery, permitted gold banking, and set up a derivatives market. The liberalization of the gold market ensured a continuous supply of gold to the jewelry sector at world market prices, which helped expand production. To be globally competitive, however, the industry had to introduce technological and design improvements as well as develop a marketing culture.
>
> The World Gold Council (WGC), an industry group, has brought in technology consultants and produced manuals on manufacturing technology, which are used in workshops and schools. Technical high schools and universities introduced classes and branches on jewelry technology and over 1,000 students are joining the workforce every year. The industry had to invest in reaching global jewelry standards, such as those for finishing and weight standardization, as well as switch from imitating foreign jewelry products to developing original designs—a clear trend of recent years, according to WGC's director in Turkey, Murat Akman. Today, the sector boasts its own brands and trademarks and employs about 750,000 people.
>
> Industry growth has been facilitated by a strong domestic demand (owing to cultural traditions) and by a thriving tourist sector. Turkish exports of precious metals and jewelry rose from a mere US$2.8 million in 1992 to US$3.7 billion in 2007 (direct exports totaled US$2.623 billion in 2007). The country exported some 135 tons of gold jewelry in 2007, including direct sales to foreign tourists that consume about 22.8 percent of Turkey's gold jewelry production.
>
> *Sources:* World Gold Council; Turkish-U.S. Business Council 2007.

countries extremely vulnerable to adverse shifts in trading patterns. Improving innovation—maximizing use of skills, upgrading and adapting technologies, and improving the technological capacities of firms—is a difficult challenge faced by all these countries.

Some countries, however, are faring well in the post–Multi-Fiber Agreement era, and some are even exhibiting signs of competitiveness and innovation in a labor-intensive industry that generally relies on a low-cost workforce. Bangladesh, for example, has maintained its garment exports even in the face of stiff competition, partly owing to investment and to government support for domestic yarn and fabric production (World Bank 2007c, 158; 2008b). From April 2009 on, Cambodia began subsidizing more than 35 percent of the amount the garment and shoe industries contribute to the National Social Security Fund "in order to secure social protection as well as the sustainability of the garment and shoe factories" during the global economic crisis.[2] In Brazil, local governments have played a key role in fostering technological innovation to increase competitiveness (see box 9.6).

> **Box 9.6 Producing Jeans in Toritama, Brazil**
>
> Toritama is the smallest *municipio* (in area) in the state of Pernamabuco in Brazil but has the state's second-highest per capita income. The local jeans cluster currently produces 15 percent of Brazil's total production. The *municipio*'s population is about 22,000, and it has 22,000 industrial sewing machines. Around 90 percent of the local economic population is engaged in jeans production. The unemployment rate is nearly zero, and migration to Toritama has soared, attracted by the jobs offered by the approximately 2,300 firms in the jeans business (1,400 of them are informal).
>
> Laundering is a crucial part of jeans production, and the 50 laundry businesses located in Toritama have always struggled to meet the demand for water, not to mention the need to properly treat effluents. Water represents up to 30 percent of the cost of laundering, not including post-use treatment. The water supply has always represented a major bottleneck, as semi-arid Toritama sits on top of a stony terrain. In 2004, the *municipio* faced a serious water crisis, and households were supplied only once a week. Businesses were forced to buy water from trucks. Yet, local jeans production did not decline that year.
>
> An innovative and cost-effective method for recycling water allowed firms to remain in business and become competitive nationwide. Based on simple technology, the used water goes through a series of containers that contain filters and special stones to purify the water. With initial assistance from a German donor agency (BFZ) and support from the state government, a local entrepreneur combined his experience with foreign technology to develop the method, which was later adapted to 40 other local laundry firms. Although the technology was not very advanced, it was cost effective, and small firms could both afford and use it. While the new method ensured business feasibility, it also helped protect the environment, as firms had previously discharged used water (containing chemicals) without treatment into the Capibaribe River, one of the state's two major rivers.
>
> *Source:* "The Impact of Innovation on Cluster Sustainability and Competitiveness" 2005.

Automobiles

The automobile industry is perhaps the most visible sector in manufacturing. Global car sales in 2008 approached nearly 60 million, with sales of more than US$1 trillion. Cars are now being manufactured in a vast number of developing countries, including Brazil, China, India, South Africa, and Thailand. Several recent trends characterize the industry as a whole, with special attention on market share, consolidation, and "greener" cars.

First, developing countries are gaining increasing market share in new car sales—passenger-vehicle sales in the BRICs (that is, Brazil, Russia, India, and China) are now similar to those in the United States, and in early 2009 China overtook the United States in annual sales. In fact, due to the saturated markets in the United States and Europe, differences in economic growth, and demographic realities, nearly all future growth in sales will come from BRICs and other developing countries. These changes have also contributed to a growing focus on vehicles designed for the conditions of local markets: two examples are the rugged small cars built for the Brazilian market and Tata Motors' recent launch of the Nano, which offers many Indians an affordable car for the first time. Second, the degree of consolidation and cooperation within the auto industry is only increasing. Tremendous changes have taken place, including new mergers and the launch of numerous joint ventures. And third, the development of "greener" cars that use the latest innovative technologies has become a global phenomenon that represents the future of the industry.

The Example of China
China has become the manufacturing hegemon of the world, and its exports have undergone unprecedented growth. Starting with less than a 0.5 percent share in world merchandise exports in 1980, Chinese exports accounted for 8 percent of the global total in 2006, which was roughly eight times the exports of Brazil or India. Early on, the Chinese government negotiated with the large multinationals that wanted access to the Chinese market, initially forcing the companies to enter into joint ventures with domestic firms. They also negotiated local content and training requirements. These requirements greatly helped the Chinese develop technological and management capability. Once the cost advantage of producing in China became apparent to multinational companies, the government relaxed the joint-venture requirement to encourage the foreign firms to bring their best technology (Dahlman 2008).

For the past two decades, the Chinese government has been able to use this approach to insist that foreign auto manufacturers enter into joint ventures with Chinese partners with the prospect of an enormous potential market as the main attraction. For example, it managed to force both Honda and Toyota to be part of joint ventures with the same Chinese manufacturer. As a result, the Chinese company was able to use the best of both systems to develop its own brand and production. Largely through the financial resources and technical expertise of such global auto behemoths as General Motors, Nissan, Toyota, and Volkswagen, their local partners (BYD Auto, ChangAn, Chery, Dongfeng, FAW, Geely, and SAIC, for example) have developed the capacity to meet domestic demand and are now poised to enter the global market ("A Global Love Affair" 2008). This capability is also evidenced by the establishment of modern research and development centers by the Chinese automakers, who are no longer content to copy the engineering and designs of their foreign partners. Another obvious competitive advantage that some of the Chinese automakers have is ownership by a government that favors their development and expansion. Still another advantage is the country's rising high-tech skill base: China is expected to produce more graduates and PhDs in science and engineering than the United States by 2010 (KPMG 2006, 4).

As further evidence of the ever-changing nature of the auto industry, Geely Automobile, one of China's largest private carmakers, recently purchased an Australian drivetrain transmission supplier. Weichai Power, one of China's largest diesel engine manufacturers, acquired a French diesel engine producer. Another Chinese company, BYD Auto, in which Warren E. Buffett has invested, launched a mass-market plug-in electric car before GM had begun marketing the much-anticipated Chevrolet Volt. Detroit's annual auto show in January 2009 was extremely low key in contrast to the very high profile show in Shanghai. Mercedes-Benz, BMW, and Porsche all unveiled new vehicle models there. "The center of gravity is moving eastward," Dieter Zetsche, chairman of Daimler, told reporters at the Shanghai show (Marr 2009). In fact, the change is even more extraordinary: in 1992, fewer than 1 million vehicles were made in China; by 2008, the figure was 9.35 million.

Even the recent global recession has not significantly slowed the growing market for cars in China, and the government has introduced several key measures that have both mitigated the effects of the economic downturn and stimulated innovation in the industry:

- Reducing retail taxes from 10 percent to 5 percent on cars with engines smaller than 1.6 liter
- Offering US$700 million in subsidies to those who trade in tractors and older vehicles for new cars and trucks
- Providing US$220 million in subsidies for upgrading automotive technologies, especially alternative energy vehicles
- Giving subsidies of up to US$8,800 to local governments and taxi companies for each hybrid vehicle purchased.

The result has been a very explicit government policy favoring small and alternative fuel–powered cars (Liu 2009, 24). The recent measures described above clearly offer concrete incentives for building and buying cars with small or hybrid engines. One result is that China is now home to the largest selection of electric and hybrid cars. In fact, the Chinese Automotive Industry Plan calls for creating capacity to produce 500,000 "new-energy" vehicles, such as battery electric cars and plug-in hybrid vehicles. The plan aims to increase sales of such new-energy cars to account for about 5 percent of China's passenger vehicle sales. In addition, the long-term objective of the government is clearly to consolidate the auto industry and move it up the value chain. Currently, 150 companies are licensed to produce motor vehicles. The goal is to winnow them down to a much smaller number, with about 10 emerging as globally competitive.

Services

Services make up about 70 percent of global GDP and consist perhaps of an even more diverse set of activities than manufacturing: banking, retail, insurance, education, media, health care, information technology, hospitality, law, tourism, and consulting are just some examples of activities within the service sector. A number of entrepreneurs in developing countries are finding a niche in different sectors, often with support from their governments, which place a priority on those sectors as an avenue for development, growth, and employment.

Tourism

Tourism is the world's fastest-growing sector. From 1950 to 2007, international tourist arrivals rose from 25 million to 903 million. The overall export income generated by these arrivals grew at a similar pace, faster than the world economy, to over US$1 trillion in 2007 (UNWTO 2008). Tourism accounts for 7.6 percent of the world's workers (more than 60 million in China alone) and generates nearly 10 percent of its income. The impact of tourism is reflected not just in these figures but also through its links to other economic sectors and industries such as construction, manufacturing, and restaurants. In a country like Kenya, where the salary of a single hotel or restaurant worker supports four other people, tourism jobs are essential. According to Geoffrey Lipman, assistant secretary general of the UN World Tourism Organization, "Tourism is a good development agent because poor countries don't have to manufacture it . . . the market comes to these countries, then wanders around depositing foreign-exchange income whenever it's directed, including poor rural areas" ("Wish You and Your Money Were Here" 2009).

In 1950, the top 15 destinations absorbed 98 percent of all international tourist arrivals; in 1970, 75 percent; and in 2007, 57 percent, as new destinations emerged, many of them in developing countries. Over the years, the tourism sector has become increasingly sophisticated, specialized, and segmented, particularly where innovative practices have flourished. Some destinations have embraced mass tourism (Jamaica, Mexico, parts of Spain), while others have focused on luxury travelers (Maldives, Tahiti). Each of the Caribbean islands has acquired a different and sometimes unique reputation among travelers. The role of the state in promoting and supporting tourism has varied but has often been extremely important. For three "types" of tourism—ecotourism, destination tourism, and medical tourism—governments in a number of countries have worked hand in hand with the private sector to establish a specific niche.

Ecotourism. Countries with noteworthy natural endowments such as jungles, diverse flora and fauna, beaches, and coral reefs, have embraced ecotourism, which is generally defined as responsible travel to natural areas that conserves the environment and improves the well-being of local people. The attraction of this market segment lies in its marketability and typically higher revenues per traveler. Ecotourism is growing at a yearly rate of 25–30 percent; it currently represents 5 percent of the international market. Countries that rely heavily on high-end tourists include Costa Rica, with its national park system; Botswana, Kenya, South Africa, and Tanzania, with their game parks; and Bhutan, with its strict travel guidelines. In Rwanda, the tourism industry is planning to attract mainly high-value ecotourists interested in the country's mountain gorillas and other primates; while still in the nascent stage, the Tourism Working Group represents a public-private partnership that serves as a forum for debate, information sharing, and collaboration.

Destination or Cultural Tourism. Petra, one of Jordan's most popular tourist attractions, became a United Nations Educational, Scientific, and Cultural Organization (UNESCO) World Heritage site in 1985 and is a central element of the country's National Tourism Strategy, which estimates that tourism receipts will rise from US$800 million in 2003 to close to US$2 billion in 2010. Other examples include the rapid rise of Dubai as a tourist destination; the popularity of other UNESCO World Heritage sites like the Taj Mahal (India), Ha Long Bay (Vietnam), the Giza Pyramids (Egypt), and the Grand Canyon (United States); or the annual carnival in Brazil, whose rapid growth as a social and economic phenomenon led to a huge expansion of business opportunities and compelled government agencies and the private sector to work together on issues related to planning, managing, supporting infrastructure and equipment, and providing important services.

Medical Tourism. The American health care system accounted for US$2.4 trillion in costs in 2007 (projected to be US$4.4 trillion by 2018). As the health care market becomes increasingly global, it is not surprising that hospitals such as Thailand's Bumrungrad, India's Wockhardt, Singapore's Parkway Health, and the Republic of Korea's Health Care Town are dispelling the idea that health care is a local service. They are attracting thousands of American and European patients, using state-of-the-art procedures at a fraction of the price at home. In fact, some American companies are now giving their employees the option of going overseas for certain procedures, and the American Medical Association has recently issued guidelines for foreign medical travel (Connell 2006; UNDP 2008). It is estimated that for Americans the cost saving of performing most medical procedures overseas is at least 60 percent (Newfarmer, Shaw, and Walkenhorst 2009, 74).

To profit from tourism, countries need to invest in infrastructure, marketing, and human resources to attract international visitors and domestic tourists. Governments are working closely with the private sector to modernize tourism infrastructure and, in times of recession, to cut visa fees and partner with hotels, airlines, and tourist sites to reduce prices. Even in very poor countries such as Mauritania and Rwanda, niche tourism markets have developed for high-end international travelers and illustrate the need for government and the private sector to work together to ensure provision of essential services.

Tourism's main comparative advantage is that visitor expenditures also have a major and catalyzing effect on production and employment creation. Through the consumption of local products in hotels, tourists stimulate the development of small businesses in the production and services sectors and generate links to agriculture, fisheries, food processing, and light manufacturing, such as the garment industry. Tourism is therefore a cross-sectoral activity, as visitors spend a substantial amount of money outside the hotel for food, transportation, guides, entertainment, shopping and handicrafts, entrance fees, and so on. Estimates of such expenditures vary according to the type of hotel and local circumstances but can range from half to nearly double the expenditures in the hotel. Therefore, tourism can also create investment opportunities for small and medium enterprises.

"Because the barriers to tourism are lower than for many traditional exports, such as sugar and textiles, developing countries perceive tourism to be one of the few global industries in which they can be successful players," says David Bridgman, an investment marketing specialist at the Multilateral Investment Guarantee Agency (MIGA). Barriers to entry tend to be much lower than for most manufactured products, while transaction costs generated by border barriers, administrative barriers, transportation, time, and distance are also usually lower.

Tourism also generates demand for infrastructure improvements in such key areas as water and sanitation, telecommunications, and financial services, all of which are integral to successful tourism. These improvements also benefit the local community. Hotels depend on local employees and therefore upgrade their employees' skills through formal and informal training; in many countries, local people are rising through the ranks to technical and senior management positions. In Damascus, Syria, for example, a new hotel project backed by MIGA has allocated US$2.5 million to local staff who will be employed by the hotel. Furthermore, since tourism is labor-intensive, requiring about two employees per hotel room in developing countries, it tends to employ a high number of entry-level and female workers whose working conditions are often healthier and safer than in other sectors (MIGA). Tourism can also be an important element of export diversification, because it reduces many of the information costs involved in earning foreign exchange. It provides a local source of foreign demand to help producers learn about consumer preferences and standards in developed country markets, as well as about more sophisticated goods and exotic cultural goods.

The Example of Costa Rica

Since the late 1980s, Costa Rica has been a very popular nature travel destination and the most popular country to visit in Central America. Its main competitive advantage is its well-established system of national parks and protected areas, which cover around 23.4 percent of the country's land area, the largest in the world as a percentage of a country's territory. Although it has only 0.03 percent of the world's landmass, it has a rich variety of flora and fauna and is estimated to contain 5 percent of the world's biodiversity. The country has plenty of beaches, both on the Pacific Ocean and on the Caribbean Sea, within short travel distances and several volcanoes that can be visited safely. These natural attributes have been combined with a national vision, good infrastructure, and a strong streak of conservation and environmental sustainability. Not surprisingly, Costa Rica has become known for its ecotourism, with tourist arrivals growing by an average annual rate of 14 percent between 1986 and 1994. Since 1999, tourism has earned more foreign exchange than bananas, pineapples, and coffee combined. The number of visitors rose from 329,000 in 1988 to 1 million in 1999, to 2 million in 2008, when the country earned more than US$2.2 billion in spite of the global recession. In 2007, tourism made up about 8 percent of the country's GDP, accounted for 22.3 percent of the foreign exchange generated by all exports, and was responsible for 13.3 percent of direct and indirect employment.

For more than two decades, the government has been working to develop and consolidate its tourism industry in an effort to diversify its export earnings base and to offset the migration of manufacturing employment to countries with cheaper labor. Since the 1980s, it has helped move the industry up the value chain in terms of market segment and perception (table 9.2).

In Costa Rica, 33 laws regulate tourist businesses and tourism activities. The Costa Rica Tourism Board is the national regulatory institution. It operates

Table 9.2 Moving Up the Value Chain in Tourism in Costa Rica, 1980s–Future

Indicator	1980s	1990s	2000–Future
Development phase	Pioneer	Growth	Evolution
Market segment	Academics; tourists with strong nature-related interests and satisfaction with basic amenities	General interest in nature; tourists with moderate interest in ecotourism but desire for more upscale amenities; sun and surf with nature experiences	Sun and surf with moderate adventure activities; intercultural tourism with nature experiences
Market perception	Hard-core ecotourism	Moderate ecotourism	Multiphase tourism with strong interest in nature

Source: Adapted from Zamora and Obando 2001.

under the Ministry of Tourism and has managerial autonomy and its own budget. It has a board of directors and an executive president appointed by the government. Civil society is organized in different ways for tourist activity. Regional chambers of tourism are integrated with the National Chamber of Tourism. In addition, there is the Costa Rican Association of Tourism Professionals and Costa Rican Association of Hoteliers and Related Concerns. Tourist microentrepreneurs are also organized under the National Chamber of Tourist Microbusinesses, and there is also a Costa Rican Association of Tour Operators. Local communities have formed associations and cooperatives in various places throughout the country with local tourism as their main activity. Private reserves, which cover 1 percent of the national territory, rely on ecotourism as their main source of revenue. They carry out activities in coordination with, and backing from, the government agencies. NGOs also carry out tourism activities, many of which emphasize ecotourism. Examples include the Tropical Scientific Center, the Biodiversity and Tourism in Costa Rica Association, and the Monteverde Conservationist Association. Furthermore, some NGOs like the Instituto Nacional de Biodiversidad and the Neotropic Foundation are dedicated to education and to building awareness of the importance of biodiversity and a sound environment, among other things. A group of national organizations organized EXPOTUR, an annual event that has positioned itself as the main venue for the commercialization of tourism.

Training for tourist activities is provided by the National Training Institute, under an agreement with the Costa Rica Tourism Board; various state and private institutions collaborate. The institute organizes short technical and specialization courses in diverse fields related to tourism, as well as courses for trainers in areas such as development of personnel, administrative skills, and efficiency of services. Training courses are also offered in the country's main tourist regions, directed to the different subsectors (hotel management, food and beverage services). All nature guides in the country must be trained and certified by the institute and be registered with the tourism board. As for formal education through public universities, the University of Costa Rica offers a bachelor's degree in ecological biodiversity and tourism in Costa Rica.

As noted earlier, the effects of tourism often spill over into other sectors of the economy and produce numerous benefits.[3] In Costa Rica, tourism has helped revitalize the arts and crafts industry through greater demand for both quantity and quality. Various national and local artisans' associations are promoting and supporting arts and crafts and help distribute products throughout the country. In addition, a growing number of artisans take advantage of sustainable biodiversity resources and exploit waste materials as part of a commitment to conservation and sustainable use of biological resources. Similarly, the food and beverage industry has benefited from demand for greater quality and variety; for example, natural fruit juices are being used more and in more creative ways in a number of beverages.

Information Technology Services

In 2006, global trade in information and communication technology (ICT) reached more than US$3.5 trillion. ICT-related foreign investment set a record the following year, when about 20 percent of all cross-border mergers and acquisitions were related to ICT (OECD 2008). About half of all ICT goods production now comes from countries that are not members of the Organisation for Economic Co-operation and Development (OECD), and some of these, such as China and India, are increasingly home to top ICT firms. IT services—application development, systems integration, IT infrastructure services, consulting, manufacturing engineering, and software development—offer enormous opportunities for many countries. It is estimated that the annual market for IT services and IT-enabled services is potentially about US$475 billion, of which less than 15 percent has been exploited (World Bank 2009). India has focused on key services such as software development (see box 9.7).[4] Other countries have achieved similar

---

Box 9.7 The Software Industry in India

The Indian software industry has been remarkably successful. It has grown more than 30 percent annually for 20 years, with 2008 exports projected at around US$60 billion. India exports services—two-thirds of which are for the United States—to more than 60 countries. Key factors behind the industry's success have been public investment in technical education, the facilitating role of the Indian diaspora, a favorable policy environment, and growing global demand for IT services.

The government's investment in technical education in the 1960s created a series of elite technical and management institutes that have provided Indian firms with a pool of English-speaking professionals, many of whom were sent to the United States to work for limited billable projects. The Indian diaspora was instrumental in facilitating links between U.S. companies and IT firms in India. Software engineers who returned to India could be hired by local firms. In recent years, Indian firms have increasingly conducted software development offshore, a trend made possible by the maturity of the industry and the development of IT infrastructure. The Indian government liberalized the telecommunications market and adopted a number of policies to promote IT exports (including loosening import rules for the necessary equipment, establishing technoparks, and providing tax incentives). To ensure a continuous supply of qualified personnel, the Ministry of Human Resources Development helped expand computer science departments, encouraged the private sector to open training institutions, and introduced quality control systems.

Indian software firms quickly moved up the value chain, from performing low-cost programming abroad to providing comprehensive software development services for overseas clients. Revenue per worker more than doubled over a decade. Furthermore, an increasing number of firms now meet international certification requirements for key quality standards, further enhancing the credibility of Indian brands in international markets.

Source: Dahlman and Utz 2005.

success, while still others have developed small but vital industries in more challenging environments (see Harabi 2009). Even a poor country like Kenya is showing nascent signs of a developing ICT industry with a core of well-educated entrepreneurs focusing on customized applications for mobile phones and new digital content (Zachary 2008b).

The Example of Vietnam

Vietnam is another country making significant strides in this sector. Its IT industry is considered one of its most dynamic sectors, and its rapid development has been actively helped by the government. The market for IT has been gradually opened for more players. Previously, the Vietnam Post and Telecommunication Corporation was the only corporation providing telecommunication services in Vietnam. Beginning in 1995, the government began to license nonstate companies to do business in this sector and has greatly liberalized the pricing of telecommunication services, which has resulted in a more competitive sector (Hong 2007).

The government of Vietnam has consistently supported the development and growth of the IT sector through policy interventions. It has made major investments in network modernization and capacity upgrading and in high-technology and software parks. A number of resolutions and laws on IT development have been approved during the past 15 years, and the development of IT is one of seven priorities in Vietnam's science and technology development strategy up to 2010 (and in the upcoming five-year plan). Intel recently received a license from the Ministry of Investment and Planning to build a US$300 million assembly and test facility to produce chips and computer parts, thus becoming the first major foreign investor in high technology in the country and adding its seventh assembly site to its global network.

The Vietnamese software industry had revenues of over US$400 million in 2007 and has been recording 40 percent annual growth. Currently, around 25,000 IT professionals are estimated to be working in software firms. For at least a decade, Vietnamese officials have expressed strong belief in information technology as a key to successful economic development and have acted accordingly. The software industry is currently the most subsidized economic sector in Vietnam: businesses involved in software production and services, both local and foreign invested, are exempt from corporate income tax (28 percent) for four years from the date they generate their first taxable income, and software products receive a zero percent value-added tax and are free from export tax. Further incentives have been recently offered to assist with training support for significant software projects. The government has plans to attract some of the 3 million overseas Vietnamese to bolster 600 existing software development firms. These companies employ 15,000 people, mainly in Ho Chi Minh City and Hanoi, up from 170 firms and 5,000 workers in 1999.

Among its other policy instruments to support the development of the IT services industry,[5] the government has instituted several specific and explicit pro-innovation policies:

- Tax exemption on business income earned from scientific research and technological development, products manufactured during test production, and products made from technology applied for the first time in Vietnam
- Expenses incurred for scientific and technological research, innovations, and initiatives allowed as a deductible expense for computation of income tax
- Intellectual property rights protection for computer programs and compilation of data, as well as layout design for semiconductors and integrated circuits
- Fifty-year copyright protection and 20-year protection for invention patents.

Policy Conclusions

Although agriculture, manufacturing, and services are very different sectors, they all share some key factors for success. What Rwandan coffee, Chinese automobiles, Costa Rican tourism, and Vietnamese IT services reveal is that success is a long-term process that usually involves direct and indirect government action in the context of an innovation system composed of diverse actors (see table 9.3). Innovations can and do emerge at

Table 9.3 Competitive Industries and Innovation Systems in China, Costa Rica, Rwanda, and Vietnam

Feature	Chinese automobiles	Costa Rican tourism	Rwandan coffee	Vietnamese IT services
Vision and leadership	National priority	National priority	National coffee policy; President Kagame's leadership	National priority
Framework conditions	Political stability; economic incentives	Political stability	Political stability	Political stability; tax and other incentives
Education and research	Growing R&D facilities; abundance of science and engineering graduates	University degrees in tourism-related subjects; training for guides	Focus on technical and vocational training; workshops for farmers	University and vocational education
Infrastructure	Support from government in many forms, including subsidies	Support from numerous NGOs and government agencies	Support services to farmers and cooperatives	Support from government agencies; IT parks
Industrial system	150 firms of different sizes and capabilities	Local SMEs	Farmers' cooperatives	New firms; joint ventures
Intermediaries	Joint ventures, government ownership of some firms	Costa Rica Tourism Board; civil society groups	NGOs, USAID (donor), National University of Rwanda	IT parks
Demand	Large domestic market; potential global market	Specialized tourism, such as ecotourism, adventure	New and growing market for specialty coffee	Global IT market

Source: Author.
Note: IT = information technology; NGOs = nongovernmental organizations; R&D = research and development; SMEs = small and medium enterprises; USAID = U.S. Aid for International Development.

any point in the system as the result of consciously mediated or coordinated interactions among different types of agents. Innovations therefore do not necessarily depend on any particular single action, because in practice innovation has multiple sources.

What specific insights emerge from the four industries examined about how innovation is promoted, especially from the perspective of public policies? Table 9.3 summarizes these industries according to the main features of the innovation system framework. Policy makers may consider a number of the issues that the studies bring out:

- *The innovation system framework—useful but "messy."* Innovative systems are inherently interactive and somewhat chaotic. Interaction of the different actors should be emphasized more than the infrastructure. In all four industries examined, the types of collaboration that have taken place between public and private sector entities have been both critical and complex.

- *Timing and serendipity.* Though impossible to estimate, the role of timing and serendipity cannot be overlooked. Rwanda entered the coffee export market in the mid-1990s at a time of growing demand for specialty coffee. Costa Rica embraced ecotourism when nascent environmental concerns made this a viable industry. China is experiencing a new era in its automobile industry as it focuses on state-of-the-art battery technologies and a rapidly growing domestic market, even as a number of traditional car companies such as General Motors and Chrysler are bankrupt or merged. But timing and luck are problematic, and it is usually very difficult to pick winners (World Bank 2005).

- *Vulnerability—a constant presence.* Catering to a global market can expose weaknesses. In 2008, the salmon industry in Chile has suffered from quality problems related to the health of its fish, and safety and quality problems have plagued Chinese exports of toys, pharmaceuticals, and pet food. Kenyan agricultural exports, owing to insufficient rain and an exponential increase in fertilizer and pesticide prices, suffered tremendously in 2008: the production of maize declined from 34 million bags to 24 million; shipments of high-value fruit and vegetables to the European Union have declined owing to high fuel prices and a growing trend on the part of customers to buy more local and seasonal produce; and Kenyan roses and other cut flowers may see a drop in demand as the European market becomes saturated. In addition, human resources are a growing challenge. Many countries face a tremendous deficit of skilled labor, especially managers and workers with IT skills.

- *The central importance of learning.* In assessing common elements of technological adaptation derived from various studies of technological leapfrogging into profitable export activities, the importance of "tacit knowledge" (information, skills, and interactions and procedures embedded in individuals or organizational structures such as firms, networks, and public institutions) should not be underestimated (Chandra 2006). This notion points to the crucial role played by organizational structures in the innovation process and illuminates the difficulties inherent in technology transfer. It also serves to highlight the importance of technological learning processes in their own right and thus the need for policy attention to this issue. In short, it is essential for a country to discover its own pragmatic innovation agenda and approach. Mexico, for example, has at least four different models and institutions. This requires self-discovery and experimentation. If, for instance, knowledge is rapidly diffused among firms in an industry, the "leading practice" can be mainstreamed, with competitive benefits for the entire industry (Dahlman 2009).[6]

- *Financing—an essential component.* Entering new markets, promoting new industries, and supporting entrepreneurs are inherently risky endeavors. It is important therefore to study policies, institutions, and initiatives that have performed well in this respect. Examples include the Ireland Linkage Program, Singapore's assistance to export-oriented firms, and Nicaraguan support to small and medium enterprises. This role can also be played by donors and other international partners: USAID and the World Bank, for example, offered support to the Rwandan coffee industry at a critical time.

- *Achievement of critical mass.* Innovation of the quality and extent that can make a significant difference in a sector or subsector depends on a critical mass of entrepreneurs, financing, supportive government policies, adequate human resources, positive market conditions, and the like. There is no magic recipe or precise calculation of the size of this critical mass and of how much of each "ingredient" is required for a subsector to flourish and achieve a "tipping point." Success depends on conditions within the country, timing, and some degree of serendipity. The innovation system framework is particularly useful here since it can illustrate effectively how key interventions can increase the odds that a tipping point will occur as a critical mass is established.

- *Public policy as a key driver of innovation and competitiveness.* Government policies can make a real difference in creating and sustaining national competitive advantage. Although globalization may appear to weaken the role of government, its role has, in fact, become stronger than ever. Public policy plays an important role in whether conditions are favorable to innovation (Porter 1990). It can influence the operating conditions and institutional structures that surround firms. Thus, government's most powerful roles are indirect rather than direct. That is, they "steer" by shaping the business environment rather than by intervening directly. While public policies alone cannot produce innovation, well-crafted policies can facilitate, push, and reinforce it as desirable behavior.[7] Conversely, poorly conceived public policies can stifle, delay, or penalize innovation. On the one hand, some public policies can be seen as difficult to embrace initially due to their imposition of new regulations and costs; on the other hand, they can eventually promote a level of innovation that would not have occurred otherwise, as has been the case particularly with environmental policies that have stimulated a new wave of technologies for cars, home construction, and packaging and even a product as simple as the incandescent light bulb.[8]

A major constraint to the competitiveness of many firms is the absence of the right kinds of institutions to support technological change. Such institutions carry out R&D, evaluative testing, quality assurance, enforcement of laws and standards, networking, and information dissemination. Since the market alone is not sufficient to promote the organizational interactions required for innovation, the state must create new institutions where they do not exist, restructure institutions in response to change, and reshape interactions among firms and organizations through the use of incentives (Oyelaran-Oyeyinka 2006). In short, one role of public policy is to change institutional actors and modify institutional rules as new circumstances arise to encourage technological learning. Not surprisingly, the most essential factors for influencing technological catch-up are an environment that nurtures the ability to learn and apply new technologies and the facilitating hand of government.[9]

It is clear that a wide range of public policies can foster (or impede) innovations that lead to productivity gains, which in turn translate into greater competitiveness. But it is also clear that the policies that are most relevant will vary from country to country and be determined by local values, institutional cultures, business conditions, and key production inputs in particular subsectors. As a result, although developing countries can often learn useful lessons from the successful experiences of other nations, they will have to design their own strategies. To do so, of course, they may find inspiration in the approaches of others. But their choices of what to use and what to discard will and should be conditioned by local values, institutional capacities, and economic conditions.

Notes

1. In fact, Starbucks' need to purchase premium coffee and support Fair Trade practices has led it to pay consistently higher prices to its producers, as described in its corporate social responsibility report.

2. Deputy Prime Minister Men Sam An, quoted in the *Cambodian Daily*, April 28, 2009.

3. See Newfarmer, Shaw, and Walkenhorst (2009) for an additional example of butterfly chrysalises for export.

4. For a more detailed treatment of the Indian IT/software industry and the facilitating role of the government in its development, see Dahlman and Utz (2005) and Chandra (2006).

5. For a comprehensive list of policy instruments, see World Bank (2008c, 86).

6. A similar result can be achieved even in the least developed countries through arrangements such as industrial clusters, where leading practice can evolve. See Zeng (2008).

7. Achieving policy coordination and coherence is a difficult challenge for governments. See OECD (2009, 17).

8. It was a law passed by the U.S. Congress in 2007 setting much higher efficiency standards for incandescent light bulbs that forced manufacturers to develop innovative light bulbs that use 50–75 percent less energy and last three to five times longer than previous models.

9. See Chandra (2006) for more information on specific government policies affecting technological adaptation and learning.

References and Other Resources

Bourne, Joel K., Jr. "The Global Food Crisis: The End of Plenty." 2009. *National Geographic,* June 9.

Chandra, Vandana. 2006. *Technology, Adaptation, and Exports: How Some Developing Countries Got It Right.* Washington, DC: World Bank.

Collier, Paul. 2007. *The Bottom Billion: Why the Poorest Countries Are Failing and What Can Be Done About It.* New York. Oxford University Press.

Connell, John. 2006. "Medical Tourism: Sea, Sun, Sand and Surgery." *Tourism Management* 27,

Dahlman, Carl. 2008. Innovation Strategies of the BRICKS (Brazil, Russia, India, China, and Korea): Different Strategies, Different Results.

———. 2009. "Different Innovation Strategies, Different Results: Brazil, Russia, India, China, and Korea (the BRICKs)." In *Innovation and Growth: Chasing a Moving Frontier,* ed. Vandana Chandra, Deniz Eröcal, Pier Carlo Paodan, and Carlos A. Primo Braga, 131–68. OECD and World Bank: Paris and Washington, DC.

Dahlman, Carl, and Anuja Utz. 2005. *India and the Knowledge Economy: Leveraging Strengths and Opportunities.* Washington, DC: World Bank.

Department of Tourism and Commerce Marketing, Dubai. http://www.dubaitourism.ae/.

Dugger, Celia W. "Ending Famine, Simply by Ignoring the Experts." *New York Times,* December 2.

Economist. 2008. September 20.

"A Global Love Affair: A Special Report on Cars in Emerging Markets." 2008. *Economist,* November 13.

Global Wind Energy Council. 2008. "Global Wind 2007 Report." http://www.gwec.net/fileadmin/documents/test2/gwec-08-update_FINAL.pdf, 50–54.

Harabi, Najib. 2009. "Knowledge Intensive Industries: Four Case Studies of Creative Industries in Arab Countries." Paper prepared for the World Bank Institute, World Bank, Washington, DC.

Henderson, Joan. 2006. "Tourism in Dubai: Overcoming Barriers to Destination Development." *International Journal of Tourism Research* 8: 87–99.

KPMG. 2006. "Globalization and Manufacturing," 4.

Larsen, Kurt, Ronald Kim, and Florian Theus. 2009. *Agribusiness and Innovation Systems in Africa.* Washington, DC: World Bank.

Li, Junfeng. 2008. "Opinion: China's Wind Power Development Exceeds Expectations." Worldwatch Institute. http://www.worldwatch.org/node/5758.

Liu, Melinda. 2009. "A Lean, Green Detroit." *Newsweek,* May 4.

Marr, Kendra. "China Emerges as World's Auto Epicenter: As Detroit Crumbles, Beijing Picks Up the Pieces—at a Bargain." 2009. Washington Post.com. MSN Web site. http://www.msnbc.msn.com/id/30802161/.

Newfarmer, Richard, William Shaw, and Peter Walkenhorst. 2009. *Breaking Into New Markets: Emerging Lessons for Export Diversification.* Washington, DC: World Bank.

Nguyet, Hong Vu Xuan. 2007. "Promoting Innovation in Vietnam: Trends and Issues." Central Institute for Economic Management (Vietnam). Presentation at the Forum on Innovation in the African Context, Dublin, March 6–8.

OECD (Organisation for Economic Co-operation and Development). 2007. *Business for Development: Fostering the Private Sector.* Paris: OECD.

———. 2008. *Information Technology Outlook.* Paris: OECD.

———. 2009. *The OECD Innovation Strategy: Draft Interim Report.* Paris: OECD.

Oyelaran-Oyeyinka, Banji. 2006. *Learning to Compete in African Industry: Institutions and Technology in Development.* Hampshire, UK: Ashgate Publishing Limited.

Porter, Michael E. 1990. *The Competitive Advantage of Nations.* New York: Free Press.

Rede NÓS Seminar. 2004. "The Impact of Innovation on Cluster Sustainability and Competitiveness." Jointly organized with the Pernambuco Secretary of Science and Technology (SECTMA), February 25.

Rukazambuga, Daniel. 2008. "Agricultural Innovation and Technology in Africa: The Rwanda Experience in the Coffee, Banana, and Dairy Commodity Chains." Paper prepared for the World Bank Institute, World Bank, Washington, DC.

Saint, William. 2007. "How National Public Policies Encourage or Impede Agribusiness Innovation: Studies of Six African Countries." Paper prepared for the World Bank Institute, World Bank, Washington, DC.

Spielman, David J., and Regina Birner. 2008. "How Innovative Is Your Agriculture? Using Innovation Indicators and Benchmarks to Strengthen National Agricultural Innovation Systems." Discussion Paper 41, Agricultural and Rural Development Department, World Bank, Washington, DC.

"Top 100 Espresso: Coffee Statistics Report—2010 Edition." 2008. http://www.top100espresso.com/coffee_consumption_statistics_report.html.

Turkish-US Business Council http://www.turkey-now.org (Turkish-US Business Council. 2007. *Turkey Brief: Turkish-US Relations,* http://www.taik.org/db/Docs/Turkey_Brief_2008.pdf pages 48–50)

UNCTAD (United Nations Conference on Trade and Development). 2007. *The Least Developed Countries Report 2007.* New York: UNCTAD.

UNDP (United Nations Development Programme). 2005. *Innovation: Applying Knowledge in Development.* London: Earthscan.

———. 2008. *Creative Economy Report 2008.* Geneva: UNDP.

UNIDO (United Nations Industrial Development Organization). 2005. *Industrial Development Report 2005: Capability Building for Catching Up.* Vienna: UNIDO.

UNWTO (United Nations World Tourism Organizations). 2008. *Tourism Highlights: 2008 Edition.* Geneva: UNWTO.

Watkins, Alfred, and Anubha Verma. 2008. *Building Science, Technology, and Innovation Capacity in Rwanda: Developing Practical Solutions to Practical Problems.* Washington, DC: World Bank.

"Wish You and Your Money Were Here." 2009. *Time Magazine,* May 4.

World Bank. 2005. *World Development Report: A Better Investment Climate for Everyone.* Washington, DC: World Bank.

———. 2007a. *World Development Report 2008: Agriculture for Development.* Washington, DC: World Bank.

———. 2007b. *Enhancing Agricultural Innovation: How to Go Beyond the Strengthening of Research Systems*. Washington, DC: World Bank.

———. 2007c. *Building Knowledge Economies: Advanced Strategies for Development*. Washington, DC: World Bank.

———. 2008a. *Global Economic Prospects: Technology Diffusion in the Developing World*. Washington, DC: World Bank.

———. 2008b. *The Global Textile and Garments Industry: The Role of Information and Communication Technologies (ICTs) in Exploiting the Value Chain*. Washington, DC: World Bank.

———. 2008c. *International Good Practice for Establishment of Sustainable IT Parks*. Washington, DC: World Bank.

———. 2009. *Information and Communications for Development: Extending Reach and Increasing Impact*. Washington, DC: World Bank.

World Gold Council: http://www.gold.org/

Zachary, G. Pascal. 2008a. "The Coming Revolution in Africa." *Wilson Quarterly,* Winter Issue.

———. 2008b. "Inside Nairobi: The Next Palo Alto?" *New York Times,* July 20.

Zamora, Natalia, and Vilma Obando. 2001. *Biodiversity and Tourism in Costa Rica*. San Jose: Instituto Nacional de Biodiversidad.

Zeng, Douglas Zhihua. 2008. *Knowledge, Technology, and Cluster-Based Growth in Africa*. Washington, DC: World Bank.

10

Building Innovative Sites

The observation that economic phenomena—and innovation in particular—are spatially polarized is not new: see, for example, Alfred Marshall's "industrial districts" at the end of the 19th century, Joseph Schumpeter's "innovation clusters," Eric Dahmen's "development blocks," Francois Perroux's "development and growth poles" in the 1950s, and, more recently, economic geographers' and economists' industrial and high-technology agglomerations and "new economic geography."

Local innovation is of particular interest both because innovation has its foundations in microeconomic (local) processes (proximity, networks, density, diversity) and because economic globalization is increasingly important. The development of global corporations has triggered the geographical division of the value-added chain, which fosters competition between local units of production, promotes the new techniques that enable the international division of production processes, increases the efficiency of transportation, and improves information and communication technology (ICT) infrastructure. At the same time, decentralization has resulted in more local power and funding.

The term *glocal* reflects the notion that links to international business activity strengthen local enterprises. And because of the connection between local innovation and success in international markets, localities have a strong rationale for developing a strategy for increasing innovation in targeted sites—special economic zones, science parks, clusters, and even cities:

- First, increased innovation in targeted sites can have a strong demonstration effect and thus lead to broader initiatives.

This chapter was prepared by Justine White.

- Second, when means are constrained, small-scale projects can be more cost-effective than larger, more ambitious efforts.
- Third, for developing countries with little experience in such policies, local innovation strategies can be a testing ground for more far-reaching strategies.

This chapter begins with a discussion of several concrete policy tools for building local innovative sites, one of which is the special economic zone. In developing countries, these zones are often associated with low-wage, low-skill production, but experience shows that they can stimulate innovation, particularly in a context of a mediocre but evolving investment climate. Science parks are also increasingly popular in developing countries, in particular for developing employment opportunities for recent tertiary graduates but also for spurring local business creation. Clusters, often considered a "silver bullet" for boosting innovation in both developed and developing countries, are networks of firms whose function may be favored by certain government policies. Finally, the chapter ends with a discussion of the way to promote innovation in cities and in the surrounding region, a topic of increasing concern all over the world.

Special Economic Zones

Although special economic zones (SEZs) have existed for several centuries in various forms, probably the first modern one was developed in 1959, near Shannon Airport, Ireland, where it still continues strong today. By the mid-1970s, SEZs numbered at least 79 in 25 countries (Jenkins, Larrain, and Esquivel 1998; FIAS 2008). Today, there are over 3,000 publicly and privately operated SEZs located in more than 135 countries, including some in Organisation for Economic Co-operation and Development (OECD) countries. According to recent estimates, SEZs in developing countries employ some 40 million people directly and 10–77 million indirectly (ILO 2003; FIAS 2008). Furthermore, although SEZ employment is negligible as a percentage of total employment,[1] the share of SEZs output in the exports of developing countries can be considerable, as in the Madagascar (80 percent), Philippines (78.2 percent), Bahrain (68.9 percent), and Morocco (61 percent) (FIAS 2008).

A special economic zone is here defined as a geographically delimited area, with a single management or administration and a separate customs area (often duty free), where streamlined business procedures are applied and where firms physically located within the zone are eligible for certain benefits (such as tax exemption for a number of years, accelerated depreciation, and investment credits) (Jenkins, Larrain, and Esquivel 1998; FIAS 2008).[2]

After defining SEZs and presenting reasons for developing them, this section then considers some of the conditions and policy tools that can be used for building the "catalyst" SEZs that drive local innovation.

Special economic zones have appealed to policy makers the world over as a way to attract multinational companies (MNCs) and create local jobs, even when the general business environment is poor. Their attractiveness comes not only from their direct economic benefits, such as employment and foreign currency generation, but also from their indirect benefits, such as technology transfer, potential backward linkages with local firms, and stimulation of local innovation. Some developing countries have succeeded in using SEZs as catalysts for transforming domestic firms and promoting innovation. Many of these zones, however, have remained "islands" isolated from their host economy and have not played a catalytic role (see figure 10.1). Remaining an island does not imply that the SEZs have not succeeded on some counts: they have often generated much-needed local employment or attracted foreign direct investment (FDI) and foreign exchange. However, they have not stimulated local enterprises to upgrade and innovate.

Host Country Benefits of an SEZ

From a host country's perspective, the benefits of an SEZ fall into two categories: *direct* benefits, which include foreign exchange earnings, attraction of FDI, increased government revenue, and export growth; and *indirect*, or dynamic, benefits, which include upgrading the skills of the workforce and management, technology transfer, demonstration effect, export diversification that enhances the trade efficiency of domestic firms, and knowledge of international markets. The latter are closely linked to the integration of SEZs into the local economy and their capacity to spark local innovation.

The creation of a "catalyst" appears to depend on whether the SEZ becomes partially integrated into the local economy through linkages with enterprises outside the SEZ. If so, the SEZ can increase the ability of domestic firms to respond to new opportunities, favor technology transfer through people, and

Figure 10.1 "Island" versus "Catalyst" Special Economic Zones

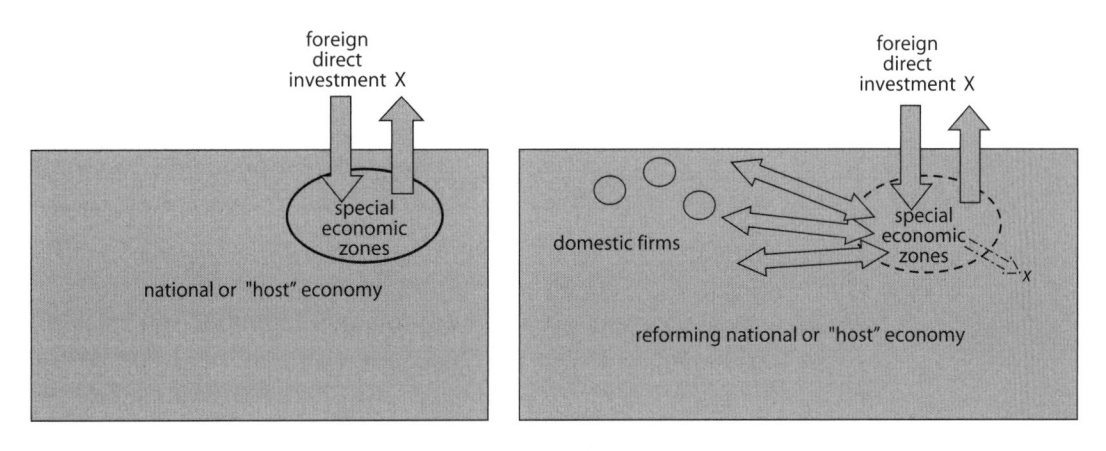

Source: Author.

stimulate the competitiveness of domestic firms (see figure 10.1) (Johansson and Nilsson 1997; Omar and Stoever 2008).

The host country can experience significant spillover effects from FDI. Horizontal spillovers, for example, include technology "leakage" from MNCs to local firms in the same industry, which occurs in various ways: first, local firms may be able to learn by observing and imitating; second, employees may leave MNCs to join local firms, bringing with them new technology and management know-how (a "domestic skilled diaspora"); and, third, MNCs may provide knowledge or know-how that can also be used by domestic firms.

Vertical spillovers, or backward links, take the form of positive externalities through the supply chain. As a channel through which information and material flow between a firm and its suppliers, these links create a network of economic interdependence. Multinationals located within the SEZ may want to transfer technology to their local suppliers outside the SEZ to achieve lower production costs, increased specialization, and better adaptation of technologies and products to local environments. Lall (1980) notes that such links can take several forms. A multinational could, for example, help prospective suppliers build production capacity, provide technical assistance or information to raise the quality of suppliers' products or facilitate innovations, or offer training and help in management and organization (see UNCTAD 2001).

Horizontal and vertical linkages benefit the local economy through increased output and employment, improved production efficiency, technological and managerial capabilities, and market diversification. Although these links can be extremely important for fostering local innovation, the onus of developing them should not be on the firms inside the SEZ; when that approach has been tried in the past, the effort has not succeeded (see box 10.1). Furthermore, imposing local content and other burdensome requirements is often impractical because of intense competition between SEZs (FIAS 2008). Host governments can, however, create attractive conditions, facilitate contacts, and provide various direct or indirect incentives that make it cost effective for foreign companies in SEZs to get supplies from local sources. The Republic of Korea's outsourcing program is one example, and, in Shenzhen, China, SEZ administrators provide individually tailored directories listing prospective domestic suppliers.

Several authors contend that the ease of establishing backward links is constrained either by prevalent local industrial development or by sectoral specialization (FIAS 2008; ILO/UNCTC 1988). Jenkins, Larrain, and Esquivel (1998) provide a statistically significant econometric connection between backward links and the country's level of industrialization, although the causality of this link is not demonstrated.[3] As for the claim of sectoral favoritism, others have argued that some sectors are more receptive to the development of backward links than others. The authors explain that, because of the very nature of manufacturing, the electronics industry should generate

> **Box 10.1 The Development of Backward Links: A Successful and a Less Successful Example**
>
> *Republic of Korea:* When the Masan zone began operations in 1971, domestic firms supplied just 3.3 percent of materials and intermediate goods to companies in the zone. Four years later, they supplied 25 percent and, eventually, 44 percent. Consequently, domestic value added increased steadily from 28 percent in 1971 to 52 percent in 1979. In all, the evidence indicates that the Korean government successfully encouraged backward links with local industries and subcontractors. Local companies supplying export processing zone (EPZ) firms had preferential access to intermediate and raw materials. The zone administration also provided technical assistance to subcontracting firms.
>
> *Dominican Republic:* During the 1980s, the share of domestic value added in total output decreased, from 40–45 percent in the early years of the decade to just 25–30 percent in the later years. And, through lack of government interest or incentives, there were few backward links between domestic firms and industries in SEZs. Until 1993, domestic firms that wanted to sell products to companies in the zones needed an export license, which was difficult to obtain. In addition, even though the legislation stated that domestic firms could recover import duties paid for materials used in products sold to EPZ firms, they were almost never able to do so.
>
> *Source:* Author.

more links to the domestic economy. Others contend that links are difficult to establish in any industry but particularly in the textile industry (Basile and Germidis 1984; ILO 1998). The accumulated experience of some countries could support these assertions. Many of the most successful SEZs in Asia, however, initially attracted labor-intensive industries with relatively unsophisticated technologies (textiles, basic electronics) that required unskilled workers and then upgraded to more technology-intensive and higher value-added sectors. These successful SEZs were extremely good at moving away from the low-skill, labor-intensive industries of their early years of operation. New garment industries were not allowed in Taiwan, China's export processing zone (EPZ) as of 1974, for example.

Some Pointers from More Successful SEZs

Many of the more successful SEZs (see table 10.1) benefited from an increasingly well-trained domestic workforce, most often the result of on-the-job training inside the SEZ and major efforts to improve the national education system. In 1968, for example, 57 percent of the SEZ workforce in Taiwan, China, had only elementary school training; in 1990, 87 percent had more than elementary training. In the 1970s, 80 percent of the workforce in the country had completed middle school; this proportion was 95 percent in 1990. In terms of gender, these figures are even more dramatic. In the 1970s, only 20 percent of women working in SEZs in Korea had completed high

Table 10.1 Training for Workers in SEZs in Selected Economies

Country	Training provided
China (Shenzhen)	Three months on-the-job training for operators (one month for class and two months for production practice); over 80 adult education institutes (1990) but weak links between needs of enterprises in the EPZ and skills provided
Republic of Korea (Masan)	Three months on-the-job training for operators; overseas training for skilled workers (mainly in Japan)
Malaysia	Three months on-the-job training for operators; quality control cycles with monetary and other incentives (gifts, medals, and commendation letters) for identifying problems and suggesting ways of solving them; little training for computer programming, technical engineering, and design work
Mauritius	Three months on-the-job training for operators (75 percent minimum salary for trainees); lack of trained intermediate workers
Philippines	One day to a few weeks on-the-job training for operators; rotation by some firms (Japanese) of operators to make them familiar with 10–18 interrelated tasks (three-month rotation)
Sri Lanka	One to three months on-the-job training for operators
Taiwan, China (Kaohsiung)	Three months on-the-job training for operators; cooperative training programs between school or college and the firm in the EPZ; provision of general education by school or college and special technology training by firms; some overseas training
Thailand (Lat Krabang)	Three months on-the-job training for operators; off-the-job training; study and experiment in the classroom and laboratory for some workers; overseas training (at parent company) for core employees in management and technology

Source: Kusago and Tzannatos 1998.

school, as compared with over 95 percent today. And it would seem that a skilled local workforce can be an important decision factor for high-technology MNCs (see box 10.2).

Some SEZ management companies have also given training to local companies outside the SEZ. For example, SFADCo provided direct training for local industry outside the SEZ during the early 1960s (Callanan 2000). The literature, however, has given little attention to the role of labor turnover from SEZs as a channel for the diffusion of technology and innovative processes to the domestic economy. It would seem, though, that labor turnover can be significant for transferring technology and managerial know-how to domestic firms. Some countries (box 10.3) have used fixed-term, nonrenewable two- to five-year contracts for local managers in SEZs.

Experience has shown that improving the general business climate of the host country is essential for developing a catalyst SEZ (FIAS 2008). Although SEZs can theoretically be located anywhere, their medium- to long-term viability and their capacity to create local dynamics seem to require undertaking domestic business reforms at the time the SEZ is designed or shortly afterward,

> **Box 10.2 Attracting High-Technology Investments in an SEZ in Costa Rica**
>
> Intel's construction of a US$300 million semiconductor assembly plant in Costa Rica came as a surprise to many, especially in view of the country's small size and the fierce competition from countries such as Brazil, Chile, Mexico, the Philippines, and Thailand over winning that investment. Success was attributed to the country's exceptional education system, commitment to openness, general political stability, high-level political support by the head of state, mobilization of the business community, and capacity to respond rapidly to requests.
>
> *Source:* Spar 1998.

> **Box 10.3 SEZs and Labor Circulation—A "Domestic Skilled Diaspora"?**
>
> - In the Masan Zone in *Korea*, it is estimated that 3,000–4,000 people received specialized training, in the zone and abroad (mainly Japan) and that half of them eventually left the zone to work in local electronics firms.
> - In *Taiwan, China*, under government guidance, personnel from firms in the zones were placed at potential suppliers' factories to offer advice on production methods and quality control.
> - *Shannon, Ireland*, had high labor turnover between the SEZ and the domestic economy, with many managers leaving to create competing firms outside the SEZ.
> - In *Shenzhen, China*, workers were appointed by the government for a three-year term and were then required to leave the zone. Many managers subsequently started their own firms, capitalizing on experience gained in the SEZ.
>
> *Source:* Jenkins, Larrain, and Esquivel 1998; Leong 2007; Callanan 2000.

so that SEZ companies can source local content competitively, and entrepreneurs can set up firms to compete with SEZ companies.[4]

The example of Shenzhen offers some interesting insights into the technological upgrading of SEZs and the stimulation of innovation in the domestic economy (box 10.4). After testing business climate reforms in the SEZ, the Chinese government launched nationwide reforms to match or emulate the business climate tested within the zone. Exports from the SEZ to the domestic economy were authorized. Furthermore, owing to a strict labor policy but also to voluntary departure, many employees left SEZs to create rival firms. This factor put competitive pressure on firms within the SEZ to innovate or disappear.

Finally, several studies have insisted on the importance of location (FIAS 2008). In fact, many governments, responding to the need to create employment and economic opportunities in rural areas, have established SEZs in

> **Box 10.4 A Tale of Two Countries—Investment Climate Reform**
>
> India's Kandla, the first SEZ in Asia, was launched in 1965, and the first SEZ in China in 1980. Being a first mover gave little advantage to Kandla, however, while China's SEZs, particularly Shenzhen, have been a phenomenal success.
>
> One major difference between the two countries was their approach to economic reform and free trade. In contrast, China's SEZs were intended to serve as both test beds and spearheads for implementing wider-ranging economic reforms and trade liberalization in the rest of the country—goals they clearly achieved. Thus, SEZs in India remained isolated enclaves, while SEZs in China were rapidly overtaken and threatened by competitive domestic firms. To remain relevant, Chinese firms became more technology intensive and more business friendly and offered better services to companies.
>
> *Source:* Leong 2007.

more remote locations. However, these have often failed to become catalysts. An SEZ in a city or in a peri-urban area has easier access to firms, capital, and labor and can integrate into the economy more easily. Table 10.2 summarizes the main policies aimed at encouraging innovation through SEZs.

Science Parks

Although science parks are popular for developing local capacity for innovation and for creating employment for tertiary and technical graduates, many have not achieved the hoped-for success. This section defines and describes science parks, details their establishment, and then explores ways to help ensure their sustainability.

Evolution of Science Parks

A science park is an organization and property development managed by specialized professionals who seek to increase the competitiveness of their city, region, or territory of influence.[5] The park does so by concentrating mature technology, science, or research-related businesses (which can be MNCs), fostering collaboration among them and knowledge-based institutions, and transferring knowledge to the market place (Sanz 2004). Two well-known examples are the Hsinchu Science Park in Taiwan, China, and the Cambridge Science Park in England. Science parks are often associated with or operated by institutions of higher education or research institutes.

Science parks, first created in the 1960s, are now found all over the world, although they continue to be concentrated in developed countries. While early parks were often simply real estate developments, recent generations have focused more on services—particularly on business services to start-up

Table 10.2 Encouraging Innovation through SEZs

	Policies	Examples
Fostering links	Attractive conditions and incentives that make it cost effective to use local content	Korea, Rep.; Taiwan, China
Increasing domestic capabilities	Investment in training and technology upgrading of domestic workforce to match; allowing or encouraging domestic firms to have same access to hardware (machines) to improve production (Korea)	Skills Development Fund (Singapore), Penang Skills Development Center (Malaysia), Satellite Relations Program (Taiwan, China), Intel Corporation (Costa Rica)
Stimulating labor circulation	Encouragement of placements in local firms by managers inside SEZs, lifting restrictions on labor circulation	Shenzhen; Taiwan, China
Accompanying investment climate reforms	Strengthening of the overall national investment climate outside the SEZ so that domestic firms can flourish	All successful, dynamic SEZs
Emphasizing location	Importance of physical location (proximity to economic hub) and infrastructure links	All successful, dynamic SEZs

Source: Author.

companies, as well as leisure services for tenants—to increase their impact and attractiveness to businesses and their employees (see figure 10.2) (Sanz 2004).

Typically, the science park of today can be defined by four "functional components" and several physical components. The functional components include the following: (a) *businesses:* established MNCs, domestic companies, and start-ups in various combinations; (b) *knowledge providers:* university research and education infrastructures, applied research labs, and facilities usually handled by public bodies; (c) *industry support services:* business incubators and enterprise development areas, usually managed by private operators; and (d) *financial support services:* venture capital, regional development agencies, or banks. The physical components include infrastructure development, office buildings, meeting rooms, transportation, power, and ICT connectivity.

The combination of functional and physical components promotes economic development and competitiveness by creating new business opportunities and adding value to mature companies, fostering entrepreneurship, incubating new innovative companies, generating knowledge-based jobs, and building attractive spaces for knowledge workers. Many developing countries have created science parks to obtain technology transfer, skills, capital, and exposure to MNC research for both universities and domestic companies (see box 10.5); to create employment for graduates with advanced degrees who often do not find employment otherwise; and to slow brain drain.

An Urban Phenomenon

Statistics from the International Association of Science Parks show that science parks are an overwhelmingly urban phenomenon: over 66 percent are in

Figure 10.2 Evolution of Science Parks over Time

early parks: stand-alone physical space	1990s: connections	2000 and beyond: economic driver for the region
• real-estate operations • campus-like environment, selling single parcels of land • focus on industrial recruitment • few, if any, ties between tenants and university or federal laboratories • little business assistance and few services provided	• anchor with R&D facilities aligned with industry focus of park • innovation centers and technology incubators more common • multitenant facilities constructed to accommodate smaller companies • some support for entrepreneurs and start-up companies provided directly	• more and more mixed-use development, including commercial and residential • increased focus and deeper service support to start-ups and enterpreneurs • less focus on recruitment • formal accelerator space and plans for technology commercialization roles emerging • greater interest on part of tenant firms in partnering with universities • universities more committed to partnering with research park tenants • amenities from day care to conference and recreational facilities added

Source: AURP/Batelle 2007.

Box 10.5 Zhongguancun Science Park in Beijing, China

Zhongguancun, located in Beijing, China, comprises seven separate science parks, with about 17,000 firms. Beijing boasts a very high concentration of skills: 37 percent of the members of the Chinese Academy of Science and the Chinese Academy of Engineering, two-thirds of the country's PhDs, and a pool of young graduates from some of the best universities in the country (39 universities and 213 research institutes as of December 2006). Beijing is undeniably a knowledge city.

Source: Kuchiki 2007.

---

a city and 27 percent near a city, in part, for the same reasons that make cities attractive, including the concentration of talent and the location of universities. A successful science park should also be *integrated* into the host city. Several relevant factors need to be taken into account:

- *Land policy.* Correlating land availability in a metropolitan area with reasonable development ambitions over a 10-year time frame is an important factor in the choice of location. Ideally, a science park should be part of an urban economic planning map (typically drawn up on a 20-year time horizon). Commercialization—that is, selling or leasing space to companies—in science parks can be slow (1–2 hectares a year), and this pace should be recognized at the outset.

- *Ease of access.* A good but hard-to-reach site will have trouble attracting tenants, as entrepreneurs need to be able to get to their clients, suppliers, and scientific or technical partners rapidly. Many otherwise excellent science parks suffer from transportation-related problems.[6] Experience shows that entrepreneurs and their partners and clients like to exchange knowledge and ideas in restaurants, cafes, and "culturally vibrant" areas, which are often in city centers.

- *Presence of universities or research labs.* Although universities or research labs are a central feature for the park, creating a science park adjacent to a university is not enough to ensure success: spatial proximity does not necessarily guarantee networking or connections (see Saxenian, Bresnahan, Gambardella, and Wallsten 2001). Among other factors, the targeted university or research centers must be open to collaboration with firms.

- *Affordable housing.* Apart from location, another influential factor is affordable housing for entrepreneurs and their families. For this reason, affordable housing is now included in many new science parks.

The idea of integrating science parks into existing local networks and city or regional development strategies is closely related to that of location. Some 65 percent of science parks worldwide are located near business clusters. Many are also an integral part of a cluster and have sometimes even spurred its development (Silicon Valley's science parks, for example). Those considering whether to create a science park should weigh, among other factors, its location and design, as well as ways to support users, link with networks, secure financing, and ensure sustainability. Science parks can also be designed as part of an overall city or territorial economic and infrastructure development strategy (see box 10.6).

Box 10.6 Greater Sfax Development Strategy and the Science Park

Sfax's science park (Technopole de Sfax) was set up in Tunisia in 2006 to promote computerization and multimedia. A number of institutions and a research center specializing in computerization and multimedia offer training and research activities in the park, which is creating much-needed employment for the city's highly skilled but unemployed or underemployed workforce. Sfax is a major university center in the central part of the country with 20 institutions of higher education and 44,000 students enrolled in various subjects (science, technology, information technology, arts, and engineering). The science park has been designed as an integral part of the Greater Sfax Development Strategy, which maps out the city's integrated development to 2016. The science park is only one of the economic development tools leveraged for its development.

Source: Djeflat, http://www.investinTunisia.com.

A science park will typically go through various phases in its design and build out:

- *Strategy.* At the local level, the integration of the science park into other aspects of the host city should be carefully considered, particularly in developing countries. Aligning the proposal with the needs of the users of the park is the most important aspect of its strategic positioning and strongly increases its chances of success. Gap analysis, technology foresight,[7] road mapping, and needs-assessment techniques are useful for positioning the science park as a tool for local development.

- *Planning.* It is important to plan and pace in detail how to develop the infrastructure and equipment and to establish the governance rules that will support the strategy's implementation.

- *Action.* The action phase includes developing land tenure policies, centers (resource centers, technology transfer centers, incubators), real estate development (for rental, for sale, business centers, possibly housing), urban integration (ICT infrastructure, transport) that will make the science park a reality.

- *Commercialization and promotion.* Service and commercialization strategies are necessary for making the science park truly functional and should not be neglected. In this respect, the Finnish umbrella organization, TEKEL, which acts as overall coordinator and promoter of science parks in Finland, is noteworthy (box 10.7).

- *Evaluation and performance indicators.* As with all projects, a science park should receive regular evaluations and make necessary adjustments. Performance indicators used to measure and benchmark science parks include attractiveness (occupation of the park), employment, training, research and development (R&D), and enterprise data (number created, enterprise growth).

Box 10.7 TEKEL

TEKEL is an umbrella organization for Finnish science parks. It acts as an expert in and promoter of science park activities, as a national coordinator of network-based cooperation, and as a facilitator, creating and maintaining connections with the public sector, the business community, the education and research sectors, and international networks in the field. The operational impact of TEKEL extends to 14,400 companies, 2,400 of which are based in science parks.

Source: TEKEL, http://www.TEKEL.fi.

Science park clients are typically technology-intensive small and medium enterprises (SMEs) or MNCs and require a range of services, particularly for SMEs in their early stages of development. Quality and cost effectiveness of services are important for attracting prospective businesses or tenants. Chapter 3 treats support to users in further detail.

Financial Sustainability and Consensus Building

Several factors are necessary for ensuring the sustainability of a successful science park:

- It must receive external support, such as the commitment of university leadership and acceptance by the local economic development community.
- It needs financing for building construction and for start-up and equity capital.
- It should enhance the ease of doing business and the harmonization of intellectual property, patenting, and licensing.
- The park should take advantage of what the local economy has to offer and should be integrated into the community.
- The science park should help build the brands and increase the international recognition of its constituent businesses and organizations.

Some of these factors were addressed earlier; financial sustainability and consensus building are discussed below.

A science park is a costly investment, requiring funding at different stages of development, from conception to operation and production. It is necessary to finance both the physical infrastructure and some or all of the funding for the projects and companies located in the science park. Funding options for infrastructure depend primarily on the institutional arrangements of the park and the respective roles of the public and private sectors. That said, there are three basic funding options for physical infrastructure in science parks:

- *Public funding.* During the start-up phase, the public sector often funds basic infrastructure such as roads, electricity, water and wastewater, real estate development (for teaching, laboratories, and office use), and green spaces.
- *Joint public-private funding.* The private sector focuses primarily on the infrastructure and tangible assets of companies, as well as on some general services (administration, communication). Through tax incentives, however, the public sector may attract private infrastructure investment in the science park.
- *Loans and guarantees.* International organizations and development finance institutions occasionally fill the gap in long-term funding for basic infrastructure investments in science parks. They can also provide guarantees for senior and subordinated debt, credit lines, bond issuance, and the like.

The financing of companies and projects within the park can include public seed money and start-up capital, as well as private provision of funds such as venture capital.

To be financially sustainable, a science park must derive a certain amount from various revenue streams:

- *Use or sale of physical infrastructure.* Use or sale of physical infrastructure refers to rental or purchase of office space or land and use of telecommunications facilities.
- *Use of technical areas.* Technical areas include meeting and training rooms and the like.
- *Use of technological facilities.* Technological facilities include testing, experimentation, and research facilities.
- *Services.* Pricing policy should take the local context into account. Careful planning of occupancy projections is necessary, as financial difficulties will lead to reduced services for current tenants and jeopardize medium-term sustainability.

A science park should also contribute to broader economic objectives, including local employment and the increased competitiveness of local firms. The broader economic objectives cannot be met, however, if the short-term financial ones are neglected.

Clearly, a successful science park depends on a wide community of support and participation. It should seek to balance the interests of all major stakeholders, including the area's industrial, scientific, and financial leaders; representatives of business associations; potential tenants; city, regional, and national government; community organizations; and educational and academic institutions. Ideally, these stakeholders should participate in the strategic positioning of the science park and develop a sense of shared ownership and responsibility in the implementation phases. Beginning with a workshop or conference, this effort can continue through involvement in research and surveys on available local and national resources.[8]

Consensus building must result in a willingness to move forward together. Although public policy has an important role to play in the development of a science park, it is an insufficient driver on its own. Experience shows that the contributions of active networks of local champions with the ability and motivation to organize themselves—along with the dynamism and leadership of science park management—are also important factors in success.[9]

Clusters

Cluster initiatives are increasingly used for economic development in both developed and developing countries and are supported by the development community at large.[10] Popularized through *The Competitive Advantage of*

Nations (Porter 1990) and ensuing publications (OECD 1999a, 2001, 2007a), clusters have been viewed as an instrument for enabling firms to join their efforts and resources with knowledge sources and government for greater regional, national, and international competitiveness. Although clusters are not necessarily innovation systems (Oyelaran-Oyeyinka and McCormick 2007) and innovative clusters are not necessarily "high-technology" clusters, cluster initiatives may be one of the most effective means for producing an environment conducive to innovation (Anderson and others 2004).

This section defines, identifies, and discusses the rationale for clusters and describes some policy measures that support clusters and evaluate them. It also describes some lessons learned from the evaluation of cluster initiatives in developing countries.

Definition and Purpose

A cluster is a geographic concentration of interconnected companies in a particular field with links to related organizations such as trade associations, government agencies, and research and educational institutions (USAID 2008). It gives rise to external economies (specialized suppliers or pools of sector-specific skills) and favors development of specialized services in technical, administrative, and financial matters (OECD 1999a, 2007a). Clusters typically differ from science parks in several ways:

- They tend to have a sectoral specialization, which science parks do not necessarily have.
- They do not have the urban development or physical infrastructure component of science parks.
- They need not have the high-technology focus often associated with science parks.
- They tend to be larger and have more stakeholders than science parks.

Entertainment in Hollywood, fashion in Milan and Paris, information technology in Bangalore, and financial services in London and New York are well-known clusters. The California wine cluster, for example, comprises grape growers and wineries, research and education providers, and state government agencies. It is closely related to the tourism, food, and agriculture clusters.

Supportive Policies

The literature includes a very vigorous debate on cluster origination and the policies, if any, that should be used to support them. Indeed, the presence of potential benefits from cluster initiatives does not in itself justify policy intervention. A yes-or-no view on the issue of public intervention can be counterproductive (Hamdouch 2007); the diversity in the emergence and support for clusters argues for pragmatism and caution. Some clusters have a science and technology logic, some a technology and manufacturing logic; some rely on path dependencies or on the catalytic effect of returnees or a diaspora of the highly skilled (Saxenian 1999, 2006; Kuznetsov 2006), all with different degrees of political or institutional "voluntarism" (see box 10.8).

In fact, clusters can emerge and develop as a legacy of the past or as a more voluntary attempt to create a new future, or both. It is difficult, however, for policy makers simply to create a successful cluster. Indeed, Mytelka (2007) points out that a spontaneous cluster may be more likely to foster new habits, learning practices, links, and continuous innovation than constructed clusters. Once identified, however, clusters can be nurtured through policy intervention (OECD 2001).

Cluster Mapping

One of the first steps before any policy intervention is to understand the local economy and map existing, potential, or dormant clusters. This effort is particularly important in developing countries. Focusing on clusters can in fact help local and national governments better understand how their local economies work. Indeed, by looking at an economy through the lens of various clusters, local governments can more accurately identify market imperfections, detect systemic failures, and better tailor policies.

There are many ways to identify and map clusters and compare their relative scales and concentrations. While analytical tools are valuable starting points, the results of even the most rigorous methodologies will be no better than the quality of the input data, which are usually quite soft. Ultimately, one must also depend on local observers to identify latent or dormant clusters (Rosenfeld 2002). Cluster identification typically relies on several specific tools or processes:

- Research studies and reports that analyze key industries, industry sectors, and related emerging and global trends
- Analysis of local assets, including the knowledge base, natural endowments, and the like

---

Box 10.8 Different Cluster Initiators

- *India, Kazakhstan, Qatar, Thailand:* Key role played by local industrial bank
- *Central America:* Private (individual) initiative as catalyst
- *Canada and Western Europe:* Important role of chambers of commerce and the government at different levels
- *United States:* Private sector.

Source: Author.

- Analysis of stakeholders to find those that identify strongly with a cluster
- Analysis of critical barriers, gaps, and opportunities for growth in existing or emerging clusters.

Once clusters are identified, more sophisticated tools can model and map them and the relations among cluster members.

Successful Clusters

There is neither a standard recipe for the success of a cluster nor a simple set of best practices. Clusters evolve, operate, and are "embedded" in specific geographic, cultural, social, regulatory,[11] spatial, and institutional environments. This complexity has led some analysts to insist on the futility of "recipes for success" based on "success stories," as they are likely to fail, as in the case of defunct Silicon Valley copycats (Brookings 2006), if they are not adapted to the local context (Saxenian, Bresnahan, Gambardella, and Wallsten 2001). It is therefore important to keep in mind that all situations are unique and that most processes involve trial and error. Furthermore, clusters are not an end in themselves but one tool among many that can promote increased competitiveness, innovation, and, ultimately, economic growth.

In certain circumstances, government policies that facilitate networking, catalyze comparative advantages, and build effective institutions, as well as nurture the more general environment for innovation, can help a cluster gain momentum and improve both its efficiency and its capacity for innovation. Policies used to support clusters fall broadly into two main categories: improving cluster dynamics and improving the cluster environment, including evaluation of cluster support mechanisms and initiatives.

Firms in clusters do not necessarily share or circulate an economically optimal amount of knowledge and information. Policy measures, called "broker policies," attempt to establish an effective framework for dialogue and cooperation between firms and between firms and relevant public sector actors (particularly in local areas and regions) or other agencies. These measures include the following, among others:

- *Creation of platforms for dialogue and networking between firms and other stakeholders.* The construction of meeting spaces, support to institutions for collaboration, and the encouragement of networking in a broad sense, including firms but also institutions, are examples.
- *Support for the creation of knowledge-enhancing partnerships between firms and other institutions.* Public-private partnerships in specific fields (health, environment) can involve cooperation and possible coinvestment of resources. An award or other incentive structure can also encourage collaboration between universities and industry. Science parks can help in this respect and are often part of a cluster.

- *Intelligence.* Information and data on cluster-specific businesses and on economic and technological trends can be disseminated.
- *Statistics and data.* Standard statistics often fail to cover many structures and links that are crucial for measuring and understanding cluster development (Anderson and others 2004). Employment, enterprise creation, growth rates, and market projections can help clusters and government plan and design programs. For a more in-depth and focused analysis, cluster benchmark studies and status reports can be useful (Rosenfeld 2002).

Specific public policies can also support the upgrading of skills in clusters:

- *Links with vocational training programs.* These include on-the-job training with a focus on practical elements. Clusters may provide a critical mass of related needs for upgrading skills. Public initiatives may prompt firms to identify their needs and broker supply arrangements with educational institutions.
- *Cluster skills centers.* Once a cluster's needs are known, there may be a case for establishing a skills center to meet them if other providers are unable to do so or could do so only in partnership with existing providers. Such centers can survey industry needs, develop new curricula, and update skill standards. Emphasis should be on industry-specific knowledge, not on job-specific skills.
- *External learning.* Clusters that focus exclusively on internal training can cut themselves off from external sources of learning. Without access to benchmark practices and markets—particularly in less-favored regions and developing countries—clusters can limit learning.

Some services to innovative firms are particularly relevant in a cluster context:

- *Cluster technology centers.* Typical functions of cluster technology centers include applied R&D, testing and quality standards, technical advice, network brokering, technician and management training, and technical studies.
- *Cluster-based incubation.* Incubators are commonly used to support new and small business ventures. Within a cluster context, business incubation can be particularly powerful as similarities or complementarities of firms justify more highly tailored services and assistance and generate intra-incubator activity.

A common strength of clusters is their ability to pool resources and efforts to reach markets effectively. Policy makers often help by making data and information on markets (and eventually technologies) available to clusters. On the demand side, public procurement can also be a powerful tool, given its volume in most countries. It has a strong potential for developing and strengthening clusters, especially when pursued consistently over an extended time. Public procurement policies, however, are now regulated by international trade agreements and may also entail important risks by making clusters overly dependent on public demand or excessively focused on meeting that demand. In addition, the public sector is not always the most forward-looking or sophisticated customer. Finally, overspending and failure to stop supporting failing projects can be a problem, as can corruption. In some cases, though—as in Silicon Valley—public procurement has spurred new forms of collaboration and generated innovative goods and services. If clusters can be encouraged to produce for demanding consumers (box 10.9), they are likely to become less dependent on public providers and reduce some risk.

Improvement in the Efficiency of Clusters

Further means of improving a cluster's efficiency include international links, different framework conditions, and evaluation of the cluster's performance. Clusters' international links can be promoted, thanks to the strengthening of international trade and the recent improvements in transportation and communications systems. Such links can improve the access of firms, notably smaller businesses, to the wealth of global product and process knowledge, which they would otherwise have difficulty in gaining. Public cluster initiatives can use the attraction of FDI to strengthen the resource base and obtain access to leading-edge technologies and skills.

Outward FDI can also be used to enter foreign markets and gain access to pools of technology and skills. Some studies have shown that expansion abroad tends to be accompanied by higher competitiveness, productivity, and R&D in home operations as well (Van Pottelsberghe de la Potterie and Lichtenberg 2001). Support of export networks and coordinated purchasing, underwriting of delegations to international trade shows, public sponsorship of joint marketing initiatives and branding (including regional and product branding), and other services can also improve the effectiveness of clusters.

Box 10.9 Demanding Local Consumers

In Sub-Saharan Africa, because the productive capacity of many clusters is suboptimal, they remain locked into low-quality, low-income markets. One way to build a cluster's productive capacity is to have it fully engaged in producing for demanding consumers. Although entering the export market would achieve that goal, very few are ready to make that leap, because of production scales and capital and metrology, standardization, testing, and quality assurance issues, among others. McCormick and Kinyanjui argue that encouraging clusters to produce for demanding, high-volume local customers such as supermarkets, hospitals, and schools can enhance productive capacity.

Source: McCormick and Kinyanjui 2007.

Framework conditions for optimal cluster functioning include macroeconomic stability; product and factor markets; educational systems; and physical, institutional, and governance structures conducive to innovation. While developing, emerging, and transition countries may display some specific weaknesses in these areas, many have launched reforms to counteract them, notably through guidance from different World Bank investment climate assessments and *Doing Business* reports. In some countries, the interplay between formal practices (contracts, structured hierarchies, and public regulations) and informal practices (norms, routines, traditional authority, and expectations) is complex. Critical issues include whether there is conflict or complementarity between formal institutions and informal value systems. In addition, because trust is particularly important to cluster initiatives, those that can reinforce social capital and the attitudes that influence trust between stakeholders are also important.

Measurement of Performance

An array of indicators can be used to evaluate clusters. Economic goals, such as employment, are valid, of course, but because they are subject to a number of influences, they can be problematic short-term indicators. Changes in the business environment, the quality and quantity of products exported, increased collaboration between firms, and training accomplished, among others—especially in those areas targeted by cluster initiatives—are candidates for evaluation more directly related to policies.

Operational performance is a direct reflection of the quality of cluster policy, although it is not a policy goal in itself. Cluster initiatives may collect performance data to demonstrate quantifiable results to government officials or donors and to make better project management decisions. Other examples of evaluation and performance measurement include the creation of databases for tracking, external and internal assessments, satisfaction surveys, demand assessments for cluster products, and impact surveys.

Through international surveys of cluster initiatives (http://www.clusterresearch.org), data are available to benchmark cluster initiatives. The creation of an online survey tool enables local and regional governments worldwide to use the survey results to take stock of how cluster stakeholders view their situation and what can be done to improve it (http://www.clustercompetitiveness.org).

Although evaluations or even surveys of cluster initiatives in developing countries are few and far between, some evaluations of clusters in developing countries and surveys offer important recommendations (USAID 2003b, 2005, 2008):

- Recruit highly committed leadership.
- Develop a strategy to ensure adequate resources throughout the process.
- Choose the right geographic level of focus—regions, cities, states.

- Find tools to sustain momentum between stages.
- Engage potential implementing institutions from the earliest stages of the process.

The evaluations point out that the cluster-based approach to economic development in developing countries can help those countries move from an often compartmentalized and isolated activity, which focuses on one project at a time, to an integrative and enduring process. These evaluation tools can help engage the region's key suppliers in a dialogue with their customers, link local education providers with workforce managers, connect technology providers with product developers, shape physical infrastructure to meet industry's operational needs, and match financial investors to new or existing enterprises. It is important, however, to take a long-term view.

Fostering Innovation in a City or Region

A city is a complex system, and fostering innovation there requires specific actions to bring out its strengths and capabilities and to address its weaknesses. Innovation policies can play an important role in cities, notably in the developing world.[12] Innovation plays an essential role in enhancing a city's attractiveness relative to others, but some cities seem better able to innovate than others.

The world's population is increasingly concentrated in urban areas: over half the entire population lives in cities today, and about 70 percent are predicted to do so by 2050, with those in Asia and Africa registering the biggest growth. Overall, the world's population is expected to increase from 6.7 billion in 2007 to 9.2 billion in 2050, and the population living in urban areas is projected to rise from 3.3 billion to 6.4 billion, overwhelmingly in developing countries. According to a United Nations report, "The urban areas of the world are expected to absorb all the population growth expected over the next four decades while at the same time drawing in some of the rural population" (UN 2008).

At the same time, globalization has led to greater possibilities for cities to integrate into the world economy, including in the developing world,[13] resulting in increasing competition among cities, both domestically and internationally (Camagni 2002; Scott 2006; World Bank 2008). In this context, a city's ability to innovate has become crucial in determining its relative dynamism and development. Beyond the static natural advantages that may characterize and differentiate cities, innovation policies can help local authorities actively foster competitiveness. Innovation or innovation-related features stand high in the "city rankings" (such as Kiplinger's) that have proliferated, highlighting the increasing competition among cities.

Innovative activities contribute to an area's attractiveness and competitiveness, and research tends to find them spatially concentrated around cities.

Process and product innovation, as estimated in some countries, seems to be as spatially concentrated as technological innovation (NESTA 2007b). In this respect, innovation is a quintessentially urban phenomenon. To conclude that urban expansion automatically leads to innovation would be misleading, however, as some urban areas are more successful at fostering innovation than others, such as Oxford and Cambridge in the United Kingdom (NESTA 2007a, 2007b).

Urban innovation is not the privilege of developed countries. Bangalore and Hyderabad in India; Beijing, Shanghai, and Shenzhen in China; and Dubai in the United Arab Emirates, to name but a few, are all testimony to the increasing capacity for technological innovation in cities in developing countries.

Cities as Greenhouses for Innovation

The connections between innovative activities and cities are difficult to pinpoint, since innovation is systemic in nature and influenced by a multiplicity of local factors. In line with the metaphor of gardening used in chapter 1, a city can be considered a greenhouse, with the *ingredients* of innovation (firms, human capital, institutions) fortified by *nutrients* (infrastructure, cultural environment) that enable the formation of veritable urban innovation systems. Cities' innovation efforts, however, may be choked by specific *weeds*.

Firms develop new ideas and adopt and adapt those of others, which they then bring to the market: these new ideas are at the heart of the innovative process. Firms are attracted to the urban environment by the presence of many public goods and positive externalities (Marshall 1920). *Human capital* is also concentrated in cities, and a critical mass of creative and talented people is particularly important for innovation. Workers come to cities for many reasons: cities are "agglomerations of consumption" (Storper and Manville 2006), with goods and services converging in a small area; workers find jobs more easily and can hope for better salaries because of the density of firms; urban areas concentrate cultural entertainment, which is particularly important for the "creative class." Finally, *institutions* support local and regional innovation, not only by supplying many of the assets that underpin it but also by actively supporting, facilitating, and shaping the public goods that are essential for firms and the development of human capital.

Furthermore, cities offer specific "nutrients" that foster innovation. *Infrastructure* and other public goods available in cities, for example, are a substantial benefit to innovative activities. Roads, hubs, airports, rail links, and other public transportation (bus, subway, rail) all help improve innovation potential while strengthening and facilitating relations among the actors in the innovation process. Good transportation and services bring firms and institutions closer in time and distance, advancing collaboration, market transactions, and networks. Such links also increase the potential size of markets for goods, services, and labor. In addition, their density of ICT infrastructure in cities is a major advantage, providing faster and larger connections to international communications networks.

Urban planning factors, such as the availability of land and property, also affect innovation: by "designing" permissible forms of development, they largely determine how firms configure production and distribution. City infrastructure is constantly evolving, however, and requires not only continuing investment to remain functional but also new infrastructure as the city develops. Local authorities must have a strong capacity to adapt, if they are to create and maintain relevant and functional infrastructure (Léautier 2006). Urban master planning is necessary for preparing for future infrastructure development, as the costs of urban infrastructure can rise exponentially as the city fabric becomes denser.

According to some, cultural vitality, ethnic diversity, and social tolerance, all found mainly in urban environments, are essential to the creativity associated with innovation. Richard Florida's "creative class" (2002b) is supposedly attracted and held by urban environments (see box 10.10), an observation that has spurred a series of highly visible "cultural" investments in cities worldwide (Beijing, Bilbao, Budapest, London, Manchester).

Urban governance, which includes *accountability* (how cities manage their finances and communicate their achievements and use of funds to citizens) and *responsiveness* (the ability of a decentralized entity to determine and respond to the needs of its constituents), is important for urban innovation (Léautier 2006). This factor indirectly points to the sharing of power between central and local government and the importance of clear boundaries and responsibilities among the different levels of government.

In addition to these ingredients and nutrients, cities offer the additional benefits of concentration, proximity, density, and diversity (Florida and Gates 2001) and engender multilayer networks. Thus, firms in cities tend to form clusters of economic activity: vertically disintegrated networks of production units tied together in relations of specialization and complementarities. Cities

---

Box 10.10 The Creative Class?

Richard Florida's best-selling book, *The Rise of the Creative Class: And How It's Transforming Work, Leisure, Community and Everyday Life* (2002), describes a highly mobile, creative class, on which a city's fortunes increasingly turn. The capacity to attract, retain, and even pamper mobile and finicky "creatives," whose aggregate efforts have become the primary drivers of innovation, is a preoccupation of cities everywhere and has become a public policy phenomenon. A notable example is the government of Singapore's move to relax its restrictions on homosexuality, and for that matter busking and bungee jumping, in the name of encouraging urban economic innovation.

Source: Author.

also facilitate the creation of face-to-face and social networks, such as business associations and professional associations, as well as virtual and other types of networks (Sassen 2006). Many studies point to the importance of building trust and relationships and of making a place for "face time," which is critical in the knowledge economy (Storper and Venables 2004).

While cities provide a particularly nurturing environment for innovative activities, however, urban development may entail costs that can be highly detrimental to innovation. Diseconomies of scale can arise when the population increases beyond a certain size. Research has shown, for example, that the relationship between income and population becomes negative for cities with a population of more than 6 million (OECD 2006). Congestion, including increased commuting times, transportation and logistics, rentals, and environmental costs (pollution, traffic, circulation, water quality, noise, lack of green space) can negatively affect firms' activities, the functioning of networks, and human capital. In many developing countries, these negative effects may exist in cities of well under 6 million.

Moreover, while city-regions are at the forefront of employment and wealth creation, they also tend to concentrate a higher number of unemployed workers. In OECD countries, for example, around 47 percent of unemployment is concentrated in urban regions, and it is above 60 percent in Japan, Korea, the Netherlands, the United Kingdom, and the United States. In fact, most large cities, including the wealthiest ones, have large pockets of population with social problems and low standards of living. A main consequence of urban inequalities may be a higher level of criminality (OECD 2006). Social exclusion is often associated with strong residential segregation between the prosperous and the populations that live in deprived neighborhoods and derelict suburbs. The UN Habitat's *Global Report on Human Settlement: Enhancing Urban Safety and Security* (2007), underscores the importance of safety for a city's economic prosperity.

From the viewpoint of innovation, large-scale social exclusion may affect the activity of firms owing to labor skill mismatches and restricted market access. Underemployment or unemployment, criminality, and spatial polarization may also tarnish the image of the city and drive away the highly skilled, often more mobile, population.

Stimulation of Innovation in Developing Country Cities

Besides the specific tools available for fostering innovation, such as SEZs, clusters, and science parks, cities can also learn from both the positive and the negative factors that affect innovation in an urban context. Such lessons are crucial in developing countries, where they may help shape the creation and implementation of dedicated innovation policies and address weaknesses in infrastructure and in urban planning, human capital, governance, and image issues.

A major bottleneck of firm activity in developing countries, for example, is the low density, multinodal structure of their cities, aggravated by the limited availability of transportation (OECD 2008). On the demand side, high transportation costs and low consumer mobility inhibit market access for firms. On the supply side, poor infrastructure hampers the constitution of interfirm networks and clusters, while intensifying the mismatch between the employment centers and the workers' housing. Poor transportation infrastructure also means that people use cars, private minibuses, tuk-tuks, motorbikes, and the like, thereby creating close to anarchy on small roads and exacerbating traffic congestion and pollution. Proactive urban planning policies can therefore be considered important to enhance for more general citywide innovation policy actions, based on principles that explicitly aim at enhancing competitiveness.

A *land market monitoring system* may ensure sustainable housing. Experience in Korea and Japan, with regard to land pooling and readjustment programs to regularize informal settlements and mark off public and private land, could be useful; however, such policies do not in themselves enhance the housing supply but can be used in support of other policies. To avoid sprawl, developers can be given incentives to build on smaller lots in central areas.

With enterprises in developing countries increasingly concentrated in cities, meeting rising labor needs will require improving the skills of the labor workforce upstream and rationalizing the local labor market. Workers' inadequate skills and insufficient training point to deficiencies in the general education system. While education policies have consequences for social development that go well beyond economic competitiveness, innovation-related concerns are outlined in chapter 6.

Human capital mismatches also result from poor diffusion of knowledge to economic activities or from insufficient integration of the labor market into the wider economy. Better links between sources of knowledge and the labor market is at the heart specific innovation policies, such as SEZs, clusters, and science parks. Whatever the tool adopted, prior identification of regional competitive advantage is important.

The overwhelming importance of the informal economy, which typically represents a sizable share of the economic activity in developing countries, constitutes a major challenge to labor market integration. One possible approach is to see informal activity as an incubator for entrepreneurship in poor areas. Formal and informal firms are also likely to interact along the production process. Local labor market policies could use formal-informal firm networks to give informal firms a way to integrate into the formal sector. Like Naples (see box 10.11), Bogotá and Istanbul have designed specific programs—ranging from market constructions for relocated vendors to zoning for informal commerce—to induce informal workers to integrate into the formal labor market.

> **Box 10.11 CUORE in Naples**
>
> In the Neapolitan economy, the informal economy represents a very high share of total economic activity—perhaps as much as one-quarter of total employment—and is plagued by organized crime and localized poverty traps. In 1999, the municipality and the University of Naples Federico II set up the Urban Operational Centers for Economic Renewal (CUORE) to develop cooperation between the state and informal enterprises. The project consists of a network of neighborhood service centers for entrepreneurs. A team of specially trained young professionals are in charge of identifying informal firms and helping those willing to change their status. They offer them interesting incentives, including marketing support (for example, participation in a trade fair), cooperation with other firms, or legal assistance. While the extent of organized crime has so far limited results, contacts by informal firms with the CUORE have helped clarify the motivation or pressures that explain the choice of the informal sector.
>
> *Source:* OECD 2008.

As for governance, city innovation policies should ideally be part of the larger framework of national strategies for innovation. While implementation may take place locally, such important measures as taxation, trade policy, public funding of research, and regulation of the business environment all take place nationally. This factor may be challenging in developing economies, where institutional, financial, and political constraints may limit the autonomy of planning agencies or local authorities.

It may be useful to put a single agency in charge of coordinating regional planning at the city-region level to articulate and implement national and subnational strategies favorable to local innovation and competitiveness: housing, land use, transportation, and the labor market (OECD 2008). The agency can comprise municipal, regional, and national representatives, the crucial point being that local government should be represented to ensure agreement on the policies to be implemented. According to the subject, trade unions and representatives of firms might also be included in the decision-making process. Box 10.12 provides a topical example of a successful approach.

Besides dedicated innovation policies, which are often decided nationally, cities have some scope for local action to increase their city's attractiveness and competitiveness. In the specific context of development, the so-called "urban entrepreneurialism" approach, notably used for rehabilitating old industrial "rust-belt" cities in the United Kingdom (OECD 2007b), offers some ideas relevant for cities whose negative image discourages business and investment. In an attempt to promote growth-oriented strategies, the entrepreneurial paradigm is concerned (among others) with enhancing the image of a city through appropriate marketing, domestically and abroad, to attract capital and skilled workers. In this sense, it takes due account of the emphasis laid on the creative

> **Box 10.12 The Vancouver Agreement**
>
> The Vancouver Agreement is an urban development initiative that has brought together the governments of Canada, British Columbia, and Vancouver since 2000. The parties are committed to working together with communities and business on a coordinated strategy to promote sustainable economic and social development, with pooled resources. A governance committee is composed of members appointed by the different levels of government, while implementation is facilitated by a working group of senior managers of public agencies. A wide range of representatives from the different partners regularly meet to tackle issues identified in common. One of the objectives is to promote innovation, while streamlining the way public agencies work, in partnership with the private and nonprofit sectors. A key component of the agreement is also the implementation of a business cluster strategy to leverage economic and employment opportunities for inner-city residents, focusing on construction, business services, and tourism.

class to foster innovation and competitiveness. One of the first steps in actually designing local competitiveness policies consists of carefully auditing the strengths and weaknesses of a city. Based on case studies, Cities Alliance has compiled appropriate methods for "city auditing" (2008).

The renovation of old town buildings or the launching of development programs can modify a city's identity and provide physical evidence of dynamism at moderate cost, especially as the private sector can be associated with the project. As part of an all-encompassing branding strategy, a cultural event can increase a city's visibility, while providing local work and new skills.

Conclusion

Developing innovative local sites and integrating them into a wider city-region innovation strategy can be rewarding for developing country governments. Such sites may boost local innovative capacities, create employment and growth, and have important nationwide demonstration effects.

Developing an innovative special economic zone potentially involves a coordinated set of policies, many of which aim at upgrading domestic conditions. These policies may focus on links and spillovers between firms in the SEZ and firms outside, in particular by building domestic capabilities in local firms and domestic labor force training to take advantage of spillovers. Fostering labor circulation from the SEZ to the domestic economy can also help create spillovers. Experience suggests that reforms of the domestic investment climate, to emulate to some extent the conditions of the SEZ, can help domestic firms grow. Finally, it is important to choose the location of the SEZ carefully.

Science parks have flourished in developing countries in recent years. A successful science park requires gaining the commitment of university leadership and acceptance by the local economic development community. Financing must be obtained (for construction, start-up, and equity capital), and conditions must be attractive (ease of doing business, harmonization of intellectual property, patenting, licensing). Other areas requiring attention include zone design and services, specialization, and branding and international recognition.

Policies that support *clusters*—including broker policies, training, support services to firms, and demand-side policies, as well as helping establish international connections and export products or services—can also be effective for fostering local innovation.

The importance of the location of these different sites, as well as their integration into the overall city-region economic area cannot be overstated. Integrating some or all of these tools into a more comprehensive city-region innovation strategy can increase the potential market for goods, as well as the labor pool of specialized workers, by taking advantage of the density and diversity of urban settings. Specific urban factors such as infrastructure improvement can also increase the effectiveness of such tools in an urban context. As cities are complex systems, however, urban features can also make it harder for firms to be innovative. A city's innovation strategy should be balanced and should, where possible, strengthen the city's advantages while recognizing and correcting its weaknesses.

Finally, both national and international policies have an important influence on what is permissible locally. Effective decentralization, a powerful facilitator of local initiatives, is ideally coordinated harmoniously with policy at higher levels of government. Different structural policies that frame the business environment (see chapter 4) and international trade policies influence the local level.

Notes

1. SEZs account for less than 1 percent of the global workforce, but significantly more in certain countries or regions. SEZs employ 4.6 percent of the active population in Honduras, 6.2 percent in the Dominican Republic, 8 percent in Tunisia, 10 percent in Fiji, 12 percent in the Seychelles, 24 percent in Mauritius, and 25 percent in the United Arab Emirates (FIAS 2008).

2. This broad definition of SEZ is in line with that of the Revised Kyoto Convention of the World Customs Organization. It encompasses free trade zones, export processing zones, free ports, and enterprise zones.

3. Some data from Taiwan, China, are interesting in this respect (see Omar and Stoever 2008, 149).

4. The World Bank's regular investment climate assessments and *Doing Business* reports can provide useful guidance on necessary national reforms.

5. This section uses the term *science park* in line with the International Association of Science Parks (IASP) definition to cover several different terms and expressions including science park, technology park, technopole, technopolis, technology precinct, research park (see http://www.iasp.org). As the IASP states, "Although there may be certain differences between them, projects under these aforementioned labels share many goals, elements and methodology, and have innovation at the core of their business."

6. For example, Paris-Sud, a science park in the southern residential outskirts of Paris, has been plagued by problems from the outset owing to poor transportation infrastructure and access.

7. The most commonly used technology forecasting model is the Delphi method. It is a systematic, interactive forecasting method that relies on a panel of independent experts.

8. Surveys have addressed the following questions: Who does what in terms of innovation and research in the region? What are the basic needs of firms? How can existing training programs contribute to development? What local resources are already in place that can be used? What type of financing is available? For a complete list of suggested questions, see the forthcoming Science Park guidebook (EIB, WBI, EuroMed).

9. The recruitment process, continuous training, and reward system of the science park management staff and their access to specialized competencies through external consultancies are potentially important factors to consider.

10. The community of international organizations and bilateral aid organizations is quite active in clustering in developing countries. The World Bank, UN Industrial Development Organization, U.S. Aid for International Development, and Sida (Swedish International Development Cooperation Authority)/VINNOVA broker between domestic actors to support cluster development. The European Union Commission has also supported cluster initiatives in new member states by allocating funds to regional actors for broad-based modernization of infrastructure and involvement in transnational R&D cooperation.

11. The regulatory environment affecting clusters covers funding organizations (banks, venture capital companies, business angels, public funding agencies); law firms (particularly those specialized in IPR); and regulatory bodies (standardization committees, ethical commissions).

12. Cities and city-regions are used interchangeably here to describe the same phenomenon, although city-region is the more appropriate term, as cities, as an administrative area, are often only a very small part of the urban area. In fact, city-regions often have several local level governments.

13. According to the GaWC report (2008), 14 developing country cities are included in the rankings: Mexico City, Moscow, and São Paolo (Beta World Cities); and Bangkok, Beijing, Budapest, Caracas, Istanbul, Jakarta, Johannesburg, Kuala Lumpur, Manila, Shanghai, and Warsaw (Gamma World Cities).

References and Other Resources

Anderson, Thomas, Sylvia Schwaag Serger, Jens Sörvik, and Emily Wise Hansson. 2004. *The Cluster Policies Whitebook 2004*. Malmo: International Organisation for Knowledge Economy and Enterprise Development (IKED).

AURP/Batelle. 2007. *Characteristics and Trends in North American Research Parks: 21st Century Directions*. http://www.aurp.net/more/FinalBattelle.pdf.

Basile, A., and D. Germidis. 1984. *Investing in Free Export Processing Zones*. Paris: Organisation for Economic Co-operation and Development.

Brookings Institution. 2006. "Making Sense of Clusters: Regional Competitiveness and Economic Development." Metropolitan Policy Programs Discussion Paper, Brookings Institution, Washington, DC.

Callanan, B. 2000. *Ireland's Shannon Story: A Case Study of Local and Regional Development*. Dublin: Irish Academic Press.

Camagni, R. 2002. "On the Concept of Territorial Competitiveness: Sound or Misleading?" European Regional Science Association (ERSA) conference papers series, ERSA, Vienna, Austria.

Cities Alliance. 2008. *Cities Alliance Annual Report*. http://www.citiesalliance.org/publications/annual-report/2008-annual-report.html.

Economist. 2000. "The Geography of Cool." April 13, 91–93.

FIAS. 2008. *Special Economic Zones: Performance, Lessons Learned, and Implications for Zone Development*. Washington, DC: World Bank.

Florida, R. 2002a. "The Economic Geography of Talent." *Annals of the Association of American Geographers* 92 (4): 743–55.

———. 2002b. *The Rise of the Creative Class.* New York: Basic Books.

———. 2005. "The World Is Spiky." *Atlantic Monthly.* October, 48–51.

Florida, R., and G. Gates. 2001. "Technology and Tolerance: The Importance of Diversity to High-Technology Growth." Center on Urban and Metropolitan Policy Survey Series, Brookings Institution, Washington, DC.

GaWC (Globalization and World Cities). 2008. *The World According to GaWC 2008.* http://www.lboro.ac.uk/gawc/world2008t.html.

Hamdouch, A. 2009. "Innovation Clusters and Networks: A Critical Review of the Recent Literature." Paper presented at the 19th EAEPE Conference, Universidade do Porto, Portugal, November 1–3.

ILO (International Labour Organization). 1998. "Labour and Social Issues Relating to Export Processing Zones." Paper prepared for the International Tripartite Meeting of Export Processing Zone-Operating Countries, Geneva, Switzerland, September 28–October 2.

———. 2003. "Employment and Social Policy in Respect of Export Processing Zones." Committee on Employment and Social Policy. GB.286/ESP/3. Geneva: ILO.

ILO (International Labour Organization)/UNCTC (United Nations Centre on Transnational Corporations). 1988. *Economic and Social Effects of Multinational Enterprises in Export Processing Zones.* Geneva: ILO.

Jenkins, Mauricio, Felipe Larrain, and Gerardo Esquivel. 1998. "Export Processing Zones in Central America." Development Discussion Paper 646, Harvard Institute for International Development, Harvard University, Cambridge, MA.

Johansson, H., and L. Nilsson. 1997. "Export Processing Zones as Catalysts." *World Development* 25 (12): 2115–28.

Ketels, C., G. Lindqvist, and Ö. Sölvell. 2006. *Cluster Initiatives in Developing and Transition Economies.* Stockholm: Center for Strategy and Competitiveness.

Kuchiki, A. 2007. "Clusters and Innovation: Beijing's High-Technology Industry Cluster and Guangzhou's Automobile Industry Cluster." Discussion Paper 89, Institute of Developing Economies, Chiba, Japan.

Kusago, T., and Z. Tzannatos. 1998. "Export Processing Zones: A Review in Need of Update." Social Protection Discussion Paper 9802, World Bank, Washington, DC.

Kuznetsov, Yevgeny. 2007. *International Migration of Talent and Home Country Development: From First Movers to a Virtuous Cycle.* Washington, DC: World Bank Institute.

———, ed. 2006. *Diaspora Networks and the International Migration of Skills.* Washington, DC: World Bank Institute.

Lall, S. 1980. "Vertical Inter-Firm Linkages in LDCs: An Empirical Study." *Oxford Bulletin of Economics and Statistics* 42: 203–6.

Léautier, F., ed. 2006. *Cities in a Globalizing World: Governance, Performance, and Sustainability.* Washington, DC: World Bank Institute.

Leong, C. 2007. "A Tale of Two Countries: Openness and Growth in China and India." DEGIT (Dynamics, Economic Growth, and International Trade) conference paper. http://www.ifw-kiel.de/VRCent/DEGIT/paper/degit_12/C012_042.pdf.

Marshall, A. 1920. *Principles of Economics.* 8th ed. London: Macmillan.

McCormick, Dorothy, and Mary Njeri Kinyanjui. 2007. "Industrializing Kenya: Building the Productive Capacity of Small Enterprise Clusters." In *Industrial Clusters and Innovation Systems in Africa: Institutions, Markets and Policy.* Tokyo: United Nations University Press.

Mytelka, Lynn K. 2007. "From Clusters to Innovation Systems in Traditional Industries." In *Industrial Clusters and Innovation Systems in Africa: Institutions, Markets and Policy.* Tokyo: United Nations University Press.

NESTA (National Endowment for Science, Techonolgy, and the Arts). 2007a. *Innovation and the City: How Innovation Has Developed in Five City-Regions.* London: NESTA.

———. 2007b. "What Role Do Cities Play in Innovation, and to What Extent Do We Need City-Based Innovation Policies and Approaches?" Working Paper 01/June 07, NESTA, London.

OECD (Organisation for Economic Co-operation and Development). 1999a. *Boosting Innovation: The Cluster Approach.* Paris: OECD.

———. 1999b. *Business Incubation: International Case Studies.* Paris: OECD.

———. 2001. *Innovative Clusters: Drivers of National Innovation Systems.* Paris: OECD.

———. 2006. *Competitive Cities in the Global Economy.* Paris: OECD.

———. 2007a. *Competitive Regional Clusters.* Paris: OECD.

———. 2007b. *Globalisation and Regional Economies: Can OECD Regions Compete in Global Industries?* Paris: OECD.

———. 2008. *Competitive Cities: A New Entrepreneurial Paradigm in Spatial Development.* Paris: OECD.

Omar, K., and W. Stoever. 2008. "The Role of Technology and Human Capital in the EPZ Life-Cycle." *UNCTAD Transnational Corporations Journal 17* (1).

Oyelaran-Oyeyinka, B., and D. McCormick, eds. 2007. *Industrial Clusters and Innovation Systems in Africa: Institutions, Markets and Policy.* Tokyo: United Nations University Press.

Piore, M., and C. Sabel. 1986. *The Second Industrial Divide: Possibilities for Prosperity.* New York: Basic Books.

Porter, Michael E. 1990. *The Competitive Advantage of Nations.* New York: Free Press.

———. 2000. "Location, Competition, and Economic Development: Local Clusters in a Global Economy." *Economic Development Quarterly* 14 (1): 15–34.

Rosenfeld, S. 2002. *Creating Smart Systems: A Guide to Cluster Strategies in Less Favored Regions.* Brussels: European Commission.

Sanz, Luis. 2004. "Survey of Science Parks Highlights Global Trends and Best Practice." *Newsbits* 3 (1): 6–20.

Sassen 2006. "Four Dynamics Shaping the Ongoing Utility of Spatial Agglomeration." Proceedings of the Cambridge Econometrics conference, "Greater Cities in a Smaller World," Cambridge, July 4–5.

Saxenian, A. 1999. *Silicon Valley's New Immigrant Entrepreneurs.* San Francisco: Public Policy Institute of California.

Saxenian, A., T. Bresnahan, A. Gambardella, and S. Wallsten. 2001. "'Old Economy' Inputs for 'New Economy' Outcomes: Cluster Formation in the New Silicon Valleys." *Industrial and Corporate Change* 10 (4): 835–60.

Scott, A. J. 2006. "The Changing Global Geography of Low-Technology, Labor-Intensive Industry: Clothing, Footwear, and Furniture." *World Development* 34 (9): 1517–36.

Smarzynska, B. 2002. "FDI Spillovers through Backward Linkages: Do Technology Gaps Matter?" Working Paper 3118, World Bank, Washington, DC.

Sölvell, Ö. 2008. *Clusters: Balancing Evolutionary and Constructive Forces.* Stockholm: Ivory Tower Publishing.

Spar, D. 1998. "Attracting High Technology Investment: Intel's Cost Rican Plant." FIAS Occasional Paper 11, World Bank, Washington D.C.

Storper, M., and M. Manville. 2006. "Behaviour, Preferences and Cities." *Urban Studies* 43 (8): 1247–74.

Storper M., and A. J. Venables. 2004. "Buzz, Face to Face Contact and the Urban Economy." *Journal of Economic Geography* 4 (4): 351–70.

UN Habitat. 2007. *Global Report on Human Settlement: Enhancing Urban Safety and Security.* Nairobi, Kenya: UN Habitat.

UNCTAD (United Nations Conference on Trade and Development). 2001. *World Investment Report 2001: Promoting Linkages.* Vienna: UNCTAD.

UNIDO (United Nations Industrial Development Organization). 2001. *Development of Clusters and Networks of SMEs: The UNIDO Programme.* Vienna: UNIDO.

USAID (U. S. Agency for International Development). 2003a. "The Economic Impact of Cluster Initiatives under the Competitiveness Initiative Project." http://pdf.usaid.gov/pdf_docs/PNADI044.pdf.

———. 2003b. *Promoting Competitiveness in Practice: An Assessment of Cluster-Based Approaches.* Washington, DC: USAID.

———. 2005. *Cluster Initiatives in Developing and Transition Countries.* Washington, DC: USAID.

———. 2008. "The Cluster Approach to Economic Development." Technical Brief 7, USAID, Washington, DC.

Van Pottelsberghe De La Potterie, B., and F. Lichtenberg. 2001. "Does Foreign Direct Investment Transfer Technology across Borders?" *Review of Economics and Statistics* 83 (3): 490–97.

Veltz, P. 1996. *Mondialisation, Villes et Territoires—L'economie d'archipel.* Paris: PUF

Von Hippel, E. 1994. "'Sticky Information' and the Locus of Problem Solving: Implications for Innovation." *Management Science* 40 (4): 429–39.

World Bank. 2008. *World Development Report, 2009: Reshaping Economic Geography.* Washington, DC: World Bank.

World Bank, EIB (European Investment Bank), and Medibtikar. Forthcoming. "Science Park Development Management Toolkit."

11

Stimulating Pro-Poor Innovations

Most discussions of innovation policy focus on improving the capacities of national research and development (R&D) institutions to address the needs of the formal economy. Yet, the majority of the world's poor—2.6 billion people or 40 percent of the world's population—live on less than US$2 a day and derive their income primarily from subsistence agriculture or work in informal enterprises (mostly in South Asia and Africa).[1] This chapter focuses on how "inclusive innovation"—policies that promote innovation for the poor and by the poor—can help improve the productivity and livelihood of those who operate mostly in the informal economy.

How to Define Inclusive Innovations, Pro-Poor Innovations

According to Anil Gupta, inclusive or harmonious development is recognized as an important goal for socioeconomic development in most developing countries, particularly in Brazil, China, India, and South Africa (Gupta 2007). Inclusion can take place by treating economically poor and disadvantaged people as consumers of public policy on assistance and aid for basic needs or as consumers of low-cost products made by large corporations or by the state or other enterprises (Prahalad 2005). Inclusion can also take place by building the capacity of the poor to produce what they already know how to and do produce, as well as building the capacity of the poor to use their innovations

---

This chapter was prepared by Anuja Utz. Background research was provided by Anna Reva, consultant, World Bank Institute.

and outstanding traditional knowledge either as is or by blending or bundling it with knowledge of others into products marketed by them or other enterprises. In addition, links with R&D institutions that can take such technologies or products and develop value-added products for eventual diffusion through commercial or noncommercial channels can also help inclusion.

Pro-poor innovation is another way of thinking about this issue. According to Berdegué, a pro-poor innovation system can be defined as a multistakeholder social learning process that generates new knowledge, puts it to use, and expands the capabilities and opportunities of the poor (Berdegué 2005). In pro-poor innovation processes, institutions play a critical role: they determine the extent to which the poor are able to participate in the innovation process and share in the potential benefits. Institutions include social norms of behavior, habits, routines, values, and aspirations, as well as laws and regulations, all of which are rooted in a given society's history and culture. The importance of institutions to innovation processes creates several challenges. The institutional framework, for example, may require substantial changes before certain pro-poor innovations can take off: laws and regulations governing intellectual property rights may have an antipoor bias; secure access to assets such as land or credit may be difficult or impossible for the poor; owing to social norms, poor women may be prevented from taking on certain roles required for innovation; social stratification may block the formation of the social networks needed for innovation; and manipulation of product markets may destroy the economic incentive to innovate. Innovation is spurred when the actors involved have reasonable assurances that they can benefit from their efforts and that free riding and other forms of opportunistic behavior will be contained. Institutions provide that needed assurance. Innovation requires cooperation, and cooperation is rooted in institutions that help build trust. A main implication is that pro-poor innovation strategies and policies cannot be "one size fits all" but need to fit the particular conditions of different social settings (Berdegué 2005; Gupta 2007).

A Snapshot of the Lives of the Poor

Understanding the needs of the poor is at the core of inclusive innovation policy. Some of their most basic needs are captured in the Millennium Development Goals (MDGs) (see table 11.1) and have received substantial attention from governments and the international community. Much of the knowledge and specific technology required to address the MDGs exists: basic nutritional information and sanitation techniques, preventive medicine, environmentally friendly technologies, cheap mobile phones, and the like. Table 11.1 tracks the needs of the poor in Latin America and the Caribbean, East Asia and the Pacific, South Asia, and Sub-Saharan Africa. Most of the world's poor are concentrated in South Asia and Sub-Saharan

Table 11.1 Progress toward Meeting Millennium Development Goals in Four Regions, 2006

Millennium development goal	Latin America and Caribbean	East Asia and Pacific	South Asia	Sub-Saharan Africa	Average for low income
Eradicate extreme poverty and hunger					
Share of poorest quintile in income	–	–	–	–	–
Percentage of underweight children under 5	5.1	12.9	41	27.0	35.3
Achieve universal primary education					
Primary completion rate (percent)	99	98	80	60	73
Promote gender equality					
Ratio of enrollments of girls to boys in primary and secondary school (percent)	101	99	90	86	89
Reduce child mortality					
Under age 5 mortality rate/1,000 births	26	29	83	157	112
Improve maternal health					
Maternal mortality ratio/100,000 births	130	150	500	900	650
Contraceptive prevalence rate/percentage of married women ages 15–49	69	79	53	22	44
Combat HIV/AIDS and other diseases					
HIV prevalence (percentage of adults)	0.6	0.2	0.7	5.8	1.7
Incidence of tuberculosis per 100,000 people	57	135	174	368	221
Ensure environmental sustainability					
Carbon dioxide emissions per capita (metric tons)	2.5	3.3	1.0	0.9	0.9
Access to improved sanitation facilities (percentage of population)	77	51	37	37	38
Fixed-line and mobile phone subscribers per 100 people	73	58	19	15	17

Sources: UNDP 2007; World Bank 2008b.
Note: Data are from 2006 or from most recent year for which they were available.

Africa. Consequently, these are the regions most in need of efforts to boost education, broaden health care, and improve livelihoods.

Globally, about 1 billion people live on less than US$1 a day; and 2.6 billion, or 40 percent of the world's population, live on less than US$2 a day. Rural areas account for three in every four people living on less than US$1 a day. Beyond low incomes and lack of physical assets, the poor are deprived of most essential services, such as health care, education, social protection, and access to infrastructure (particularly, roads, water, and electricity); and they face crime, corruption, and burdensome regulations (for example, it is harder for them to register a business or obtain title to their land).

Poor people's needs go beyond those tracked by the MDGs, however. The key issue is to get existing knowledge to the poor and to provide the means (supporting institutions, education and skills, finance) to help them use it. Moreover, long-term sustainability requires higher income-earning opportunities and increasing the productivity of existing micro- and small enterprises (MSEs). Much of the knowledge and technology for upgrading informal enterprises already exists. The challenge is to transfer it effectively by enhancing education and skills, increasing access to finance, and linking MSEs to markets. The potential of information and communication technologies (ICTs) can be harnessed to address some of these needs. Overall, pro-poor innovation policy requires thinking about the roles of the various actors (government, private sector, nongovernmental organizations), the package of assistance measures, and effective means of delivering them as close as possible to the communities that need to be reached. Most needed is the development of effective networks to help disseminate existing knowledge and to provide advice and support for the necessary initiatives.

Equally important to strengthening the capabilities of the poor is strengthening incentives, policies, and institutions. Top-down, supply-driven initiatives have often proved ineffective for addressing the needs of the poor. Inclusive innovation policy presupposes a change in institutional culture and mandates the involvement of the poor in identifying their development priorities and in providing incentives for various actors to serve their needs more effectively. This change will entail closer collaboration among public R&D entities, industry, universities, nongovernmental organizations (NGOs), donors, and global networks. The poor can also gain by organizing themselves. In the Indian state of Andhra Pradesh, for example, community-based development initiatives have led self-help groups to develop mutual insurance schemes, lending and savings operations, and marketing strategies for new agricultural products.

Organization of the Chapter

This chapter provides an overview of inclusive innovation and, in particular, highlights the mechanisms that the formal sector can use to address the needs of the poor. It analyzes current experience with the promotion of grassroots innovation and indigenous or traditional knowledge. Finally, it discusses means of helping the informal sector absorb existing knowledge and technology.

Harnessing Formal Innovation Efforts for the Poor

A first approach to promoting inclusive innovation is to harness, increase, and redirect formal efforts, especially in agriculture (which employs the majority of the poor in South Asia and Sub-Saharan Africa). The actors involved include public R&D institutions, such as those focused on agriculture, and universities; the private sector, including corporate social responsibility (CSR) initiatives; and global networks and NGOs. This section also considers the potential use of ICTs for reaching the poor.

The Power of Agriculture to Reduce Poverty

As mentioned, three out of four of the world's poorest people live in rural areas, and that represents 2.1 billion earning under US$2 a day. Most make their living, directly or indirectly, from farming. Agriculture is the main livelihood of about 2.5 billion people, including 1.3 billion smallholders and landless workers. Increased agricultural productivity can reduce poverty directly by raising farm incomes and indirectly through labor markets, to the extent that it creates employment opportunities for the poor. In South Asia and Latin America, 25 percent of working-age men—usually the poorest—are employed as farm laborers. If farm production expands, they will benefit. Increased productivity of nontradable staple foods also reduces domestic food prices for poor consumers. In addition to the urban poor, who spend a large share of their incomes on food, more than half of poor rural households are typically net food buyers that stand to benefit from lower prices. Studies from India show that, in the long term, food prices have a major influence on whether people can rise out of poverty. Agriculture has special powers that, properly tapped, can offer a way out of poverty for millions. Agriculture's contribution to poverty reduction varies depending on the country:

- In agriculture-based countries, a productivity revolution in smallholder farming can raise incomes and reduce poverty, as in the case of cocoa in Ghana.
- Transforming countries where urban dwellers' rapidly rising incomes leave many of the rural poor behind, as in China and India, requires a comprehensive approach that offers rural populations multiple pathways out of poverty: encouraging shifts to high-value agriculture, decentralizing non-farm economic activities to rural areas, and providing assistance to help people move out of poverty.
- In urbanized countries, where agriculture has a smaller share of the economy but where deep pockets of rural poverty remain, agriculture can help reduce rural poverty if smallholders become direct suppliers to modern food markets (Savanti and Sadoulet 2008).

Investing in efforts to harness agriculture to reduce poverty is essential. Agricultural R&D is especially important for generating additional income and employment for the poor. India's "green revolution" offers an example of such efforts with a view to achieving self-sufficiency in food grains. It involved a package of investments in technology, comprising largely high-yield varieties first of wheat and subsequently of rice, chemical fertilizers, and agricultural research and extension, aided by public investments in infrastructure (irrigation, roads, market institutions) and price incentives that encouraged production. Agricultural R&D and innovations and applications of science in agriculture were critical in generating additional income and employment for the poor. The National Agricultural Innovation Project promotes joint efforts between research entities and the private sector to accelerate collaboration among public research organizations, farmers, the private sector, and other stakeholders, using agricultural innovations as a vehicle for a more market-oriented path toward poverty alleviation. It is an effort worthy of emulation and scaling up.[2]

Agriculture also plays a dominant role in nearly all the countries of eastern and central Africa, and many face similar agro-ecological, climatic, and development challenges. Significant scale economies can be realized through the regionalization of R&D, using networks such as the Association for Strengthening Agricultural Research in Eastern and Central Africa.[3] The challenge for such networks is to determine the regional and national research priorities with the highest potential rates of economic return. A recent study indicates significant potential for agricultural technology spillovers in the region (You and Johnson 2008). Therefore, if countries pool their resources and pursue regional initiatives in their search for technology solutions, they can hope to reap greater economic benefits.

The concept of agricultural innovation systems is also attracting interest, especially in Africa.[4] This "systems" perspective aims at developing stronger coordination and collaboration in agricultural education, research, extension, and farmer organizations. It involves exploring ways of sharing or reducing the financial risks of investing in technology and innovation, tapping more effectively into scientific and technological resources nationally and internationally, promoting public-private partnerships, creating demand-responsive research and educational institutions, and building capacity for technological learning.

Building on Public R&D and University Initiatives

Universities and public research centers in developing countries have the potential to become central actors in pro-poor innovation by using their considerable capabilities to address the needs of the poor. Most institutions of higher education in these countries, however, focus on education and training and devote few resources to research. Research bodies also tend to be isolated from local communities, and they do not direct their efforts toward finding solutions to problems in agriculture, health, or industry. Some public research centers and universities, though, have attempted to change their orientation as a result of government policies and incentives, a shortage of funds, and the need to generate income by developing commercial products for local communities or as a result of a decision by the institution's management. India has been somewhat successful in harnessing the resources of the large public research system, which still focuses primarily on defense, space, and energy, to address infrastructure needs in poor communities (box 11.1).

Similarly, several universities in Africa have attempted to revise their curricula or establish new programs and research centers to meet social needs. Successful examples include the Kigali Institute of Science and Technology, which has developed a number of pro-poor technologies, and the Malaria Research and Training Center in Bamako, Mali, which has gained international

> **Box 11.1 Adapting Public Research Systems for Development Needs in India**
>
> *Use of space technology for development:* Advances in space-based Earth observation technology and its applications have the potential to provide economic security and better living standards. For example, Sujala, a watershed development project in Karnataka, relies on high levels of community participation in five districts and scientific planning tools such as satellite remote sensing, geographic information systems, and information technology. Similarly, under the Rajiv Gandhi National Drinking Water Mission, more than 2,000 groundwater maps covering about 45 percent of the country (mainly problem zones) have been prepared, and more than 24,000 wells have been drilled.
>
> *Technology applications for rural India:* The mission of the Council for Scientific and Industrial Research (CSIR), India's largest public R&D infrastructure, is to provide scientific and industrial R&D that maximizes economic, environmental, and social benefits for the people of India. CSIR takes a people-oriented development and delivery approach. Its labs have been instrumental in the revival of India's world-famous handmade blue pottery. Its research led to product and quality improvements and product diversification and thus enabled this ailing traditional industry to grow and extend its markets beyond India. Another example is the technology to desalinate water using reverse osmosis. CSIR labs have worked on designing a multichannel ceramic membrane with optimum channel configuration for upscaling technology for purification of arsenic contamination in groundwater. In addition, CSIR has been working on herbal products, especially oil-yielding mint plants. Nearly 400,000 hectares of land are used to cultivate the Kosi, Himalaya, and Sambhav varieties (*Menthol sinesis*) developed by CSIR. These pest-resistant and high–oil-yielding varieties have been adopted by 20,000 farmers and have generated 40 million man-days of employment. India is now the largest exporter of menthol mint and its oil, displacing China to second position. CSIR labs have also been working on food-processing and leather-processing technology.
>
> *Source:* Dutz 2007.

recognition for the quality of its research and its links with groups of traditional healers at the local level (box 11.2).

Apart from research on socially relevant topics, a number of universities are also introducing community service initiatives, implemented jointly by professors and students, either on a volunteer basis or as part of the requirements for obtaining a degree. University-led community service programs have obvious benefits. They provide students with opportunities to obtain practical and problem-solving skills and learn relevant indigenous practices while contributing to the social and economic development of impoverished and marginalized communities. The Dominican Republic, Mexico, Nicaragua, and South Africa have introduced mandatory community service as a requirement for obtaining a degree, for all or some disciplines. The results have been mixed, however, as meaningful and systematic engagement with

Box 11.2 Pro-Poor Innovations at University Research Centers in Africa

Kigali Institute of Science and Technology: The Kigali Institute of Science and Technology (KIST) was founded in 1997 as Rwanda's first higher education institution of technology. From the outset, the university management aimed to focus the curricula and research on community-related problems and to combine conventional teaching with technology transfer initiatives. The university's Centre for Innovations and Technology Transfer has developed an impressive list of pro-poor technologies, including low-cost hand- and foot-powered water pumps capable of lifting water as much as 9.5 meters without electricity, for irrigation purposes in rural areas; rainwater-harvesting systems for areas without adequate piped water; a dual crop dryer that uses either sunshine or biomass (such as rice husks, sawdust, or firewood); solar water-heating systems; more efficient cooking stoves; and biomass plants in which the resulting methane gas replaces almost two-thirds of the firewood otherwise needed for cooking or heating water. KIST seeks frequent feedback from communities to further improve its products. For instance, recommendations from the institution's community development officers, many of whom are women, led to the design of lighter oil presses for easier use by women. The officers also work with rural women's groups, helping them improve their businesses with the aid of simple technologies. As a result, some groups have started supplying restaurants with fruit juices, dried mushrooms, tomato concentrates, jams, and honey by introducing better food-processing devices and techniques that guarantee more consistent quality. Marketing of technologies developed by the university facilitates economic development in communities and also helps KIST generate additional resources to supplement its budget.[1]

The Malaria Research and Training Center of Bamako University in Mali: Created in 1992, Bamako University's Malaria Research and Training Center (MRTC) is internationally recognized for its contributions to research on malaria and improvement of public health standards. MRTC has a clear strategy, broad local and international partnerships, and government support. It collaborates with local authorities and doctors who know local needs and behavior. Field research is conducted with the cooperation of local people. MRTC's international activities include cooperation with universities, research centers, and international agencies. Overall, the importance of the goals pursued and the results it has obtained have helped MRTC raise funds and boost its research capacities. MRTC has opened its doors to students from other African countries and collaborates with units at various African universities, thus encouraging the diffusion of research excellence through partnership networks and the development of local capacity. The center has published more than 200 articles in international scientific journals since 1992. It has been successful in international grant competitions and has been hailed as a center of excellence by the Agence Universitaire de la Francophonie and the U.S. National Institutes of Health. MRTC is also certified by the U.S. Food and Drug Administration to conduct clinical tests according to international standards, for example, tests of antimalaria vaccines. MRTC researchers have organized a network with traditional doctors for immediate care of infected persons in the Bandiagara region, an initiative that has significantly reduced malaria mortality. In 1997, prior to the program, the mortality rate among children under five years of age was 20–30 percent. By 2005, it had been reduced to 5–7 percent.[2]

Sources: 1. Bollag 2004. 2. World Bank 2007a.

poor communities requires additional staff, financial resources, and support from local governments and NGOs. In addition, in most countries that have incorporated community service into the curriculum, there has been lack of monitoring and evaluation of compliance or results of such initiatives.

Stronger Incentives and Funding Needed. Stronger incentives and more funding are needed if existing public R&D and university-enabled initiatives are to unleash their potential. Mechanisms for increasing the focus on inclusive innovation include institutional mandates—that is, competitive research grants and targeted funding for research teams and institutes that produce relevant innovations—as well as prizes and public awards. As a policy thrust, government should explicitly direct research institutes, universities, and other publicly funded institutions of learning to do more to address the needs of the poor.[5]

To underscore the high priority placed by government on the reorientation of the public R&D infrastructure, a special pro-poor innovation fund could be set up to provide matching grants for R&D with a pro-poor orientation, in addition to earmarking part of research institutes' budgets for that purpose. The fund should cover not only R&D in the public institutions but also joint R&D with universities, NGOs, and private enterprises; it should also cover scaling-up, pilot plant, testing, and market testing.

Notably, the marketing, dissemination, and commercialization of the developed products are weaker than the development of the pro-poor solutions themselves. Many public research institutions have no experience with entrepreneurial activities. Mechanisms are therefore needed to demonstrate, scale up, and disseminate innovations. The precise nature of the mechanisms depends on the innovation and its potential applicability. Those that are in the nature of public goods should be widely disseminated to the target population. Those that can be commercialized should be licensed to the producers or organizations that can do so. One possibility would be to create a professional body entrusted with field trials and demonstration for diffusion, adaptation, and assimilation of technologies for the poor. Such an entity would hire professionals trained in market research and related professions and offer competitive compensation.

University Engagement with the Poor. Measures are needed to encourage universities to work on the technological, economic, and social needs of the poor, including by adapting university programs and curricula to the realities of their communities. Universities' civic engagement should be encouraged and widely popularized, including through media campaigns. This approach would help demystify the image of universities as ivory towers and build trust and cooperation between disadvantaged groups and university communities. The results of the community service programs introduced by universities in the developing world should be evaluated and the lessons learned disseminated to educational institutions generally. Universities in developing countries can learn a lot about structuring, managing, and financing community service initiatives from Western institutions where such practices have long been common. They can also obtain expertise from international networks, such as the Talloires Network, which unites institutions from around the world committed to promoting the civic role and social responsibilities of higher education.[6] The network provides participating members with expertise to support the process of building civic engagement, the possibility of twinning with another university, and public relations and media links. Special funds should be allocated by the government for research at universities that addresses the needs of the poor. As in the case of public labs, financial incentives, awards, prizes, and special recognition should go to researchers and research teams that develop relevant innovations. Universities should also have access to matching grant funds to work with research institutes, NGOs, and private firms to undertake R&D, scale up, and market testing of these innovations.

Encouraging the Private Sector to Serve the Needs of the Poor
The private sector can play a significant role in improving the lives of the poor in at least three ways: by developing affordable products and services tailored to the needs of low-income consumers, by creating job opportunities and increasing the productivity of the poor, and by addressing some of their needs through CSR initiatives.

The Poor as Consumers and Producers. Generally, the private sector does not focus on developing products and services for the informal sector because of a wide perception that there are no profits to be made in low-income markets. Throughout the developing world, however, the poor pay much more for basic products and services than the better off. For instance, urban slum dwellers pay between 4 and 100 times as much for drinking water as do middle- and upper-middle-class households. Food costs 20–30 percent more in the poorest communities, while annual interest rates charged by moneylenders can be as high as 2,000 percent (Prahalad and Hammond 2002). As a result, there is a real opportunity for private companies to establish profitable operations in this segment of the market, while bringing lower-cost and better-quality goods to poor consumers.

A recent report by the World Resources Institute and the International Finance Corporation (2007) looks at market size and business strategy at the base of the economic pyramid (BOP), which encompasses some 4 billion people. The report gives empirical measures of their aggregate purchasing power and behavior as consumers, which suggest significant opportunities for market-based approaches to increasing their productivity and incomes and to empowering their entry into the formal economy (box 11.3).

Box 11.3 The 4 Billion at the Base of the Economic Pyramid

Four billion low-income people, a majority of the world's population, form the base of the economic pyramid. Yet these 4 billion—with annual incomes below US$3,000 in local purchasing power—have substantial purchasing power: the BOP constitutes a US$5 trillion global consumer market. These markets are often rural, especially in rapidly growing Asia, very poorly served, dominated by the informal economy, and relatively inefficient and uncompetitive. These population segments for the most part are not integrated into the global market economy and do not benefit from it. The BOP population has other characteristics as well:

- *Significant unmet needs:* Most people in the BOP have no bank account and no access to modern financial services. Most do not own a phone. Many live in informal settlements, with no formal title to their dwelling. Many lack access to water and sanitation services, electricity, and basic health care.
- *Dependence on informal or subsistence livelihoods:* Most of the BOP lack good access to markets for selling their labor, handicrafts, or crops and have no choice but to sell to local employers or to middlemen who exploit them. As subsistence and small-scale farmers and fishermen, they are uniquely vulnerable to destruction of the natural resources they depend on.
- *Affected by a BOP penalty:* Poorer people often have goods and services that are more expensive, of low quality, or difficult or impossible to access.

Addressing the unmet needs of the base of the pyramid is therefore essential for raising welfare, productivity, and income and for rising above poverty. Analysis of BOP markets can help businesses, governments, and the development community:

- For businesses, it is an important first step toward identifying opportunities, considering business models, developing products, and expanding investment in these markets.
- For governments, it can help focus attention on reforms needed in the business environment to allow the private sector a larger role.
- For the development community, a successful market-based approach would bring significant new private sector resources into play, allowing development assistance to be more targeted to segments and sectors for which no viable market solutions are at present found.

Source: World Resources Institute and IFC 2007.

As Prahalad (2005) persuasively argues, large companies can use their considerable technological, organizational, and marketing capabilities to create and deliver products and services to those at the bottom of the pyramid and make a profit doing so. This process requires mobilizing the investment capacity of large firms, the knowledge and commitment of NGOs, and the communities that need help to join forces to find solutions. Such approaches have been successfully adopted by a number of corporations throughout the world (box 11.4).

Box 11.4 The Private Sector as Provider of Products and Services for the Poor

Patrimonio Hoy, an affordable housing program: CEMEX, a global building materials company, has developed an effective mechanism for serving low-income Mexican families through its Patrimonio Hoy program. Launched in 1998, the program provides individual technical assistance to families wishing to improve their housing conditions. The building plan is organized into packages of materials that are ordered according to need. The weekly charge per family is US$15, which covers the costs of the materials and a package of services, including free access to technical consultants, guaranteed prices for 70 weeks (typical project duration), one year of storage of materials, and home delivery of construction materials, as well as access to credit to cover up to 80 percent of the value of materials received. By the end of 2007, a total of 185,000 families had benefited from the program. Credits of US$83 million had been granted, with on-time payment of more than 99 percent. The customers have been able to decrease the average time for construction of a room from five years to just over one year and to lower the cost of construction by approximately 20 percent.[1]

Bringing reliable water supply to poor communities in Manila: The Manila Water Company, owned by Ayala Corporation in the Philippines, received a concession to provide services to the East Zone of the Manila Metropolitan Area in 1997.[a] The company prioritized delivery of service to low-income consumers and developed innovative solutions to address their challenges: lack of water facilities (toilets and faucets), as well as in-house piping, inability to cover installation costs, and widespread belief among the poor that running water is more expensive than water sold by vendors in a can. Manila Water lets communities decide if they want individual or collective installation, metering, and billing. The company offers three options: one meter per household, one meter for 3 or 4 households, and a bulk meter for 40 to 50 households. Where households band together, the connection fee (ordinarily ₱7,000 a household) can fall by as much as 60 percent, depending on the number of customers who shoulder the cost of pipes, meter, and installation. Submeters measure water use in each household, and each member of the group meter takes responsibility for paying the total bill, an arrangement that in effect gives consumers (and Manila Water) group insurance coverage on payment. About 30 percent of the urban poor served by Manila Water now pool their bills, and in communities using this technique the company collects 100 percent of the money owed. Consumers recognize both the savings from collective installation and the fact that ensuring sustained service depends upon their own actions. Manila Water has provided jobs to more than 10,000 people, either as couriers who deliver bills or as contractors who help lay pipelines; the company also fosters the development of small supplier businesses and cooperatives, such as printing outfits. It has also made small loans, in partnership with the Bank of the Philippine Islands and the International Finance Corporation, to organized groups operating microenterprises such as street stalls and food services. So far, directly measurable benefits include serving 5.1 million residents, training more than 1,000 engineers, and disbursing a US$16 million annual payroll in impoverished east Manila.[2]

Sources: 1. Rangan and others, 2007, 156–67; WBI 2008. 2. Beshouri 2006; Rangan and others 2007, 213–21.
a. See http://www.manilawater.com/.

Productivity Capacities of the Poor. Low-income households, however, should not be viewed merely as consumers. The private sector can also encourage the productive capacities of the poor by helping them transfer appropriate knowledge and technology and including them in value chains. Such initiatives not only help increase the incomes of the poor but also help create additional business, value for corporations. One example comes from encouraging entrepreneurship among rural women. Hindustan Unilever, one of India's leading businesses, has developed a new mechanism to reach out to rural consumers while empowering poor women. Through Project Shakti (*strength* in Sanskrit), the company recruits disadvantaged women to sell the company's products door to door. The women are linked to microfinance banks, where they can obtain funds for an initial investment of about US$220, and are provided with the necessary training to become competent business operators. Most women generate a net profit of about US$150 annually, a substantial addition to the average rural household's income of US$250. Launched in 2000, by 2007 the project had trained 46,000 entrepreneurs covering 100,000 villages and had reached 3 million households in rural India, making it the world's largest sustained home-to-home operation. Hindustan Lever has also created a four-week training program for all participants and employs some of the company's leading entrepreneurs as trainers.[7]

A McKinsey analysis has tried to bring together exportable lessons from the experiences of Manila Water, CEMEX, and Hindustan Lever (Beshouri 2006). It finds that communities are frequently in a better position than companies to resolve issues that make it uneconomical to serve low-income groups. Three business models are emerging (table 11.2). The first, "collective accountability," focuses on collection problems associated with direct lending or postpaid services. It involves developing small groups, such as CEMEX's family clusters or Manila Water's collective billing units, whose members substitute for the company's monitoring efforts and provide "social insurance" to one another. The second business model, "scalable, embedded distribution," reduces costs and promotes a company's reputation by enlisting trusted community members (Cemex's Patrimonio Hoy monitors and Hindustan Lever's entrepreneurial women) to provide the distribution infrastructure for goods and services. The third, "livelihood partnerships," offers additional benefits to a core product or service. Rather than treating communities purely as collections of consumers, companies that take this approach provide low-cost, productivity-enhancing assistance, such as Manila Water's training and cooperative business programs. These initiatives bridge cultural gaps between company and community, create positive associations with the company's brand, raise switching costs, and promote micromarket activity, with positive consequences for both the community and the companies doing business there.

Table 11.2 Three Models for Enabling Businesses to Serve the Poor Economically

Business model	Core issue addressed	Community-based intervention	Relevant industries	Sample businesses
Collective accountability	Problems with collection; pilferage	Small groups monitor use, promote compliance, provide social insurance.	Utilities (water, electricity); finance	Manila Water (Philippines); ICICI Bank (India)
Scalable, embedded distribution	Traditional delivery too costly relative to purchase size and density of consumers	Low-cost community-based distribution points employ key workers in low-income areas.	Fast-moving consumer goods; telecoms; low-value consumer goods	Indofood (Indonesia); Hindustan Lever (India); Kodak Brazil
Livelihood partnership		Business offers additional services around core products and services that promote primary demand while providing training or cooperative business programs to community.	Telephony services; utilities (water, electricity); agriculture	Globe Telecom (Philippines); Manila Water (Philippines); ITC e-Choupal (India)

Source: Beshouri 2006.

Investment, Jobs, and Opportunities. National and local governments can do more to encourage the private sector to invest, create jobs, and improve productivity, not only to promote growth but also to expand opportunities for poor people. Governments should remove regulatory impediments that prohibit the private sector from serving the poor. For instance, utility companies may not be able to serve urban slum dwellers because they do not have formal rights to their dwellings. In this case, addressing that issue is a first necessary step. Streamlining bureaucratic procedures is equally important. For example, AES-EDC, an electricity provider in Venezuela, has developed a prepaid electricity card to extend its services to and collect payments from the urban poor. The innovation was welcomed by the poor communities but has not been introduced owing to delays in government approval (Rangan and others 2007, 197–204).

Government agencies can also encourage the private sector to serve disadvantaged communities and population groups through special procurement initiatives and pro-poor public-private partnerships. These arrangements can become an effective mechanism for providing infrastructure as well as health, education, and telecommunications services for the impoverished communities as long as roles and responsibilities are clearly delineated and quality is continuously monitored (box 11.5).

Governments should consider allocating more funds to encourage efforts that focus on the challenges facing the poor. They might establish a pilot inclusive innovation fund to support R&D by public R&D entities, the private sector, universities, and NGOs aimed at meeting the needs of the informal

> **Box 11.5 Pro-Poor Public-Private Partnerships**
>
> *Expanding access to secondary education in the Philippines:* The government of the Philippines uses the Educational Service Contracting (ESC) scheme to support enrollment of low-income students in private schools in areas with a shortage of public high schools. Eligible schools must be certified and charge relatively low fees. Family income for eligible students cannot exceed US$1,280. The per student payment to private schools is set at US$71 and cannot exceed the unit cost of delivery in public high schools. Schools cannot charge the students any additional fees. The number of ESC-funded students grew from 4,300 in 158 schools in 1986 to 280,216 in 1,517 schools in 2003. In 2002, ESC contracts covered 22 percent of students in private high schools (equal to 13 percent of all private school enrollments). An assessment of the certification procedure in one region showed that less than 10 percent of schools were below standard.[1]
>
> *Public-private partnership for hand washing:* Hand washing with soap is one of the most effective and inexpensive ways to prevent diarrhea and pneumonia, which together are responsible for approximately 3.5 million child deaths every year. The Global Handwashing Day is an initiative of the Public-Private Partnership for Handwashing of the Water and Sanitation Program, the United Nations Children's Fund (UNICEF), the World Bank, and a variety of other partners.[8] Studies by Safeguard, an antibacterial soap made by Proctor and Gamble, show that hand washing with soap is an effective way to save children's lives. In China, Safeguard is partnering with the Red Cross, UNICEF, and the World Bank to establish a Safeguard–Red Cross Health Great Wall Foundation and build sanitation and water facilities in schools in Sichuan. On October 15, 2008, Safeguard initiated the distribution of hand-washing education leaflets to 8 million students around the country. In addition, it posted Global Handwashing Day information at nearly 1,500 bus stops in the cities of Beijing, Guangzhou, and Shanghai, thus reaching over 90 percent of the population in these cities.[2]
>
> *Sources:* 1. Patrinos 2006. 2. Global Handwashing Day, http:// www.globalhandwashingday.org/; Proctor and Gamble, http://www.pg.com/company/our_commitment/globalhandwashingday.shtml.

sector, on a matching-grant basis. Such projects should undergo continuous monitoring and evaluation. If the pilot proves successful, the government should earmark a small percentage of the public R&D budget to support an inclusive innovation fund on a continuing basis. The fund should cover scaling up, piloting, testing, and commercialization. Competition for scarce funds would be based on transparent eligibility and evaluation criteria.

It is crucial for large domestic firms, multinational corporations, government agencies, NGOs, and, most important, the poor to come together to solve problems. The business community and society need to be made more aware of potential opportunities for win-win solutions. Public recognition and awards should be given for the most successful business initiatives aimed at the poor. By crafting community-based strategies that reflect the specific characteristics of the low-income segment, companies can tap into a huge growth opportunity for themselves and achieve competitive rates of return while also delivering important developmental benefits to the communities they serve.

Socially Driven Pro-Poor Initiatives. Companies or organizations may also look beyond profit to help deal with some specific needs of the poor, such as basic literacy, preventive medicine, and health-related initiatives, in a spirit of corporate social responsibility.[9] CSR is an increasingly important part of the business operations of transnational corporations and of many national companies in developing countries. Bill Gates has recently referred to "creative capitalism" as an "attempt to stretch the reach of market forces so that more companies can benefit from doing work that makes more people better off" (Gates 2008).

Among the BRICs (Brazil, Russia, India, and China), Russian companies do not fare very well on CSR, but Brazil is quite active in this area: some 1,300 companies are members of Instituto Ethos, a network of businesses committed to social responsibility. Ethos tries to influence public policy and corporate behavior "to establish a socially responsible market." Brazilian firms such as Natura, a cosmetics company, and Aracruz, a pulp and paper producer, are widely known for their CSR efforts. India also has a long tradition of paternalistic philanthropy; large family-owned firms such as Tata are active in providing basic services, such as schools and health care, for local communities. It is still early days for CSR in China, but pressure to take CSR more seriously is growing. In Shanghai in October 2007, 13 foreign and domestic companies launched the Chinese Federation for Corporate Social Responsibility. Among international organizations, the United Nations promotes CSR around the world through the Global Compact. Many NGOs are also cooperating with big companies on joint projects: examples include BP's arrangements with NGOs to distribute stoves in rural India and ABN AMRO's collaboration on microfinance in Latin America with ACCION International (*Economist* 2008c).

The private sector engages in development-related activities in various ways, including through core business practices: for example, the development of new business products and innovative ways to deliver affordable goods and services, public-private partnerships, corporate philanthropy, and transparent and responsible engagement in public policy dialogue, rule making, and institution building (WEF 2006). Throughout the developing world, companies are reaching the poor, particularly in education, health, and job creation (box 11.6).

Many private companies are supporting sustainable development through their ethical business practices as well as through a trickle-down effect on the community around them. The mass media and governments should draw attention to this social consciousness. In some countries, the media create annual lists of the most socially responsible companies. Public recognition as

> **Box 11.6 Examples of Socially Driven Pro-Poor Initiatives**
>
> ***Promoting basic education for disadvantaged children:*** Since 1997, the Coca-Cola's Little Red Schoolhouse program in the Philippines has given disadvantaged children in remote areas access to basic education by building schools and training educators. With more than 60 schools built, over 30,000 students, 750 teachers, and 3,100 parents and community members have benefited. By 2009, 19 more schools would be built. This program received the Support and Improvement of Education Merit Award during the 2006 Asian Corporate Social Responsibility Awards ceremony.[1]
>
> ***Strengthening the national health system of Tanzania, Abbott Laboratories:*** The Abbott Fund and the government of Tanzania have formed a public-private partnership, one of the most comprehensive initiatives in Africa, to strengthen the country's health care system. Key areas of focus include modernizing facilities, training staff, improving hospital and patient management, and expanding capacity for testing and teaching. Centered at Muhimbili National Hospital in Dar es Salaam, the Abbott Fund initiative also includes support for more than 80 hospitals and rural health centers across the country. The fund has invested more than US$50 million, and, to date, more than 7,800 health workers have been trained in effective HIV care, HIV testing has been provided to more than 180,000 people, and the state-of-the-art clinical laboratories serve hundreds of patients each day.[2]
>
> ***Creating employment in the tourism sector, Serena Hotels:*** Serena Hotels employ 3,000 people in East Africa and has a policy of using local, national, or regional suppliers wherever possible to boost local economic activity. Most food and beverage items are sourced locally. The hotels advise and train local suppliers to meet their quality standards. For instance, in Tanzania, Serena has worked with bottled water manufacturers and local fruit and vegetable growers, so that these items can be sourced locally. Where the hotels work with larger companies, preference is given to those that work with smallholders. Similarly, the hotels promote the cultural heritage of numerous local communities and ethnic groups by purchasing a range of products from them for the furnishing and decoration of their properties as well as for sale to guests.[3]
>
> *Sources:* 1. Coca-Cola, http:// www.thecoca-colacompany.com/citizenship/pacific.html#3. 2. Abbott Laboratories, http://abbottglobalcare.org/sections/Strengthening/default-2.html? 3. Ashley, de Brine, Lehr, and Wilde 2007.

well as various awards to the companies with the most prominent CSR initiatives could also be effective; however, the companies should be evaluated according to clear benchmarks.

Much more can be done to make CSR an established business practice. Clear metrics should be developed for evaluating the impact of such initiatives on social conditions. CSR initiatives also need to be better understood so that companies can allocate their funding appropriately and stakeholders, notably the communities concerned, can influence decision making. For transparency, it may be opportune to develop a set of common CSR indicators that companies can use to communicate their performance, both internally and externally. These might be established by working with leading international institutions, such as the Global Reporting Initiative. Finally, if CSR is to be pursued sustainably, investors will need to appreciate the links with financial performance and understand the challenges of delivering long-term social returns in a context of shrinking financial horizons. A dialogue between the business and the financial communities on social responsibility is essential and should help provide a stronger analytical case for CSR.

The public sector can facilitate CSR initiatives by developing frameworks for assessing local or national CSR priorities, by engaging the private sector in the public policy process (for example, for national sustainable development or poverty reduction strategies), and by building a stable and transparent environment for pro-CSR investment, including norms for strengthening social, environmental, and economic governance and means of enforcement. Government agencies can also develop or support CSR management tools and mechanisms, including voluntary product labeling schemes and benchmarks and guidelines for company management systems or reporting. In South Africa, the government has gone a step further by creating fiscal incentives and by leveraging public procurement or investment.

Plugging into Global Networks and Supporting Local NGO Initiatives

Global R&D networks can also be harnessed to meet the needs of the poor. They may be particularly useful for low-income countries with capacity constraints and small states that cannot achieve economies of scale in a certain research field. Some of the best-known R&D efforts on international public goods are the Consultative Group on International Agricultural Research (CGIAR), which was behind the green revolution, and the Global Research Alliance, which unites over 50,000 scientists working on health, transportation, and climate change. There are also major initiatives in health and pharmaceuticals, such as the International AIDS Vaccine Initiative; the Global Fund to Fight AIDS, Tuberculosis, and Malaria; the Global Alliance for Vaccines and Immunization (GAVI); and the Bill and Melinda Gates Foundation. Democratization has allowed the growth of local NGOs in many developing countries. These grassroots organizations are often able to reach the most remote and underserved communities, and local governments should collaborate with them more systematically in establishing local regulations, policy dialogue, and the design and implementation of community development programs.

The Consultative Group on International Agricultural Research. CGIAR was established in 1971 to help achieve sustainable food security and reduce poverty in developing countries through scientific research and research-related activities in agriculture, forestry, fisheries, policy, and the environment. Today, more than 8,000 CGIAR scientists and staff are active in over 100 countries. CGIAR's major achievements include planting quality protein maize on over 600,000 hectares in 25 countries; adoption of "zero-till" technology on more than 3.2 million hectares in South Asia, which resulted in higher productivity; and promotion of new high-yielding rice varieties for Africa, which are currently planted on 100,000 hectares. It is estimated that for every dollar invested in CGIAR, US$9 worth of additional food is produced in the developing world (CGIAR, http://www.cgiar.org/).

The Global Alliance for Vaccines and Immunizations. GAVI is a public-private partnership with a single focus: to improve child health in the poorest countries by extending the reach and quality of immunization coverage as part of better health services. Countries with a gross national income per capita below US$1,000 in 2003 can qualify for GAVI support, which is provided in response to country proposals. Currently, 72 countries receive GAVI support. Since GAVI's creation in 2000, its support has prevented 2.9 million future deaths, protected 36.8 million children with basic vaccines (against diphtheria, tetanus, and pertussis), and protected 176 million with new and underused vaccines (hepatitis B, haemophilus influenza type b, and yellow fever). With the support of GAVI, spending on children's vaccines in the poorest countries more than doubled between 2000 and 2005. GAVI has also developed innovative funding schemes. Apart from direct donations, GAVI piloted the Advanced Market Commitments to facilitate the development of new vaccines for the developing world. Donors commit money to guarantee the price of vaccines once they have been developed, thus creating the potential for a viable future market. Decisions on which diseases to target, criteria for effectiveness, and long-term availability are made in advance. This approach provides vaccine makers with the incentive they need to invest in research and manufacturing. The first such pilot is for a vaccine to prevent Pneumococcal disease, which is expected to prevent 1.6 million deaths a year (GAVI, http://www.gavialliance.org).

Engineers without Borders–International. EWB–I is an international association of national EWB groups. Its mission is to partner with disadvantaged communities to improve their quality of life through education and implementation of sustainable engineering projects and to offer engineers and engineering students new experiences. Projects range from the construction of sustainable systems that developing communities can own and operate without external assistance to empowering such communities by enhancing local, technical, managerial, and entrepreneurial skills. These projects are proposed by, and completed with, contributions from the host community working with the EWB's teams. Most initiatives are small (examples include development of water supply systems, drip irrigation, off-grid electricity, and small-scale construction in poor communities of the developing world) and do not compete with projects implemented by private consulting firms (EWB-I, http://www.ewb-international.org/).

Mobilizing the Power of ICTs to Reach the Poor

Modern information and communication technologies have enormous potential for extending access to essential social services and providing the poor with new economic opportunities. The past 15 years have seen an explosion in the penetration of mobile phones and wider use of the Internet in the developing world. Even the poorest people have access to mobile telephony (thanks to sharing), and this has changed their lives not only by saving travel time and connecting them to markets but also by helping them access essential services, such as mobile banking or health care. By November 2007, the total number of mobile phone subscriptions in the world had reached 3.3 billion; while this number represents about half the human population, the number of users is hard to calculate since some users have multiple subscriptions and some are inactive (Reuters 2007). Nonetheless, the International Telecommunications Union estimates that 24 percent of the population in developing countries does not have access to mobile telephony and that rural populations are particularly disadvantaged (ITU 2008).

One of the steps that developing countries can take to enhance access to mobile telephony is to reduce or cut taxes on mobile handsets or connection fees: new mobile subscriptions in Bangladesh fell from 11 percent to 7 percent after a US$14 connection tax was imposed, whereas India saw penetration increase from 1 percent to more than 5 percent over three years following a reduction in handset import duties (World Bank 2007a). Providing access to the Internet is also important, as is the availability of locally relevant content.

A number of traditional and new technologies can be used to reach the poor. For instance, radio and television can be used for educational purposes and for providing communities with information that is important for their livelihood, for example, on health and environmental issues, agriculture, or prices for major crops. Walkie-talkies and personal digital assistants have proven effective in helping primary health workers who often work in isolated rural environments exchange experiences with colleagues and improve their practices and the outcomes for their patients. Likewise, introduction of automatic teller machines (ATMs), which can accept, store, and dispense cash, and point-of-sale (POS) terminals can significantly expand access to financial services for the poor. Banks in Brazil, for example, use POS terminals, such as bankcard readers, at retail and postal outlets for bill payment, savings, credit, insurance, and money transfer in nearly every municipality in the country. These terminals can be established at a cost of less than 0.5 percent of the cost of establishing a typical bank branch (CGAP 2006). Owing to advances in technology, many banks are piloting ATMs and POS terminals that are suited to the rural infrastructure and the needs of the poor. For instance, the U.S.-based NCR Corporation has developed biometric ATMs, which use fingerprint authentication and offer a simplified menu and local language voice instructions that can be used by the illiterate population of rural India (Murali and Jaishankar 2007). Similarly, in cooperation with hardware manufacturers, VISA International developed a battery-powered wireless POS device suitable for rural areas. The device costs US$125, while most POS devices in developed countries cost about US$700.

E-government. Andhra Pradesh has pioneered e-seva, a network of public Internet offices where citizens can pay bills online. Those who have computers and credit cards can go to http://esevaonline.com; those without can visit an e-seva center. In Andhra Pradesh, e-seva now processes 110,000 transactions a day, worth Rs110 million (US$2.8 million), and is growing by 25 percent a year. Some 60 percent of all payments for public services in the state are made electronically. The state government wants to extend the network of e-seva centers from the current 119 to 4,600 across the state, one for every six villages. The plan is to use existing post offices. The business is outsourced; a private contractor recruits the staff, provides the computers and premises, and, in return, receives a small commission on each payment. The next stage will be to widen the scope of the system: a pilot project will allow people to apply for driving licenses online instead of queuing. But the most important move is to make the mobile phone, rather than the computer, the platform for payment. E-seva services will also be provided by "m-banking." Customers will be able to pay bills by sending an SMS (Short Message Service) and a security code (*Economist* 2008b).

Improvement in Access to Secondary Education. In 1968, Mexico developed Telesecundaria without external financing. Its main objective was to solve the problem of access to technology in rural areas. It targeted students in the 200,000 rural communities with populations of less than 2,500. In 1998, 15 percent of Mexico's lower-secondary students were educated through the program (World Bank 2005a).

Use of Walkie-Talkies for Maternal Health. The Rural Extended Services and Care for Ultimate Emergency Relief program, launched in March 1996 in the Iganga District of eastern Uganda, was designed to link traditional rural community health providers with the formal health system in a cost-effective way. Traditional health providers were given walkie-talkies to contact nurses and physicians if they encountered complications in a delivery. During the initial three-year period, the increased number of deliveries under trained personnel and increased referrals to health units led to a reduction of about 50 percent in maternal mortality rate in the district (World Bank 2007a).

Enhancement of Economic Opportunities. In rural villages in Bangladesh, where no telecommunications service previously existed, the Grameen Village Phone program provides mobile phones to very poor women who use them to operate a business. These microentrepreneurs purchase the phone with a loan from a Grameen Bank and then sell use on a per call basis. The typical "village phone lady" has an average income three times the national average. The most obvious benefit is the economic impact on the entire community, as phone users can bypass middlemen and connect directly to buyers and get better prices for their produce. Following its success in Bangladesh, the Village Phone program is being replicated in Uganda and Rwanda (Grameen 2005).

Access to Financial Services. Smart Communications, Inc., is a leading national telecommunications provider in the Philippines. It launched its mobile banking program "Smart Money" in 2000. Customers must sign up for Smart Money accounts at Smart stores. Thereafter, they can deposit and withdraw cash at Smart stores and thousands of retail outlets, ranging from supermarkets to individual kiosks and roadside stands. The money is held by Banco de Oro, a traditional bank, thus giving customers what are often their first bank accounts. Customers' mobile phones are their primary means of access. Clients can also receive remittances from family members through a Smart Padala program. The beneficiaries exchange electronic money into cash at any Smart Padala Center or at over 10,000 locations of partner organizations, such as pharmacies or rural banks (Ganchero 2007; Kramer, Jenkins, and Katz 2007).

Promoting Grassroots Innovation and Knowledge Initiatives

An important means of encouraging inclusive innovation is to support grassroots innovation networks and indigenous and traditional knowledge initiatives and to promote and diffuse their innovations.

Defining Grassroots Innovation Networks and Indigenous and Traditional Knowledge

The poor are valuable sources of informal innovation. They make extensive use of traditional or indigenous knowledge and experiment to produce valuable solutions to the challenges faced by their communities. The results of their efforts (in crafts, agriculture, or health care) are poorly documented and usually limited in their application to the innovator or the community in which he or she lives. Wider diffusion and scaling up of these innovations could help reduce poverty and generate income opportunities for the poor:

- *Grassroots innovation networks.* Grassroots innovation networks support individual or collective efforts that result in innovative products based on traditional or indigenous knowledge. Grassroots innovation programs focus on alleviating poverty through local knowledge, innovations, and practices, largely produced and maintained at the grassroots level. In some cases, value may be added by the science and technology sector, but the traditional knowledge and lead ideas emerge locally. In India, the largest and best-known NGO programs are the Honey Bee Network and the Society for Research and Initiatives for Sustainable Technologies and Institutions; the two largest government programs are the Grassroots Innovation Augmentation Network and the National Innovation Foundation.[10] The government has also set up the Traditional Knowledge Digital Library to prepare a computerized database of indigenous knowledge on medicinal plants.

- *Indigenous knowledge.* Indigenous knowledge is also referred to as traditional or local knowledge and encompasses people's skills, experience, and insights used to maintain or improve their livelihood (Subba Rao 2006). Indigenous knowledge is a key element of the social capital of the poor and constitutes their main asset for gaining control of their lives. Its special features include the fact that it is *local,* in that it is rooted in a particular community and situated within broader cultural traditions and based on the experience of those who live in that community; *tacit* and therefore not easily codifiable; *transmitted orally* or through imitation and demonstration; *experiential* rather than theoretical; *learned through repetition,* a defining characteristic even when new knowledge is added, as repetition helps retain and reinforce it; and *constantly changing,* produced as well as reproduced, discovered as well as lost, though external observers often see it as somewhat static (World Bank 1998). The development process should encourage the potential contribution of indigenous knowledge to locally managed, sustainable, and cost-effective survival strategies (box 11.7) (Gorjestani 2000).

- *Traditional knowledge.* Traditional knowledge is a form of knowledge that is traditionally linked to a certain community. It is knowledge developed, maintained, and passed on from generation to generation, sometimes through specific customary systems of knowledge transmission. This type of knowledge is created every day and evolves as individuals and communities respond to the challenges posed by their social environment. Some traditional knowledge is closely associated with plants and other biological resources, such as medicinal plants, traditional agricultural crops, and animal breeds. For example, traditional Chinese medicine (TCM) is an important part of China's cultural heritage and dates back thousands of years. Chinese herbal medicine includes many compounds that are not used in Western medicine. Advanced TCM practitioners in China are interested in statistical and experimental techniques that better distinguish medicines that work from those that do not, and TCM practitioners have recently cooperated with Western medical practitioners. For instance, at the Shanghai cancer hospital, a patient may be seen by a

> **Box 11.7 Using Indigenous Knowledge to Improve Health and Raise Agricultural Productivity**
>
> *Improving the life expectancy of AIDS patients:* In the coastal region of Tanga in Tanzania, traditional healers have treated the opportunistic diseases of over 4,000 HIV/AIDS patients with herb-based medicines. Patients report that the medicines help increase appetite and weight gain, stop diarrhea, reduce fever, and treat skin diseases. Most see results within 7–30 days of beginning treatment. Furthermore, many of these patients have lived five to seven years longer than if they had not been treated. The regional hospital has also given a floor to traditional healers who have established close collaboration with modern doctors. The healers have been trained as HIV/AIDS counselors, peer educators, condom distributors, and health care providers. Thus, leveraging traditional and modern knowledge has been crucial for increasing the effectiveness of HIV/AIDS prevention and treatment activities.[1]
>
> *Improving soil fertility in Burkina Faso:* Sahelian farmers have experimented with various soil and water conservation techniques to restore, maintain, or improve soil fertility. One technique, the plant-pit system, or Zaï, originated in Mali and was adopted and improved by farmers in northern Burkina Faso. Zaï is a planting pit with a diameter of 20–40 centimeters and a depth of 10–20 centimeters (depending on soil type), to which organic matter is added. After the first rainfall, the matter is covered with a thin layer of soil, and seeds are placed in the middle of the pit. Zaï conserves soil and water and controls the erosion of encrusted soils. The advantages of the technique are that it (a) captures rain and surface and run-off water; (b) protects seeds and organic matter from being washed away; (c) concentrates nutrient and water availability at the beginning of the rainy season; (d) increases yields; and (e) reactivates biological activities in the soil and eventually leads to an improvement in soil structure. The technique can reportedly increase production by about 500 percent if properly executed. The World Bank partnered with a local NGO to facilitate dissemination and scaling up of the Zaï technique from 2002 to 2004: 32 villages adopted the practice, and on average farmers have achieved a surplus production on one hectare of more than half a ton.[2]
>
> *Sources:* 1. World Bank Indigenous Knowledge Program. 2. World Bank 2005b.

multidisciplinary team and treated concurrently with radiation, surgery, Western drugs, and traditional herbal medicines. One result of this collaboration has been the creation of peer-reviewed scientific journals and medical databases on TCM.

Defining Challenges to Grassroots Innovation and Indigenous and Traditional Knowledge

Grassroots innovators and those who possess indigenous or traditional knowledge face various difficulties. In some cases, the culture of their communities is threatened, and the very survival of the knowledge is at risk. Another difficulty is the lack of respect and appreciation for knowledge that has not been scientifically validated. Research on grassroots innovations, especially in India, points to several challenges: high transaction costs for scouting and documenting innovations, the need to add value, and difficulties for commercialization and financing (Dutz 2007).

Grassroots and indigenous innovators face uncertainty because of a lack of organizing frameworks: they do not have information about the need for their innovations and how to find users. As a consequence, innovators are mostly indifferent to diffusing their knowledge and do not seek potential scale effects, efficiency, or productivity gains from their innovations. They can be said to be caught in an "indifference trap" and to hold back productive innovations and discoveries that they could share.

Another obstacle to sharing indigenous knowledge, especially in Africa's low-income agriculture sector, is the absence of an effective knowledge-sharing mechanism. To increase efficiency and productivity continually, producers need the support and advice of others and a cohesive learning and sharing network.

The lack of an intellectual property rights regime is another challenging issue for several reasons:

- First, in many cases of traditional knowledge, it is not clear who "owns" the intellectual property. Since many individuals and several communities can often claim to "possess" the knowledge, it is not clear who should control its dissemination or benefit from the revenues it generates.
- Second, there is little if any tradition of intellectual property rights in the informal economy. Individuals may be unaware that they possess valuable or patentable knowledge, and they may value secrecy over productive exploitation.
- Third, most useful innovations of this type do not meet the patent laws' technical requirements of novelty.
- Fourth, the costs of patenting are typically beyond the limited means of the informal sector's innovators.
- Fifth, advocates of traditional knowledge and volunteers who work with traditional communities often disagree about the nature of intellectual property rights, the balance between the needs of the communities at large and their individual members, and the best ways to leverage the knowledge into revenue-generating commercialization.

The issue of legal protection is therefore central, as it concerns the commercial exploitation of indigenous and traditional knowledge by others and thus raises questions of legal protection against misuse, the role of prior informed consent, and the need for equitable benefit sharing. A comprehensive strategy for protecting such knowledge should therefore consider the community, national, regional, and international dimensions. While the options and technicalities of protection systems are diverse, a common thread is that protection should principally benefit the holders of the knowledge, and in this case the indigenous and traditional communities and peoples that develop, maintain, and identify culturally with this knowledge and seek to pass it from generation to generation (box 11.8).

Box 11.8 Benefit-Sharing Arrangements and Intellectual Property Protection for Indigenous and Traditional Knowledge

Improving the livelihood of the Kani tribe: The medicinal knowledge of the Kani tribe, an ethnic group of some 16,000 people in southwestern India, was used in the development of the antistress and antifatigue drug, Jeevani. The scientists of the Tropical Botanic Garden and Research Institute used tribal knowledge about the properties of the wild plant arogyapaacha to develop the drug. The institute transferred the manufacturing rights to Aryavaidya Pharmacy Coimbatore Ltd., with the agreement to share the license and royalty income 50–50 with the Kani. A trust fund was established to manage the income from the drug's commercialization. In 2001 alone, the Trust Society, fully managed by the Kani, received INR 1.35 million (about US$30,000) in royalties and fees, which were invested in an interest-bearing account. The funds have been used to finance various self-employment schemes for Kani youth. As sales of Jeevani have grown, so has demand for the raw material. The Forest Department agreed to permit the Kani to cultivate the plant and sell the raw drugs in semiprocessed form to the manufacturer. This project, coordinated by the Trust Society, will provide additional income to the Kani.[1]

Sustaining economic opportunities in the Amazon: AmazonLife is a Brazilian fair-trade company that makes bags from a cotton-based fabric; the fabric is rubberized with natural latex sustainably harvested from wild rubber trees in the Amazon Rainforest for sale on the international market. The raw rubber is pressed into the cotton backing by the Seringeros (rubber tappers) and Indians that live deep in the Amazon Rainforest. The process for making the vegetal leather was protected with a patent, and the Seringeros are co-owners of the patented process. The Treetap brand was registered as a trademark. Only AmazonLife can use the vegetal leather process or market imitation leather as Treetap. Today, Amazon communities work in 32 production units in the forest and have a guarantee that they can produce and sell 40,000 sheets of wild rubber laminates per year at 10 times the previous price. Everyone involved in production is guaranteed a decent wage by the company's fair-trade policy. The product has received Forest Stewardship Council certification, ensuring the long-term sustainable production of the wild rubber. With some 200 families—approximately 1,000 persons—involved, the product's success has created elevated and sustainable economic opportunities for the people of the Amazon.[2]

Developing an anti-HIV compound through traditional methods: Traditional healers in the Falealupo village of Samoa have for centuries used a tea made by

> **Box 11.8 continued**
>
> steeping ground-up stems from the mamala tree to treat yellow fever virus and hepatitis. The Samoan healers introduced Western research scientists to the plant's healing capacity. The National Institutes of Health and the AIDS Research Alliance used the plant to isolate a compound called prostratin, which is thought to have high potential as an HIV retroviral. In 2004, the University of California at Berkeley and the Samoan government signed an agreement allowing the university's researchers to use the mamala tree to develop an anti-AIDS drug. The university will share any royalties from the sale of a gene-derived drug with the people of Samoa.[3]
>
> ***Creating varieties of blight-resistant rice:*** *Oryza longistaminata* is a wild rice that grows in Mali. Local farmers considered it a weed, but the migrant Bela community developed detailed knowledge of its agricultural value and recognized that *Oryza longistaminata* has stronger resistance to diseases such as rice blight than many other kinds of local rice. Guided by their traditional knowledge, researchers subsequently isolated and cloned a gene that confers resistance in rice plants.[4]
>
> *Sources:* 1. Finger 2004, 16; WIPO n.d. 2. http:// www.lightyearsip.net/ip_brazil_treetap.shtml. 3. http://www.lightyearsip.net/ip_samoa_mamala.shtml. 4. WIPO n.d.

Addressing the Challenges

Several policy measures can be taken to make a better use of indigenous knowledge:

- In spite of the activity surrounding grassroots innovation, as in India, not much has been done to assess or quantify its contribution to improving the livelihood of people in the informal sector. There is virtually no information on the costs or impacts of innovations, although there have been many and some have been licensed in India and elsewhere. What is needed is good monitoring and evaluation to support grassroots innovations that appear to be making a positive contribution. A pilot-inclusive innovation fund might be an appropriate mechanism.

- It is important to create local knowledge-sharing networks to help innovators share their inventions with potential users and other innovators both to gain recognition for their work and to increase knowledge generation for further innovation. Such networks call for public support. The greater the number of adopters of an innovation in the network, the greater the probability that users will continue to innovate. The policy objective of a local knowledge-sharing network should be to find workable strategies to increase allocative efficiencies and scale effects. It should also provide for knowledge "connections" to enable innovators, adopters, and intermediaries to interact; for innovators to enhance the innovation process; for adopters to find solutions to their problems; and for intermediaries to help connect and support interactions or improve the knowledge-sharing environment.

- Legal protection requires the participation of communities and countries from all regions to produce effective and equitable outcomes that are acceptable to all stakeholders. The challenges are diverse and far-reaching and involve many areas of law and policy. They go well beyond even the broadest view of intellectual property.

- Many international agencies and processes are concerned with these and related issues. However, responses should be coordinated and consistent. The preservation and protection against loss and degradation of traditional knowledge should go hand in hand with its protection against misuse and misappropriation. Thus, when traditional knowledge is recorded or documented with a view to preserving it for future generations, care needs to be taken to ensure that this effort does not inadvertently facilitate misappropriation or illegitimate use of the knowledge.

- National laws are also currently the prime mechanism for achieving protection and practical benefits for holders of traditional knowledge. Brazil, Costa Rica, India, Panama, Peru, the Philippines, Portugal, Thailand, and the United States have all adopted *sui generis* laws that protect at least some aspect of traditional knowledge.[11]

Enabling the Informal Sector to Absorb Knowledge and Technology

Another way to promote inclusive innovation is to help those in the informal sector, including enterprises, better absorb existing knowledge and technology. Most of the world's poor find work in subsistence agriculture or in the informal sector. Informal employment comprises half to three quarters of nonagricultural employment in developing countries: 48 percent in North Africa, 51 percent in Latin America, 65 percent in Asia, and 72 percent in Sub-Saharan Africa (WBI 2005). Therefore, raising the productivity of smallholder farms and informal enterprises can play a major role in alleviating poverty. For example, China's success in developing rural nonfarm opportunities is based on providing a flexible, demand-driven package of services—not just technology, but also information, technical assistance, marketing, supply networks, and supply chains.

Helping informal enterprises better absorb existing knowledge requires a multipronged strategy for addressing a variety of constraints faced by microentrepreneurs (such as low skills; lack of access to credit, modern technologies, and market information; and lack of links to potential buyers). The key issue is how to get existing knowledge to the poor and provide them with the means (supporting institutions, education, finance, and the like) to use it. An illustration is provided by the plethora of roadside motor mechanic shops all over India. Millions of mechanics do all kinds of repairs, and their problem-solving abilities and novel solutions show that their talents can be harnessed to increase productivity and achieve business ends. Most mechanics lack basic education, however, and have no access to formal engineering or science training. Creating a network of such entrepreneurs and giving them better access to modern training, knowledge, quality assurance, and finance could help them provide high value to customers, while increasing their productivity and their incomes. This effort would require the involvement and collaboration of different actors: research bodies, educational institutions, banks, NGOs, and large corporations, along with a strong government coordinating and supportive role. A recent study on nurturing entrepreneurship in India's villages points out that India should do more to empower its villagers, nurture entrepreneurial activity, and take advantage of its strengths in the private sector (Khanna 2008). Corporations need a seat at the table of village reform, even multinational corporations. Agreements such as that between Bharti Enterprises and Wal-Mart Stores should be encouraged, as such businesses, together with local ones, can lay the foundation for a modern agricultural supply chain linking farmers with the urban market.

Research Bodies and Academia

Public research and academic institutions do not usually have incentives to provide knowledge services to informal workers and subsistence farmers. However, there are examples of innovative approaches and successful partnerships aimed at bringing new technology and skills to the poor. For instance, scientists from the Central Leather Research Institute in India reached out to the village of Athaoni, where until recently Kalhapuri sandals were made using traditional techniques. Scientists helped reduce the time it takes to produce the sandals: the stamping process was standardized, and certain changes were made in the design, based on computer-aided technologies. This was not a top-down initiative, as the villagers were consulted during the development process. The institute's training of several hundred artisans has raised family incomes and changed views on science and development (Dahlman and Utz 2005). Similarly, CSIR's Crops Research Institute in Ghana developed improved varieties of groundnuts (peanuts) and thus helped double or even triple the yields of some 10,000 farmers (box 11. 9).

The above examples demonstrate that successful initiatives are demand driven and rely on the participation of beneficiaries at all stages of technology development and testing. They also show that while development of pro-poor technologies and new production processes is usually the domain of researchers, collaboration with other actors is necessary to ensure the absorption and effective use of innovations by the poor.

> **Box 11.9 Participatory Development of Improved Groundnut Varieties in Ghana**
>
> The Council for Scientific and Industrial Research–Crops Research Institute (CSIR-CRI) has developed improved groundnut varieties that are high yield and tolerant of drought and groundnut rosette disease and have fresh seed dormancy and consumer-preferred characteristics. The initiative came about after a series of planning workshops involving farmers, agricultural extension agents, researchers, policy makers, agro-processors, and seed and input sellers. About 70 percent of farmers identified groundnut rosette disease as widespread and devastating, often forcing them to abandon groundnut cultivation. With the support of the International Centre for Research in Semi-Arid Tropics (Mali), CSIR-CRI developed and screened a number of breeding lines for resistance to the rosette virus. Farmers were brought to the research stations to make selections based on their own criteria and later on were provided with technical support to conduct trials on their own land. As a result, four improved groundnut varieties were developed. The project led to higher yields and incomes for 10,000 farmers (primarily women as they are the major producers of this crop).
>
> The involvement of the various stakeholders was crucial for the success of the project. The Export Development Investment Fund helped link farmers to processors and groundnut exporters and also provided some loans, which enabled farmers to expand production. The Ministry of Food and Agriculture provided technical support to farmers at the trial stage and later trained certain farmers in best practices of seed production and contracted them to produce certified seed. NGOs also played a role by participating in the formulation of the project's objectives and facilitating the organization and transportation of farmers.
>
> *Source:* Essegbey 2008.

Networks to Foster Collaboration between Research Bodies and Producers

Public research centers and educational institutions generally do not have direct links with informal enterprises and farmers. The latter are usually marginalized and rarely initiate communication with formal institutions. Furthermore, poor producers have limited education and formal technical skills and often lack a common language with researchers. Therefore, special arrangements and intermediary organizations are often needed to foster collaboration between research bodies and poor producers.

In many cases, this role is played by informal sector associations and cooperatives. These bodies can be effective in addressing the needs of their members for skills development and technology transfer, lobbying for their interests, and offering internal credit schemes. Governments and donors can strengthen these associations by involving them in the identification of skills and technology needs and channeling technical assistance through them. NGOs can also build on proximity and their understanding of the needs of the informal sector to transfer knowledge, skills, and technologies that can improve productivity and the livelihood of the poor.

Ashoka. Ashoka is the global association of the world's leading social entrepreneurs. It works on three levels: (a) it supports individual social entrepreneurs, financially and professionally, throughout their life cycle; (b) it brings communities of social entrepreneurs together to help leverage their impact, scale their ideas, and capture and disseminate their best practices; and (c) it helps build the infrastructure and financial systems needed to support the growth of the citizen sector and facilitate the spread of social innovation globally. Since 1981, it has elected over 2,000 leading social entrepreneurs as Ashoka fellows, providing them with living stipends, professional support, and access to a global network of peers in more than 60 countries (http://www.ashoka.org).

Aid to Artisans. Aid to Artisans (ATA) works to create economic opportunities for artisan groups around the world where livelihoods, communities, and craft traditions are marginal or at risk. The real impact of ATA's work shows in generating new sales and linking to new markets. ATA works with partners in the entire distribution channel to ensure sustainability. Designing market-driven products and business training is only part of its equation; ATA connects artisans with exporters, importers, and retailers and ensures that they are working together so that each business becomes profitable. Aid to Artisans has spent 33 years creating economic opportunities for over 100,000 artisans in more than 110 countries. Over the past 10 years, ATA's efforts have leveraged nearly US$230 million in retail sales. About 70 percent of the artisans it works with are women (http://www.aidtoartisans.org).

TechnoServe. TechnoServe is an international NGO that helps create economic opportunities for poor people around the world. It has helped Jorge Salazar, a cooperative of poor farmers who produce low-quality coffee in the highland areas of Nicaragua, diversify into more profitable crops. TechnoServe also identified export demand for root crops as an opportunity for farmers. It helped them obtain better planting material and improve their production techniques to achieve consistent quality and a sixfold increase in yields. TechnoServe also linked them to exporters such as TecnoAgro and Hortifruti (a Wal-Mart subsidiary) and negotiated for cooperative members to sell their best-quality quequisque (a local crop similar to cassava) for five times what it would fetch in local markets. Furthermore, the cooperative's bookkeeping systems were upgraded, hygiene practices were improved, and help was provided to get loans for working capital. These improvements created 80 new full-time processing and packing jobs and an additional 200 seasonal jobs (http://www.technoserve.org).

Public-Private Organizations as Bridges

Last, specially created public-private bodies can also serve as a bridge between national and global research centers and informal enterprises or subsistence farmers. These professional institutions should have a mandate for the diffusion and adaptation of technologies for informal enterprises. The recently released recommendations package on building science, technology, and innovation capacity for Rwanda provides a detailed description of the establishment of the public-private Technology Information Service, whose primary function will be to search for and adapt technologies for local needs (box 11.10). Many of the recommendations designed for Rwanda could be replicated in other low-income countries.

Private Sector

Large companies can also serve as mentors to informal enterprises and subsistence farmers. For instance, in countries where the agricultural sector is dominated by small farms, supermarkets and processors already play an important role in enhancing smallholder productivity. They often sign production contracts with farmers that include extension services, credit, supply of inputs, and a market for final products. Similarly, multinational corporations and local companies transfer skills and technologies to informal workers to expand market share or as part of their CSR strategy, as in the case of Sumitomo, a Japanese firm, which has developed modern insecticide-treated mosquito nets, Olyset Nets, which are certified by the World Health Organization. The nets are easy to use and of high quality: their insecticidal efficacy is guaranteed to last for five years, and their properties are unaffected by washing. Sumitomo initiated a joint venture, Vector Health International Ltd., with a local company in Tanzania and licensed its manufacturing expertise for Olyset at no charge. Vector Health started production in January 2007 with an annual output capacity of 4 million nets; the company plans to double annual production to 8 million nets in the future. The manufacturing process is labor-intensive and employs 1,200 local workers. Sumitomo has sent its engineers to Tanzania to train local employees in production management, quality control, and worker safety. Its aim is not to profit by collecting licensing fees for this technology; its goal is to expand local production by providing education to local staff (WBI 2008).

Role of Banks and Microfinance Institutions

Lack of access to financial services is one of the main reasons why the poor are unable to improve their skills through formal education and acquire new technologies. Banks and microfinance institutions can help address this problem by extending their services to impoverished urban communities and remote villages. Traditionally, formal financial institutions did not wish to serve the poor because of the higher transaction costs and higher perceived risks. However, this situation started to change as a result of the success of the global microfinance movement. The establishment of the Grameen Bank in Bangladesh in 1983 was a milestone: its experience has demonstrated that small loans enable the poor to run and grow simple businesses and that

> **Box 11.10 Public-Private Technology Information Service in Rwanda**
>
> Many firms in Rwanda have limited information about technological options for improving their production processes. Establishment of the Technology Information Service will help enterprises acquire and adapt off-the-shelf technologies. More specifically, it will help them find cost-effective answers to such questions as, What technologies best meet the needs of the enterprise? How does one acquire a technology from local or foreign suppliers? How can the acquired technology be adapted to local needs? How is the technology to be used and maintained? How can a technology be built locally (using blueprints and designs from abroad)? Making such information easily available can boost the local machinery industry. Starting a technology information service will involve a partnership between industry and public technology institutions. Its tasks will include cataloguing every technology hardware product (domestic and imported) for sale in Rwanda and providing contact details for the source, technical information on the product, and, if possible, prices. To begin with, the service can build an inventory of all agricultural and industrial equipment used in Rwanda. This effort will help enterprises planning to invest in a technology locate a working example through the database, see it in operation, and discuss performance with the operator. It will also allow end users at every level to make informed judgments when choosing and purchasing equipment. Furthermore, it can offer training on various aspects pertaining to technology acquisition, management, and use. The service can also establish links with similar technology information services outside of Rwanda, such as the Agricultural Engineering Services Directorate, the Ministry of Food and Agriculture in Ghana, the International Network for Technical Information, and Practical Action to enable Rwandans to stay abreast of up-to-date developments.[a]
>
> A matching grant facility should be established to complement the Technology Information Service to ease access to finance, often a key constraint for small enterprises in upgrading technology. A public-private technology acquisition fund could encourage an enterprise to build its savings to buy technology. Later, when the enterprise decides to invest its savings in productive hardware, the fund could provide proportional matching grants (for example, a small-scale investor or cooperative might qualify for matching funding of 50 percent, while a larger enterprise might be eligible for a smaller percentage of matching funding). The fund can operate as a savings bank, paying interest on money invested by individuals, cooperatives, or enterprises, with money deposited and withdrawn for any purpose at any time, without restriction.
>
> *Source:* Watkins and Verma 2008, 42–43.
> a. See http://practicalaction.org.

poor borrowers repay loans reliably, even though they possess no collateral to guarantee the loans. Today, the Grameen Bank has 7.53 million borrowers (97 percent of whom are women) and operates in over 82,000 villages. The bank has a recovery rate of 98 percent and has made a profit in all but three years of operation. Grameen Bank's internal survey has shown that 65 percent of borrowers' families have crossed the poverty line.

There is ample evidence of the positive impact of microfinance institutions on job creation and poverty reduction (box 11.11). Apart from extending access to essential financial services to the poor, microfinance can significantly enhance the market power of the banks and even transform state-subsidized institutions into viable commercial establishments, as in the case of Bank Rakyat, Indonesia.

Governments can facilitate the expansion of financial services to the poor by establishing a friendly regulatory framework (most importantly through liberalization of interest rates) and by providing banks with technical assistance on lending to the poor. Partnering with donors in this regard may be useful as a number of development agencies have established training programs for banks on pro-poor financial products and are experienced in providing partial credit guarantees to encourage banks to lend to this segment. It is also important to

---

Box 11.11 Financial Institutions That Serve the Poor

SKS Microfinance: Launched in 1998, SKS Microfinance is one of the world's fastest-growing microfinance organizations, having provided over US$831 million in loans to more than 2.5 million women in poor regions of India. Borrowers take loans for a range of income-generating activities, including livestock, agriculture, trade (such as vegetable vending), production (from basket weaving to pottery), and other businesses (from beauty parlors to photography). SKS also offers interest-free loans for emergencies as well as life insurance to its members. SKS statistics show that borrowers increase their annual income by 11 percent more than nonborrowers. SKS currently has 1,166 branches in 16 states across India and aims to reach 4 million borrowers by 2009. In 2008 alone, SKS Microfinance achieved growth of nearly 170 percent, with a 99 percent on-time repayment rate.[1]

Transformation of the state bank through microfinance: Bank Rakyat Indonesia (BRI) is a state bank run on commercial principles. It has received worldwide fame for its success in developing a nationwide microfinance portfolio, which in 2004 served 31.3 million savers with average saving accounts of US$108 and 3.2 million borrowers with average outstanding balances of US$540. BRI is particularly notable for having transformed itself in three years from a large, subsidized, state-owned financial institution to a profitable bank by providing products that are in demand: small nontargeted loans, simple passbook savings accounts, and time and demand deposits. It turned each of the 3,600 branches in its nationwide network into profit centers. BRI's risk management techniques rely on sticks and carrots: cutting off nonperforming clients from future access to finance, making site visits to clients that coincide with repayment schedules, and providing incentives for timely repayment in the form of a refund of 25 percent of the interest payment on the loan.[2]

Sources: 1. WBI 2007. 2. SKS Microfinance, http://www.sksindia.com/.

consider the need for finance when developing skills-building programs for the informal sector. Training programs will be much more effective if they are linked with loans to enable the creation or expansion of businesses based on the new skills. Therefore, government agencies, NGOs, donors, and private corporations organizing training for their suppliers should attempt to establish partnership programs with the banks at the inception of such initiatives.

A number of organizations (national and international research institutions, informal and formal enterprises, various ministries, NGOs, and private companies) have unrealized synergies. Governments in developing countries should aim to foster links among such actors both to facilitate creation and commercialization of pro-poor innovations and to ensure effective knowledge transfer to the poor. In addition to establishing a business-friendly regulatory framework, governments can support collaboration among these actors by improving access to information on the needs of the poor and on the technologies available to address them, allocating a percentage of the national budget to support collective pro-poor R&D efforts by different organizations, and developing sectoral programs on knowledge transfer to the poor.

Notes

1. According to Collier (2007), global poverty is falling quite rapidly for about 80 percent of the world. The real crisis involves a group of about 50 failing states, the "bottom billion," whose problems defy traditional approaches to alleviating poverty.

2. For more information on the National Agricultural Innovation Project, visit http://www.naip.icar.org.in/.

3. For more information, see the Association for Strengthening Agricultural Research in Eastern and Central Africa, http://www.asareca.org/.

4. See World Bank (2006); Rajalahti, Janssen, and Pehu (2007); World Bank (2008a); World Bank (2007a); Davis and others (2007); and Chandra (2006).

5. From the point of view of India's CSIR, creative programs are needed to attract innovators to address issues of importance to people living in rural communities. These include a "new idea scheme" for funding innovations at CSIR for applications in rural India; another for funding innovations in non-CSIR laboratories for applications in rural India through extramural research grants, and periodic meetings of a peer group with National Award and Young Scientist Award winners for discussions of real-life problems facing rural India. http://www.csir.res.in/External/Utilities/Frames/achievements/main_page.asp?a=topframe.htm&b=leftcon.htm&c=../../../Heads/achievements/major_achievements.htm).

6. See the Talloires Network, http://www.tufts.edu/talloiresnetwork/.

7. See WBI (2008) and Beshouri (2006).

8. See Global Handwashing Day, http://www.globalhandwashingday.org/.

9. The World Business Council for Sustainable Development defines CSR as "the continuing commitment of business to behave ethically and contribute to economic development while improving the quality of life of the workforce and their families as well as of the local community and society at large" (Holme and Watts 2000).

10. For more information, see the Honey Bee Network, http://www.sristi.org/honeybee.html; Society for Research and Initiatives for Sustainable Technologies and Institutions (SRISTI),

http://www.sristi.org/cms/; Grassroots Innovation Augmentation Network (GIAN), http://www.gian.org; and the National Innovation Foundation, http://www.nifindia.org.

11. *Sui generis* measures are specialized measures aimed exclusively at addressing the characteristics of specific subject matter, such as traditional knowledge.

References and Other Resources

Abbott Laboratories. http://abbottglobalcare.org/sections/Strengthening/default-2.html?

Abramson, Bruce. 2007. "India's Journey toward an Effective Patent System." Policy Research Working Paper 4301, South Asia Finance and Private Sector Development, World Bank, Washington, DC.

Adu-Dapaah, Hans. 2007. "Farmer Participatory Selection of Rosette Resistant Groundnut Varieties in Ghana." Council for Scientific and Industrial Research–Crops Research Institute, Ghana. Unpublished.

Arredondo, Victor, and Mario Fernandez de La Garza. 2006. *Higher Education, Community Service and Local Development.* Paper presented at CHE-HEQC/JET-CHESP conference, "Community Engagement in Higher Education," Cape Town, South Africa, September 3–6.

Ashley, Caroline, Peter de Brine, Amy Lehr, and Hannah Wilde. 2007. "The Role of the Tourism Sector in Expanding Economic Opportunity." Corporate Social Responsibility Initiative Report 23. Kennedy School of Government, Harvard University, Cambridge, MA.

Ashoka. http://www.ashoka.org.

ATA (Aid to Artisans). http://www.aidtoartisans.org.

Banerjee, Parthasarathi. 2006. "Innovation in Informal, Small and Tiny Industries." Background paper for South Asia Finance and Private Sector Development, World Bank, Washington, DC.

Barcelo, Israel. 2008. "Patrimonio Hoy." *Development Outreach* (June): 27–29.

Berdegué, Julio A. 2005. "Pro-Poor Innovation Systems." Rome: IFAD (International Fund for Agricultural Development). http://www.ifad.org/events/gc/29/panel/e/julio.pdf.

Beshouri, Christopher. 2006. "A Grassroots Approach to Emerging-Market Consumers." *McKinsey Quarterly* 4 (November): 61–71.

Boechat, Claudio, and Roberta Paro. 2007. "Votorantim Celulose e Papel (VCP), Brazil: Planting Eucalyptus in Partnership with the Rural Poor." Growing Inclusive Markets Initiative, UNDP, New York. http://www.growinginclusivemarkets.org/images/pdf/english/Brazil_VCP percent20FINAL.pdf.

Bollag, Burton. 2004. "Improving Tertiary Education in Sub-Saharan Africa: Things That Work." Africa Region Human Development Working Paper, World Bank, Washington, DC.

CGAP (Consultative Group to Assist the Poor). 2006. "Using Technology to Build Inclusive Financial Systems." Focus Note 32, CGAP, Washington, DC.

CGIAR (Consultative Group on International Agricultural Research). http://www.cgiar.org/.

Chandra, Vandana, ed. 2006. *Technology, Adaptation and Exports: How Some Developing Countries Got It Right.* Washington, DC: World Bank.

Christiansen, Niels. 2008. "Creating Shared Value through Basic Business Strategy." *Development Outreach* 10 (2): 10–12.

Cisco Systems. http://www.cisco.com/.

Coca-Cola. http://www.thecoca-colacompany.com/citizenship/pacific.html#3.

Collier, Paul. 2007. *The Bottom Billion: Why the Poorest Countries Are Failing and What Can Be Done about It.* London: Oxford University Press.

Dahlman, Carl, and Anuja Utz. 2005. *India and the Knowledge Economy: Leveraging Strengths and Opportunities.* Washington, DC: World Bank.

Davis, Kristin, Javier Ekboir, Wendmsyamregne Mekasha, Cosmas M.O. Ochieng, David J. Spielman, Elias Zerfu, and others. 2007. *Strengthening Agricultural Education and Training in Sub-Saharan Africa from an Innovation Systems Perspective: Case Studies of Ethiopia and Mozambique.* Washington, DC: International Food Policy Research Institute.

Dutz, Mark, ed. 2007. *Unleashing India's Innovation Potential.* Washington, DC: World Bank.

Economist. 2008a. "Africa Calling." June 5. http://www.economist.com/people/displaystory.cfm?story_id=11488505).

———. 2008b. "The Electronic Bureaucrat: A Special Report on Technology and Government." February 16.

———. 2008c. "Just Good Business: A Special Report on Corporate Social Responsibility." January 19.

Essegbey, George Owusu. 2008. "Agribusiness Innovation Study—The Ghana Experience." Unpublished World Bank country report prepared for the forum on Practicing Agricultural Innovation in Africa: A Platform for Action, Dar-es-Salaam, Tanzania, May 12–14. http://web.worldbank.org/WBSITE/EXTERNAL/WBI/WBIPROGRAMS/KFDLP/0,,contentMDK:21608036~menuPK:461215~pagePK:64156158~piPK:64152884~theSitePK:461198~isCURL:Y,00.html.

EWB-I (Engineers without Borders–International). http://www.ewb-international.org/.

Finger, Michael. 2004. "Poor People's Knowledge: Helping Poor People to Earn from Their Knowledge." World Bank Policy Research Working Paper 3205, World Bank, Washington, DC.

Fox, Tom, Halina Ward, and Bruce Howard. 2002. *Public Sector Roles in Strengthening Corporate Social Responsibility: A Baseline Study.* Washington, DC: World Bank.

Ganchero, Elvie. 2007. "Case Study: Smart Communications: Low-cost Money Transfers for Overseas Filipino Workers." Growing Inclusive Markets Initiative. New York: UNDP.

Gates, Bill. 2008. "How to Fix Capitalism." *Time Magazine,* August 11.

GAVI (Global Alliance for Vaccines and Immunization). http://www.gavialliance.org.

Global Handwashing Day. http:// www.globalhandwashingday.org/.

Gorjestani, Nicholas. 2000. "Indigenous Knowledge for Development: Opportunities and Challenges." http://www.worldbank.org/afr/ik/ikpaper_0102.pdf.

Grameen Technology Center. 2005. *Village Phone Replication Manual.* Washington, DC: World Bank.

Grassroots Innovation Augmentation Network, http://www.gian.org.

Gupta, Anil K. 2007. "Towards an Inclusive Innovation Model for Sustainable Development." Paper presented at the Global Business Policy Council of A. T. Kearney, Dubai, United Arab Emirates, December 9–11. http://www.sristi.org/.../Towards percent20an percent20inclusive percent20innovation percent20model percent20for percent20sustainable percent20development.doc.

Haan, Hans, and Nicolas Serriere. 2002. "Training for Work in the Informal Sector: Fresh Evidence from West and Central Africa." Occasional Paper, International Training Centre, International Labour Organization, Turin, Italy.

Healthstore Foundation. http://www.cfwshops.com/.

Holme, Richard, and Phil Watts. 2000. *Corporate Social Responsibility: Making Good Business Sense.* Geneva: World Business Council for Sustainable Development.

Honey Bee Network, http://www.sristi.org/honeybee.html.

IFC Grassroots Business Initiative. http://www.ifc.org/gbi.

InfoDev. 2006. "Improving Health, Connecting People: The Role of ICTs in the Health Sector of Developing Countries." Working Paper 7, World Bank, Washington, DC.

ITU (International Telecommunications Union). 2008. *Measuring Information and Communication Technology Availability in Villages and Rural Areas.* Geneva: ITU.

Jenkins, Beth. 2007. "Expanding Economic Opportunity: The Role of Large Firms." Corporate Social Responsibility Initiative Report 17, Kennedy School of Government, Harvard University, Cambridge, MA.

Johanson, Richard, and Arvil Adams. 2004. *Skills Development in Sub-Saharan Africa.* Washington, DC: World Bank.

Johnson & Johnson. 2007. "2007 Worldwide Contributions Program Annual Report." http://www.jnj.com.

Khanna, Tarun. 2008. "Nurturing Entrepreneurship in India's Villages." *McKinsey Quarterly.* November.

Kogiso, Mari, Mia Matsuo, and Tokutaro Hiramoto. 2008. "Social Issue-Oriented BoP Business and Japanese Companies." *Development Outreach* 10 (2): 17–20.

Kramer, William, Beth Jenkins, and Robert Katz. 2007. "The Role of Information and Communications Technology Sector in Expanding Economic Opportunity." Corporate Social Responsibility Report 22, Kennedy School of Government, Harvard University, Cambridge, MA.

Light Years IP. http://www.lightyearsip.net/ip_brazil_treetap.shtml.

———. http://www.lightyearsip.net/ip_samoa_mamala.shtml.

Lund, Frances, and Caroline Skinner. 2005. "Local Government Innovations for the Informal Economy." *Development Outreach.* March.

Malhotra, Mohini, Yanni Chen, Alberto Criscuolo, Qimiao Fan, Iva Hamel, and Yevgeniya Savchenko. 2007. *Expanding Access to Finance: Good Practices and Policies for Micro, Small and Medium Enterprises.* Washington, DC: WBI.

Mohandas, Palat, and P. R. Reddy. 2004. "Earth Observation Support for National Drinking Water Mission in India." India-United States Conference on Space Science, "Applications and Commerce," Bangalore, June 21–25. http://www.aiaa.org/indiaus2004/Earth-observation.pdf.

Muniyappa, N. C., B. K. Ranganath, and P. G. Diwakar. 2004. "Remote Sensing and GIS in Participatory Watershed Development in Rural Karnataka." India-United States Conference on Space Science, "Applications and Commerce," Bangalore, June 21–25. http://www.aiaa.org/indiaus2004/Earth-observation.pdf.

Murali, D., and P. Jaishankar. 2007. "Financial Inclusion through Biometric ATMs." *The Hindu.* October 15. http://www.hindu.com/thehindu/holnus/006200710092001.htm.

National Innovation Foundation. http://www.nifindia.org.

Neath, Gavin, and Vijay Sharma. 2008. "The Shakti Revolution." *Development Outreach* 10 (2): 13–16.

Nelson, Jane. 2006. *Business as a Partner in Strengthening Public Health Systems in Developing Countries: An Agenda for Action.* International Business Leaders Forum (Prince of Wales International Business Leaders Forum), London. http://www.hks.harvard.edu/m-rcbg/CSRI/publications/report_13_HEALTH percent20FINAL.pdf.

Osei, Robert. 2007. "Integrated Tamale Fruit Company: Organic Mangoes Improving Livelihoods for the Poor." Growing Inclusive Markets Initiative. New York: UNDP. http://www.growinginclusivemarkets.org/images/pdf/english/Ghana_ITFC percent20FINAL.pdf.

Patrinos, Harry. 2006. *Public-Private Partnerships: Contracting Education in Latin America.* Washington, DC: World Bank.

Pfitzer, Marc, and Ramya Krishnaswamy. 2007. "The Role of the Food and Beverage Sector in Expanding Economic Opportunity." Corporate Social Responsibility Report 20, Kennedy School of Government, Harvard University, Cambridge, MA.

Prahalad, C. K. 2005. *The Fortune at the Bottom of the Pyramid: Eradicating Poverty through Profits.* Philadelphia, PA: Wharton School Publishing.

Prahalad, C. K., and Allen Hammond. 2002. "Serving the World's Poor Profitably." *Harvard Business Review* 80 (9): 4–11.

Pratham. http://www.pratham.org/.

Proctor and Gamble. http://www.pg.com/company/our_commitment/globalhandwashingday.shtml.

Rajalahti, R., W. Janssen, and E. Pehu. 2007. "Agricultural Innovation Systems: From Diagnostics toward Operational Practices." Agriculture and Rural Development Discussion Paper 38, World Bank, Washington, DC.

Rangan, Kasturi, John Quelch, Gustavo Herrero, and Barton Brooke, eds. 2007. *Business Solutions for the Global Poor: Creating Social and Economic Value.* San Francisco: Jossey-Bass.

Reuters. 2007. "Global Cellphone Penetration Reaches 50 Pct." November 29. http://investing.reuters.co.uk/news/articleinvesting.aspx?type=media&storyID=nL29172095.

Satellife. 2005. *Handhelds for Health: Satellife's Experiences in Africa and Asia.* Watertown, MA.

Savanti, Paula, and Elisabeth Sadoulet. 2008. "Agriculture's Special Powers in Reducing Poverty." *Development Outreach* 10 (3): 16–19.

SKS Microfinance. http://www.sksindia.com/.

Society for Research and Initiatives for Sustainable Technologies and Institutions. http://www.sristi.org/cms/.

Subba Rao, Siriginidi. 2006. "Indigenous Knowledge Organization: An Indian Scenario." *International Journal of Information Management* 26 (3): 224–33.

Sulton, Christopher, and Beth Jenkins. 2007. "The Role of the Financial Services in Expanding Economic Opportunity." Corporate Social Responsibility Initiative Report 19, Kennedy School of Government, Harvard University, Cambridge, MA.

TechnoServe. http://www.technoserve.org.

UNDP (United Nations Development Programme). 2007. *Human Development Report 2007/2008: Fighting Climate Change.* New York: Palgrave Macmillan.

United Nations. 2005. *Improving the Lives of the Urban Poor: Case Studies on the Provision of Basic Services through Partnerships.* New York: United Nations.

———. 2007. *The Millennium Development Goals Report 2007.* New York: United Nations.

Watkins, Alfred, and Anubha Verma, eds. 2008. *Building Science, Technology and Innovation Capacity in Rwanda.* Washington DC: World Bank.

WBI. (World Bank Institute). 2005. *Development Outreach: A Better Investment Climate for Everyone.* Washington, DC: World Bank.

———. 2007. Expanding Access to Finance. Washington, DC: World Bank.

———. 2008. *Development Outreach: Business and Poverty: Opening Markets to the Poor.* Washington, DC: World Bank.

WEF (World Economic Forum). 2006. *Harnessing Private Sector Capabilities to Meet Public Needs.* Geneva: WEF.

WIPO (World Intellectual Property Organization). N.d. *Intellectual Property and Traditional Knowledge.* Booklet No. 2, p. 9. www.wipo.int/freepublications/en/tk/920/wipo_pub_920.pdf.

World Bank. 1998. *World Development Report 1998/99: Knowledge for Development.* Washington, DC: World Bank.

———. 2005a. *Expanding Opportunities and Building Competencies for Young People: A New Agenda for Secondary Education.* Washington, DC: World Bank.

———. 2005b. "Burkina Faso: The Zaï Technique and Enhanced Agricultural Productivity." IK [Indigenous Knowledge] Notes 80, World Bank, Washington, DC.

———. 2006. *Enhancing Agricultural Innovation: How to Go beyond the Strengthening of Research Systems.* Washington, DC: World Bank.

———. 2007a. *Building Knowledge Economies: Advanced Strategies for Development.* Washington, DC: World Bank.

———. 2007b. *Cultivating Knowledge and Skills to Grow African Agriculture: A Synthesis of an Institutional, Regional, and International Review.* Washington, DC: World Bank.

———. 2008a. *Agriculture for Development: World Development Report 2008.* Washington, DC: World Bank.

———. 2008b. *World Development Indicators 2008.* Washington, DC: World Bank.

———. Indigenous Knowledge Program. http://web.worldbank.org/WBSITE/EXTERNAL/COUNTRIES/AFRICAEXT/EXTINDKNOWLEDGE/0,,contentMDK:20672365~pagePK:64168445~piPK:64168309~theSitePK:825547,00.html.

World Resources Institute and IFC. 2007. *The Next 4 Billion.* Washington, DC: World Resources Institute and International Finance Corporation. http://rru.worldbank.org/thenext4billion.

You, Liangzhi, and Michael Johnson. 2008. "Exploring Strategic Priorities for Regional Agricultural R&D Investments in East and Central Africa," IFPRI Discussion Paper 776, International Food Policy Research Institute, Washington, DC.

Index

Boxes, figures, notes, and tables are indicated by *b, f, n,* and *t,* respectively.

A

Abbott Laboratories, 351*b*
ABN AMRO, 350
ACCION International, 350
Advanced Technology Program (ATP), U.S., 221–22
AES-EDC, 348
Africa. *See* Middle East and North Africa; Sub-Saharan Africa
African Growth and Opportunity Act (AGOA), U.S., 111
agriculture, 275–83
    agribusiness, 277–78, 279*b*
    AIS approach, 277–78, 278*f*
    high-value commodities, 278–80, 280*t*, 281*b*
    historical revolutions in, 6–7, 7*f*, 32–34
    important opportunities provided by, 273–74
    indigenous and traditional knowledge, using, 358*b*, 361*b*
    pro-poor innovations in, 338–40, 358*b*, 361*b*
    R&D in, 339–40
Aid to Artisans (ATA), 365

AIDS/HIV
    indigenous and traditional knowledge used to treat, 358*b*, 360–61*b*
    IPR in Brazil and, 113*b*
AIDS Research Alliance, 361*b*
Airbus, 271
AIS (agricultural innovation systems), 277–78, 278*f*
Akman, Murat, 284*b*
Albania, 195*n*2
AmazonLife, 360*b*
American Medical Association, 289
Angel Capital Electronic Network, U.S., 93
angel investment, 90*b*, 93
apprenticeship training, 189–90
Arab Republic of Egypt.
    See Egypt, Arab Republic of
Aracruz, 350
Aravind Eye Hospital, India, 5*b*, 25
Arcelor, 272
Archibugi, Daniele, 202
Arco Technology Index (ATI), 202

Argentina
    education and training in, 187
    implementation of innovation
        policy in, 237
    implementation of policy in, 242,
        244, 245, 247
    innovation surveys, 44
Armenia, 247, 249, 251
artisans
    ATA, 365
    tourist demand for crafts, 292
Aryavaidya Pharmacy Coimbatore
    Ltd., 360*b*
Ashesi project, Ghana, 250*b*, 256
Ashoka, 365
Asia. *See also* East Asia and Pacific;
        Eastern Europe and
        Central Asia; South Asia;
        specific countries
    brain drain/diaspora, 192
    informal sector in, 362
    old and new tigers of, 42,
        244, 245, 251
    societal specificities, adapting
        innovation to, 67*f*
assessment. *See* evaluation
Astrakhan, Irina, 132*n*24
ATA (Aid to Artisans), 365
ATI (Arco Technology Index), 202
ATMs (automatic teller machines),
    354–55
ATP (Advanced Technology Program),
    U.S., 221–22
Aubert, Jean-Eric, 53*n*
Australia
    business services in, 77
    Cooperative Research
        Centers, 220
    EIS SII, 203
    natural resources in, 247
    NQFs, 185
    Oslo Manual innovation
        survey, 206–7

railways as network industries
    in, 121*b*
R&D in, 158
regulatory framework in, 117
Austria, 93, 104*n*8, 138, 187
automatic teller machines
    (ATMs), 354–55
automobile industry, 25, 272,
    285–87, 296
Awuah, Patrick, 250*b*
Ayala Corporation, 346*b*

B

backward links or vertical spillovers
    from SEZs, 306–7, 307*b*
Bahrain, 88, 92, 104*n*9, 304
Bahrain Business Incubator, 88
Banco de Oro, Philippines, 356
Bangalore IT services cluster, 63, 94*b*,
    172, 253, 259, 317, 324
Bangladesh
    education and training in, 171
    electrical supply in, 131*n*13
    garment industry in, 283
    implementation of policy in, 251
    microcredit in, 54, 366–67
    pro-poor innovations in, 354,
        356, 366–67
Bank of the Philippine Islands, 346*b*
Bank Rakyat Indonesia (BRI), 368*b*
banks and banking, pro-poor, 354–55
base of the economic pyramid (BOP),
    market size and business
    strategy at, 344–45, 345*b*
Bayh-Dole Act, U.S., 154–55
Belarus, 237, 244, 247–48, 260
benchmarking. *See* evaluation
Benin, 111
Berdegué, Julio A., 336
Berkshire Plastic Networks, western
    Massachusetts, 99
Bharti Enterprises, 363
Bhutan, 231*b*, 289

Bibliotheca Alexandrina, Arab
 Republic of Egypt, 5*b*, 25–26
Bill and Melinda Gates
 Foundation, 352
Biodiversity and Tourism in Costa Rica
 Association, 292
BMW, 287
Boeing, 271
Bogotá Manual, 208
Bolar exception, 131*n*6
bootstrapping, 268
Booz Allen Hamilton, 209*b*
BOP (base of the economic pyramid),
 market size and business
 strategy at, 344–45, 345*b*
born global firms, 248, 260
Boston Consulting Group, 209*b*
Botswana, 43, 185, 289
BP, 350
brain drain/diaspora, 190–94, 192*f*,
 194*f*, 250*b*, 257–58*b*, 309*b*
Brazil
 automobile industry in, 285–86
 clusters in, 94, 95
 education and training in, 186,
  187, 189, 195*n*2
 implementation of policy in, 242,
  244, 245, 246
 indigenous and traditional
  knowledge, laws
  protecting, 362
 innovation surveys, 44
 IPR and HIV/AIDS in, 113*b*
 jeans production in, 285*b*
 metrology, standards, testing, and
  quality control, 81–82, 103*n*3
 priorities for, 49–50
 pro-poor innovations in, 335,
  350, 354, 360*b*, 362
 productivity dispersion
  in, 45–46, 45*t*
 R&D in, 137, 140*f*, 151, 153
 small firms, support for, 85

supporting key industries in, 271,
 284–86, 285*b*, 289
tourism in, 289
transition from developing to
 developed economy in, 43
BRI (Bank Rakyat Indonesia), 368*b*
BRICs, 140*f*, 285, 350. *See also* Brazil;
 China; India; Russian
 Federation
Bridgman, David, 290
broad perspective, importance of,
 9–11, 10*f*, 54, 55*f*
broker policies, 319, 330
BRS (Business Reporting System), 222
Budweiser, 272
Buffett, Warren E., 287
building innovative sites, 21, 303–34.
 See also clusters; science parks;
 special economic zones
 cities and regions, fostering
  innovation in, 323–29,
  325*b*, 328*b*, 329*b*
 OECD, evolution of innovation
  policy in, 57*b*
 as principle of innovation
  policy, 63–64
Bumrungrad, Thailand, 289
bureaucracy, as obstacle, 62
Burkina Faso, 111, 171, 358*b*
business angels, 90*b*, 93
business environment, promoting. *See
 also* regulatory frameworks;
 supporting competitive and
 innovative industries;
 supporting innovators
 domestic regulatory framework
  and, 116–24, 117*t*, 119*b*, 121*b*
 as principle of innovation policy,
  60–65, 60*f*
business incubators, 86–88, 86*b*, 88*b*,
 89*b*, 104*n*6, 260, 320
business models for pro-poor
 innovations, 347, 348*t*

Business Reporting System (BRS), 222
business services, 74–83
    basic industrial services (promotion, marketing, internationalization), 12*b*, 75
    ICT services, 12*b*, 75, 83–84
    incubators, 104*n*6
    investment promotion services, 78–79
    metrology, standards, testing, and quality control, 12*b*, 75, 80–82
    one-stop shops, 76–78, 103
    organizational/management services, 12*b*, 75
    productivity centers, 82
    public and private provision of, 76–78, 78*f*
    specialized service infrastructure, 78–83
    technology extension services, 12*b*, 75, 79–80
    types and examples of, 12*b*, 75
BYD Auto, 286, 287

C

Cambodia, 283, 284
Cambridge Science Park, U.K., 310
Cameroon, 98*b*, 171
Canada
    brain drain/diaspora, 192
    competition policy in, 131*n*11
    EIS SII, 203
    natural resources in, 247
    Networks of Centers of Excellence, 220
    Oslo Manual innovation survey, 206–7
    R&D in, 138, 158
    tariff regimes, 110
    Vancouver Agreement, 329*b*
Capability Maturity Model (CMM), 259
capacity for innovation
    factors affecting, 7, 8*f*
    implementation of policy driven by, 18–19, 19*t*
    as innovation-related indicator, 209*b*
    strategies for building, 225–27
carbon dioxide sequestration, 159
Caribbean. *See* Latin America and Caribbean
catalytic procurement, 126, 128–30
Celtel, 272
CEMEX, 245, 346*b*, 347
Central America. *See* Latin America and Caribbean
Central Asia. *See* Eastern Europe and Central Asia
Central Leather Research Institute, India, 363
CGIAR (Consultative Group on International Agricultural Research), 159, 352–53
champions, networks of, 264, 266–67
ChangAn, 286
Charple Report, 56*b*
Chen, Derek, 199*n*
Chery, 286
Chevrolet Volt, 287
Chile
    brain drain/diaspora, 194, 257–58*b*
    education and training in, 187, 195*n*4
    implementation of policy in, 245, 246, 257–58*b*
    innovation policy review in, 228
    innovation surveys, 44
    networks in, 100–101
    OECD innovation policy review of, 57*b*, 225
    R&D in, 153, 158
    salmon industry, 296
    technology extension services in, 80
    value chain, use of innovation policy to move up, 227

Chilean Economic Development
    Agency (CORFO), 80
ChileGlobal, 257–58*b*
China
    automobile industry in, 285,
        286, 287, 295*t*, 296
    brain drain, 191, 192–93, 194
    business incubators in, 87, 89*b*
    business services in, 77, 78*f*
    cities and regions, fostering
        innovation in, 324
    clusters in, 95, 98–99
    education and training in, 166, 186
    guanxi, 67
    hand washing scheme in, 349*b*
    ICT industry in, 293
    implementation of policy in, 238,
        244, 245, 255, 261–62, 263*b*
    Industrial Revolution missed by, 38
    as new economic tiger, 42
    OECD innovation policy review
        of, 57*b*, 225
    pilot programs, 132*n*20
    precommercial procurement in, 127
    priorities for, 49–50
    pro-poor innovations in, 335,
        339, 349*b*, 350, 362
    R&D in, 136, 137, 138, 139–40,
        140*f*, 142, 150, 151, 153,
        155, 161*n*6, 261–62
    science parks in, 98–99, 104*n*14
    SEZs in, 21, 309*b*, 310*b*
    small firms, support for, 85, 92
    supporting key industries in, 272,
        273*b*, 283, 285, 286–87,
        288, 293, 295*t*, 296
    TCM, 357–58
    transition from developing to
        developed economy in, 43
    wind power in, 272, 273*b*
Chinese Federation for Corporate
    Social Responsibility, 350
Chrysler, 296

Ciba-Geigy, 61–62
CIMO (Integral Quality and
    Modernization Program),
    Mexico, 188*b*
CIS (Community Innovation Survey),
    EU, 44, 206, 207–8, 213
cities
    fostering innovation in, 323–29,
        325*b*, 328*b*, 329*b*
    integration of science parks with,
        311–13, 313*b*
Cities Alliance, 329
clean water, access to, 46–47, 47*t*
climate change, 36–37, 37–38*b*
clusters, 316–23
    characteristics of successful
        clusters, 319–21, 321*b*
    consumer production, orientation
        toward, 321*b*
    in critical mass agenda, 244–46
    definition and purpose of,
        317–18, 318*b*
    education and training associated
        with, 320
    efficiency, improving, 321–22
    evaluating, 322–23
    incubators based in, 320
    mapping, 318–19
    regional-level innovation policy
        reviews, 229–30
    supporting innovators, 93–99,
        94*b*, 97*b*
    supportive policies for, 317–18,
        319, 320, 330
CMM (Capability Maturity
    Model), 259
CNA (Confederazione Nazionale
    Artiglianato), Bologna,
    Italy, 97*b*
Coca-Cola, 351*b*
Coco, Alberto, 202
coffee industry, 271, 273, 280–83,
    295*t*, 296

cognitive revolution, 34–36, 36f
collective accountability model for
    pro-poor innovations,
    347, 348t
collective associations, 96–97, 97b, 98b
Collier, Paul, 256
Colombia, 158, 195n2, 229, 327
Community Innovation Survey (CIS),
    EU, 44, 206, 207–8, 213
community service initiatives,
    university-based, 341–44
competence-based education and
    training systems, 185–86
competition policy, 118–21
The Competitive Advantage of Nations
    (Porter), 316–17
competitive industries. *See* supporting
    competitive and innovative
    industries
CONACYT (National Council for
    Science and Technology),
    Mexico, 77–78
concepts of innovation policy,
    5–11. *See also* innovation
    policy; principles of
    innovation policy
Confederazione Nazionale
    Artiglianato (CNA),
    Bologna, Italy, 97b
Consultative Group on International
    Agricultural Research
    (CGIAR), 159, 352–53
consumers, poor as, 344–45, 345b, 346b
converging technologies, 35, 36f
Cooperative Research Centers,
    Australia, 220
copyright. *See* intellectual
    property rights
CORFO (Chilean Economic
    Development Agency), 80
corporate social responsibility
    (CSR) initiatives, 338,
    350–52, 369n5

Costa Rica
    indigenous and traditional
        knowledge, laws
        protecting, 362
    Intel semiconductor assembly plant,
        5b, 26, 309b
    tourism in, 273, 289, 291–92,
        291t, 295t, 296
Côte d'Ivoire, 171, 185
cotton subsidies in Europe and
    U.S., 111
Council on Scientific and Industrial
    Research (CSIR), 149b,
    341b, 363, 364b, 369n5
creative capitalism, 350
"creative class," 325b
creativity and innovation, 168
crisis and innovation, 1–2, 6, 23
critical mass
    as implementation agenda,
        244–49, 248b
    as implementation problem, 241–42
    support for key industries in
        achieving, 297
Croatia, 203, 271
Crops Research Institute, CSIR,
    Ghana, 363, 364b
cross-border trade, 121–23
CSIC, Spain, 219
CSIR (Council on Scientific and
    Industrial Research), 149b,
    341b, 363, 364b, 369n5
CSR (corporate social responsibility)
    initiatives, 338, 350–52, 369n5
cultural issues
    adapting innovation to societal
        specificities, 67–68, 67f, 69
    technical culture, promoting, 61
CUORE (Urban Operational Centers
    for Economic Renewal),
    Italy, 328b
customs administration of
    cross-border trade, 121–23

D

Dahlman, Carl, 31*n,* 53*n,* 135*n*
Dahmen, Eric, 303
Daimler, 287
Danish Network Program, 99–100*b*
DDT, 61–62
Dehoff, Kevin, 138
Dell, 241
Delphi method, 331*n*7
Denmark, 99–100*b,* 121*b,* 253
Desarrolladora Homex, 272
developing countries, innovation policy for. *See* innovation policy
diaspora of skilled workers, 190–94, 192*f,* 194*f,* 250*b,* 257–58*b,* 309*b*
dissemination of technology, 46–48, 49*t*
Docquier, F., 191
Doing Business surveys, World Bank, 13, 132*n*21, 322, 330*n*4
Dominican Republic, 307*b,* 330*n*1, 341
Dongfeng, 286
double transformation, 267–68
Dr Reddy, 151
drinking water, access to, 46–47, 47*t*
dual models of education and training, 184–85
Dubai, 289, 324
Dubarle, Patrick, 73*n*
Duflo, E., 222–24

E

e-government, 355
e-seva, 355
EAO (Economic Assessment Office), U.S., 221–22
early working exception, 131*n*6
East Asia and Pacific. *See also specific countries*
    education and training in, 190
    MDGs in, 336, 337*t*
    old and new tigers of, 42, 244, 245, 251
    R&D, government support for, 150
    tariff regimes, 110
East-West societal contrasts, 67–68, 67*f*
Eastern Europe and Central Asia. *See also specific countries*
    brain drain, 191
    education and training in, 180
    EU, countries joining, 244
    implementation of policy in, 242, 244, 245, 249
Economic Assessment Office (EAO), U.S., 221–22
economic development incubators, 86*b*
economic growth from innovation policy, 6–7, 31–52
    comparative effects, 6*f*
    dissemination of technology, 46–48, 49*t*
    divergence between countries and regions, 38–40, 39*b,* 39*f*
    in emerging economies, 43–46, 45*t*
    historical examples of, 6–7, 7*f,* 32–36*f,* 32–37, 36*t*
    priorities for developing nations, 48–50
    professional economic analysis of innovation, 40–43, 42*t*
    simple technologies with major impact on welfare, 46–47, 47*t,* 48*t*
    systemic changes, 34–35, 34*f*
    TFP, 41–43, 42*f*
economic policy and performance, as innovation-related indicator, 209*b*
ecotourism, 288–89
Edison, Thomas Alva, 40
Edler, Jakob, 125
Edquist, Charles, 125

education and training, 15–17, 165–98
　apprenticeships, 189–90
　brain drain/diaspora workers, 190–94, 192f, 194f
　clusters associated with, 320
　competence-based systems, 185–86
　curriculum, adapting and updating, 179–81, 184
　dual models of, 184–85
　generic skill requirements, 167–69, 173–74
　ICT sector requirements, 165, 166, 172, 174, 175
　in-service, 186–89, 188b
　informal sector, skills development for, 189, 327, 328b, 369
　learning and teaching models, 175–83, 175t, 176b, 177–78b, 182b, 184
　lifelong learning, importance of, 173, 174–75, 175t, 176
　literacy rates, 170, 171, 174–75, 195n2
　NQFs, 185
　pro-poor innovations in, 349b, 351b, 355, 363, 369
　quality of, 170–71
　quantity of, 170
　R&D associated with universities, 137–38, 137t, 155–57, 156t, 158t
　relevance of, 171–73, 173t
　SEZs, 307–8, 308t
　skills required for knowledge-driven and innovative economies, 165–73, 173t
　specific skills for innovation, 169
　supply of skilled workers, problem of, 169–70
　supporting key industries, 296–97
　VET, 183–86, 320
　youth entrepreneurship programs, 180–81, 182b

Educational Service Contracting (ESC) scheme, Philippines, 349b
Egypt, Arab Republic of
　Bibliotheca Alexandria, 5b, 25–26
　education and training in, 195n2
　equity fundraising in emerging markets, 92
　supporting key industries in, 272
　tourism in, 289
EIS. See European Innovation Scoreboard
El Salvado, 283
electrical supply, 120
Embraer, 151, 271
emerging innovation leaders agenda, 249
emigration of skilled workers, 190–94, 192f, 194f, 250b, 257–58b, 309b
Engineering Research Centers, U.S. National Science Foundation, 220
Engineers Without Borders - International (EWB-I), 353–54
entrepreneurs, supporting. See supporting innovators
entry barriers, 117t, 118–21
environmental issues
　artisanal crafts, demand for, 292
　demands on limited resources, 36
　ecotourism, 288–89
　global warming, 36–37, 37–38b
　green technology, importance of, 23
　R&D in, 159
EPAM, 247
EPZs (export processing zones), 21, 251
ERVET (Territorial Development Agency of the Emilia Romagna Region), Italy, 100
ESC (Educational Service Contracting) scheme, Philippines, 349b
Esquivel, Gerardo, 306

Estonia, 5*b*, 26
Ethiopia, 276
Europe/European Union (EU).
 See also Eastern Europe
 and Central Asia;
 specific countries
 brain drain, 192
 business incubators in, 87
 CIS, 44, 206, 207–8, 213
 clusters, 318*b*, 331*n*10
 competition policy in, 131*n*11
 Eastern European countries
  joining, 244
 evolution of innovation policy
  in, 56–57*b*
 foresight approach in, 265
 international policy learning,
  commitment to, 215*b*
 mutual recognition agreements,
  use of, 112
 procurement policy in, 125
 R&D in, 136, 137, 138
 RTD FP, 221
 Rwandan coffee industry, support
  for, 281
 Six Countries Program, 61, 69*n*1
 trade/tariff regimes, 110, 111, 112
European Commission, 126
European Innobarometer
 Survey, 169–70
European Innovation Scoreboard (EIS)
 innovation-related indicators, 209*b*
 program evaluation,
  designing, 215*b*
 SII, 202–3
Eurostat, 208
evaluation, 17, 199–234
 clusters, 322–23
 country-level benchmarking
  indexes, 200–203
 country rankings across indexes,
  203–6, 204–5*t*
 GDP, alternatives to, 231–32*b*

microlevel innovation surveys,
 44–45, 206–13
 adapting, 211–12
 Bogotá Manual, 208
 in developed countries, 207–8
 in developing countries, 208–12
 examples of indicators from, 207*b*
 innovation-related indicators,
  208, 209*b*
 methodological issues, 212–13
 Oslo Manual, 206–8, 210, 211
 OECD innovation system and
  policy reviews, 57*b*
 policy review, 224–30, 226*b*
 program evaluation. *See* program
  evaluation
 regulatory reforms, 123–24
 strategic incrementalism, as element
  of, 265
 subnational implementation of
  innovation policy,
  benchmarking for, 254
EWB-I (Engineers Without Borders -
 International), 353–54
Export Development Investment
 Fund, 364*b*
export processing zones
 (EPZs), 21, 251
exports as springboard agenda, 251
EXPOTUR, 292

F
fair trade, 13–14, 109–12, 271, 288
favorable conditions for innovation,
 creating, 58*f*
FAW, 286
FDI. *See* foreign direct investment
field experiments, 222–24
Fiji, 330*n*1
financial costs of regulatory
 framework, 117*t*
financial services, access to, 354–55,
 356, 366–69, 368*b*

financial support as innovation policy
   education and training, 187–88
   incentives and instruments, 65–66
   industries, support for, 297
   for innovators, 89–93, 90*f,* 92*t*
   pro-poor innovations, 343
   R&D, incentives for, 215–18
   science parks, 315–16
   subnational implementation efforts, 254–55

Finland
   education and training in, 170, 180
   Estonia and, 5*b,* 26
   evolution of innovation policy in, 57*b*
   implementation of policy in, 253, 256
   natural resources in, 247
   railways as network industries in, 121*b*
   Science and Technology Policy Council, 57*b,* 65
   science parks in, 314*b*
   Tekes, 57*b,* 65, 247

first-mover institutions, creating, 250*b,* 251–52
first-mover problem, 241
Florida, Richard, 325*b*
foreign direct investment (FDI)
   competition policy and, 118
   from diaspora populations, 192
   domestic regulatory framework and, 116
   investment promotion services for innovators, 78–79
   IPR regimes and, 115
   science parks encouraging, 98
   technology transfer, 109

foresight approach, 265–67
Foxley, Alejandro, 257*b*
France
   angel investment in, 93
   business incubators in, 87
   business services in, 76
   education and training in, 187
   evolution of innovation policy in, 56*b*
   ICT business services in, 83
   INSERM, 219
   OSEO, 65
   railways as network industries in, 121*b*

Fraunhofer system, 56*b,* 76
free trade, importance of, 13–14, 109–12
functions of innovation policy, 11–17. *See also* education and training; evaluation; regulatory framework; research and development; supporting innovators
Fundación Chile, 257–58*b*

G

García, Ramón L., 257–58*b*
gardener metaphor
   cities as greenhouses for innovation, 324
   government as gardener, xv, 2–3, 8–9, 9*f,* 60*f*

garment industry, 274*b,* 283–84, 285*b*
Gates, Bill, 350, 352
Gaudin, Thierry, 31*n,* 53*n*
GAVI (Global Alliance for Vaccines and Immunization), 352, 353
GCI (Global Competitiveness Index), WEF, 203, 204–5*t,* 205, 215, 232*n*2
GDP. *See* gross domestic product
Geely Automobile, 286, 287
gender issues
   pro-poor innovations, 347
   SEZs employing women, 307
General Motors (GM), 137, 286, 287, 296
Georghiou, Luke, 125, 222

Germany
  business services in, 76
  evolving innovation policy in, 56*b*
  manufacturing in, 283
  Max Planck *Gesellschaft* model, 219
  public venture capital funds, 91
  railways as network industries in, 121*b*
  R&D in, 142
  VET in, 184–85
Ghana
  Ashesi project, 250*b*, 256
  Crops Research Institute, 363, 364*b*
  economic development, effect of innovation on, 6*f*
  education and training in, 172, 189, 195*n*2
GII (Global Innovation Index), INSEAD/World Business, 203, 204–5*t*, 205
Global Alliance for Vaccines and Immunization (GAVI), 352, 353
Global Competitiveness Index (GCI), WEF, 203, 204–5*t*, 205, 215, 232*n*2
global economic crisis, 1–2, 22–23
Global Fund to Fight AIDS, Tuberculosis, and Malaria, 352
Global Handwashing Day, 349*b*
Global Innovation Index (GII), INSEAD/World Business, 203, 204–5*t*, 205
global networks, 338, 352–54
Global Positioning System (GPS), 25, 127
Global Reporting Initiative, 352
Global Research Alliance, 352
global warming, 36–37, 37–38*b*
globalization
  acceleration and change in, 272
  local innovation and, 303
  of R&D, 138–39
  of value chains, 80–81, 101*b*, 303
glocal, concept of, 303
GM (General Motors), 137, 286, 287, 296
GNH (Gross National Happiness Index), Bhutan, 231*b*
Going for Growth studies, OECD, 215*b*
Government Performance and Results Act and Program Assessment Rating Tool, U.S., 222
government, role of. *See also* regulatory framework
  business services, providing, 76
  crisis and innovation, 23
  as gardener, xv, 2–3, 8–9, 9*f*, 60*f*
  importance of, 3–4, 7, 8–9
  in-service training, 187–88
  pro-poor innovations. *See under* pro-poor innovations
  procurement policies, 122*b*, 125–30, 128*f*, 129*b*
  R&D, support for. *See under* research and development
  science parks, funding, 315
  supporting competitive and innovative industries, 297–98
  supporting innovators, 73–74
  venture capital funds, public, 91
  "whole of government" approach, 9–11, 54–60, 58*f*, 59*f*
GPS (Global Positioning System), 25, 127
Grameen Bank, Bangladesh, 54, 366–67
Grameen Village Phone program, Bangladesh, 356
Granada, 191
grants
  for in-service training, 187
  for R&D, 153
  for subnational implementation of innovation policy, 254
grassroots innovation networks, 356–62

green revolution, 339, 352
green technology, 23
gross domestic product (GDP)
    alternatives to, 231–32*b*
    divergence between countries and regions, 38, 39*f*
    emerging economies' share in, 43
    innovation-related indicator, expenditure on R&D as, 209*b*
    population growth and, 33*f*
    R&D as percentage of, 135–36, 136*f*, 140–42, 141*f*
    TFP and, 42*t*, 43
Gross National Happiness Index (GNH), Bhutan, 231*b*
groundnuts (peanuts) research in Ghana, 363, 364*b*
Groupement Interprofessionnel des Artisans, Cameroon, 98*b*
groups, innovative, 64–65
Growth Commission Report of 2008, 43
guanxi, 67
Guatemala, 176*b*
Guellec, D., 217
guillotine approach, 119*b*, 124, 132*n*24
Gulf Venture Capital Association, 93
Gupta, Anil, 335
Guyana, 191

H

Haier, 272
hand washing scheme, 349*b*
"hard" versus "soft" networks, 99
HDI (Human Development Index), UN, 231*b*
health care
    medical tourism, 289–90
    Rural Extended Services and Care for Ultimate Emergency Relief program, Uganda, 355
    TCM, 357–58
Health Care Town, Korea, 289

High Tech Double Equity program, Austria, 104*n*8
high-value agricultural commodities, 278–80, 280*t*, 281*b*
highway variable message signs in U.K., 127*b*
Hindustan Unilever, 347
HIV/AIDS
    indigenous and traditional knowledge used to treat, 358*b*, 360–61*b*
    IPR in Brazil and, 113*b*
Hommen, Leif, 125
Honduras, 330*n*1
Hong Kong, China, 42, 43, 193, 219
Hong Kong Productivity Council, 82
horizontal spillovers from SEZs, 306
Hortifruti, 365
housing
    cities and regions, fostering innovation in, 327
    pro-poor innovations, 346*b*, 347
    science park locations, 313
H&Q Asia Pacific, 258*b*
Hsinchu Science Park, Taiwan, China, 310
Hulten, Charles, 42
human capital. *See also* education and training
    brain drain/diaspora workers, 190–94, 192*f*, 194*f*, 250*b*, 257–58*b*, 309*b*
    cities and regions, fostering innovation in, 326, 327
    as innovation-related indicator, 209*b*
    poor as consumers and producers, 344–50, 345*b*, 346*b*
    supply of, 169–70
    urban concentration of, 324
Human Capital Index, UNICI, 201
Human Development Index (HDI), UN, 231*b*

Hummer, 272
Hungary, 121*b,* 132*n*24, 153, 227, 245

I
I-Phone, 25
Ibero-American Network on Science and Technology Indicators (RICYT), 208
IBM, 155, 272
Iceland, 203, 207
ICT. *See* information and communications technology
immaturity of invention, 61–62
implementation of policy, 17–22, 237–69
  adapting best practices to local needs, importance of, 237–38
  building innovative sites, 21, 303–34. *See also* building innovative sites
  champions, networks of, 264, 266–67
  country contexts and capacities driving, 18–19, 19*t*
  critical mass agenda, 244–49, 248*b*
  critical mass problem, 241–42
  emerging innovation leaders agenda, 249
  evaluation as part of, 265
  exports as springboard agenda, 251
  first-mover institutions, creating, 250*b,* 251–52
  first-mover problem, 241
  foresight approach, 265–67
  institutional context agenda, 249–51, 250*b*
  institutional framework, creating, 255–62, 257–58*b,* 263*b*
  knowledge endowments agenda, 251–52, 252*t*
  microreforms, 19, 20*f,* 259–63
  pilot programs, 265, 267, 268
  principles of, 267–68
  pro-poor innovations, 21–22. *See also* pro-poor innovations
  R&D and, 244, 246, 261–62
  scaling up from micro- to macroreforms, 19, 20*f*
  science parks, 248*b,* 260
  self-discovery process, 238–42, 239–40*b*
  strategic incrementalism, 262–67, 264*f*
  at subnational level, 252–55
  supporting key industries, 20–21, 271–301. *See also* supporting competitive and innovative industries
  taxonomy of national agendas, 242–52
    critical mass agenda, 244–49, 248*b*
    emerging innovation leaders agenda, 249
    exports as springboard agenda, 251
    factors and stages involved in, 242–44, 243*t*
    institutional context agenda, 249–51, 250*b*
    knowledge endowments agenda, 251–52, 252*t*
  virtuous cycle, creating, 242–43, 255–62
in-service training, 186–89, 188*b*
Inbev, 272
incandescent light bulbs, 298, 299*n*8
inclusive innovations. *See* pro-poor innovations
incubators, 86–88, 86*b,* 88*b,* 89*b,* 104*n*6, 260, 320
India
  Aravind Eye Hospital, 5*b,* 25
  automobile industry in, 285, 286
  brain drain, 191, 192, 193, 194

cities and regions, fostering
  innovation in, 324
clusters in, 94*b*, 318*b*
competition policy and entry
  barriers in, 118–19
early technological innovation
  in, 39*b*
education and training in, 166,
  172, 186
ICT services in, 63, 94*b*, 172, 253,
  259, 293*b*, 317, 324
implementation of policy in, 245,
  249, 253, 259
indigenous and traditional
  knowledge, laws
  protecting, 362
Industrial Revolution missed
  by, 38, 39*b*
medical tourism in, 289
microcredit in, 368*b*
priorities for, 49–50
pro-poor innovations in, 335, 338,
  339–40, 341*b*, 347, 350, 354,
  355, 360*b*, 362, 363, 368*b*
productivity dispersion in, 45–46
R&D in, 136, 138–39, 140*f*,
  145, 148, 149*b*, 151,
  153, 340, 341*b*
roadside motor mechanic shops, 363
supporting key industries in, 272,
  285, 286, 293*b*
tourism in, 289
indigenous and traditional knowledge
  defined, 357–58
  IPR issues, 114, 359–60, 360–61*b*
  pro-poor innovations using,
    356–62, 358*b*, 360–61*b*
Indonesia, 42, 43, 115*b*, 195*n*2, 368*b*
Industrial Revolution, 6–7, 7*f*,
  33*f*, 34–35, 35*f*, 36*t*
industries, supporting. *See* supporting
  competitive and innovative
  industries

informal sector, 362–69
  association networks, 364–65
  economic role of, 335
  financial services, access to,
    366–69, 368*b*
  needs of poor and, 337
  skills development for, 189,
    327, 328*b*, 369
information and communications
  technology (ICT)
  business services, providing,
    12*b*, 75, 83–84
  education and training
    requirements, 165,
    166, 172, 174, 175
  innovation surveys and,
    209*b*, 211
  Internet access as basic right
    in Estonia, 5*b*, 26
  Internet Protocol, development
    of, 127
  as key services industry, 293–95,
    293*b*, 295*t*
  pro-poor innovations, 338, 354–56
  small firms, importance
    of, 84
  standardization gap, closing, 81
Information Technology and
  Innovation Foundation
  (ITIF), 209*b*
Infosys, 245
Inmetro, Brazil, 81–82
innovation activities, measurement
  of, 212
innovation and innovation policy
  distinguished, 11
Innovation Microdata Project,
  OECD, 215*b*
Innovation Norway, 228
innovation policy, 1–24
  building innovative sites, 21,
    303–34. *See also* building
    innovative sites

concepts of, 5–11. *See also* innovation policy; principles of innovation policy
crisis and innovation, 1–2, 6, 23
defining, 4, 5*b*, 25–27
economic growth from, 6–7, 31–52. *See also* economic growth from innovation policy
education and training, 15–17, 165–98. *See also* education and training
evaluating, 17, 199–234. *See also* evaluation
functions of, 11–17. *See also* education and training; evaluation; regulatory framework; research and development; supporting innovators
government. *See* government, role of
implementation of, 17–22, 237–69. *See also* building innovative sites; implementation of policy; pro-poor innovations; promoting competitive and innovative industries
importance of, xv–xvi, 22–23
models for, 9–11, 11*f*
practical approach to means of, 3–4, 7–11
principles of, 7–11, 53–70. *See also* principles of innovation policy
pro-poor innovations, 21–22, 335–70. *See also* pro-poor innovations
R&D, 14–15, 135–63. *See also* research and development
reasons for developing, 1–2, 6–7
regulatory framework, 13–14, 107–34. *See also* regulatory framework

supporting innovators, 12–13, 73–105. *See also* supporting innovators
supporting key industries, 20–21, 271–301. *See also* supporting competitive and innovative industries
innovation surveys. *See under* evaluation
innovation systems, 8, 58–59, 59*f*, 296
innovators, supporting. *See* supporting innovators
inoculations
  access to, 46, 47*t*
  GAVI, 352, 353
INSEAD/World Business Global Innovation Index (GII), 203, 204–5*t*, 205
INSERM, France, 219
institutions
  implementation of policy and institutional context agenda, 249–51, 250*b*
  institutional framework, creating, 255–62, 257–58*b*, 263*b*
  key role of, 65, 66*f*, 68
  supporting key industries, 298
Instituto Ethos, 350
Instituto Nacional de Biodiversidad, Costa Rica, 292
Instituto Poblano para la Productividad Competitiva, Mexico, 82
Integral Quality and Modernization Program (CIMO), Mexico, 188*b*
Intel, 267, 309*b*
Intel semiconductor assembly plant, Costa Rica, 5*b*, 26
intellectual property rights (IPR), 145–48
  definition of patents, 145–46
  as incentive for R&D, 147–48

indigenous and traditional
    knowledge, 114,
    359–60, 360–61*b*
patent numbers as indicator of
    R&D, 142–45, 144*t*, 151, 152*t*
patent protection, importance
    of, 146–47
regulatory framework for, 14,
    112–16, 113*b*, 115*b*, 131*n*6
Intergovernmental Panel on Climate
    Change (IPCC), 36–37
InterLink Biotechnologies, 257–58*b*
International AIDS Vaccine
    Initiative, 352
International Centre for Research
    in Semi-Arid Tropics,
    Mali, 364*b*
International Finance Corporation,
    344, 346*b*
International Fund for Agricultural
    Development, 281
International Organization for
    Migration, 191
international R&D networks, 157–59
International Telecommunications
    Union, 354
internationalization services, 12*b*, 75
Internet. *See* information and
    communications technology
investment climate
    promotion services for
        innovators, 78–79
    SEZs affected by, 310*b*
    World Bank assessments and
        surveys, 13, 130, 322, 330*n*4
IPCC (Intergovernmental Panel on
    Climate Change), 36–37
IPR. *See* intellectual property rights
Iran, Islamic Republic of
    implementation of policy in, 237,
        244, 247, 249, 261
    UNCTAD innovation policy
        review for, 229

Ireland
    business services in, 77
    implementation of policy in,
        238–41, 239–40*b*, 245,
        246, 249, 253, 260
    National Linkage Promotion
        program, 239–40*b*,
        246, 260, 297
    SEZs in, 304, 309*b*
Isaksson, Anders, 41, 42
Islamic Republic of Iran. *See* Iran,
    Islamic Republic of
"island" SEZs, 305*f*
Israel, 92, 138, 161*n*8, 203, 253
Italy
    business incubators in, 87
    cities and regions, fostering
        innovation in, 327, 328*b*
    clusters in, 104*n*10
    education and training in, 180
    implementation of policy in, 249
    industrial districts in, 64
    investment promotion services for
        innovators, 79
    networks in, 101
    railways as network industries
        in, 121*b*
    R&D in, 138
    regional-level innovation policy
        reviews, 229
    trade associations in, 97*b*
ITIF (Information Technology and
    Innovation Foundation), 209*b*

J

Jacobs, Scott, 124, 132*n*24
Jaguar/Land Rover, 272
Jamaica, 229, 288
Japan
    business services in, 77
    EIS SII, 203
    evolution of innovation policy in,
        56*b*, 57*b*

First and Second Basic Plans, 222
implementation of policy in, 251
IPR in, 131*n*9
land market monitoring systems, 327
precommercial procurement in, 127
R&D in, 136, 138, 139, 158
small firms, financial support for, 92
technology extension services in, 80
transition from developing to developed economy in, 43
urban environments, unemployment concentrated in, 326
Japan External Trade Organization, 89*b*
Japan Productivity Center, 82
Jaruzelsky, Barry, 138
Jaumotte, F., 217, 218
jeans production in Brazil, 285*b*
Jeevani, 360*b*
Jenkins, Mauricio, 306
Jhai Coffee Farmer Cooperative, Lao PDR, 271
Jiao Tong University, Shanghai, China, 153, 155
John Henry effect, 223
Jordan, 92, 132*n*20, 289
Jorge Salazar (cooperative), 365

K

KAM (Knowledge Assessment Methodology), World Bank, 17, 200–201, 203, 204–5*t*, 205, 215*b*
Kandla, India, 310*b*
Kani tribe, India, 360*b*
Kazakhstan, 318*b*
Kenya
    clusters in, 95
    education and training in, 187
    guillotine approach in, 119*b*, 132*n*24
    licensing reforms in, 119*b*, 131–32*n*18
    supporting key industries in, 271–72, 276, 288, 289, 294, 296

Kigali Institute of Science and Technology (KIST), Rwanda, 340, 342*b*
Kim, Ronald, 237*n*, 279*b*
Kinyanjui, Mary Njeri, 321*b*
KIST (Kigali Institute of Science and Technology), Rwanda, 340, 342*b*
Knowledge Assessment Methodology (KAM), World Bank, 17, 200–201, 203, 204–5*t*, 205, 215*b*
knowledge endowments agenda, 251–52, 252*t*
knowledge, indigenous and traditional. *See* indigenous and traditional knowledge
knowledge vouchers, 85*b*
Korea, Republic of
    clusters in, 94
    developing countries, grouped with, 161*n*8
    economic development, effect of innovation on, 6*f*
    education and training in, 170, 184, 186, 187
    guillotine approach in, 132*n*24
    imitation with local improvements in, 115*b*
    implementation of policy in, 243, 249, 253
    innovation policy review in, 228
    land market monitoring systems, 327
    medical tourism in, 289
    OECD innovation policy review of, 225
    as old economic tiger, 42
    Oslo Manual innovation survey, 206–7
    precommercial procurement in, 127
    R&D in, 140–42, 150, 151, 153
    regulatory reform in, 132*n*23

SEZs in, 307–8, 307b, 309b
small firms, support for, 85, 92
transition from developing to developed economy in, 43
urban environments, unemployment concentrated in, 326
kosehtsushi, 80
Kreditanstalt für Wiederaufbau, 91
Kuwait, 88
Kuznetsov, Yevgeny, 73n, 237n

L
lablets, 267
Ladegaard, Peter, 124
Lall, S., 306
land market monitoring systems, 327
land policy, coordination of science parks with, 312
Land Rover/Jaguar, 272
Länder (state) programs, Germany, 91
Lao People's Democratic Republic, 271
Large Hadron Collider (LHC), 63
large-scale programs, 63
Larrain, Felipe, 306
Larsen, Kurt, 165n, 279b
Latin America and Caribbean. *See also specific countries*
   agriculture in, 339
   brain drain, 191
   implementation of policy in, 242, 243, 244, 245
   informal sector in, 362
   innovation surveys, 44, 206–7, 208–10
   MDGs in, 336, 337t
   pro-poor innovations in, 339, 350
learning, new models of, 17b, 175–83, 175t, 177–78b, 182b, 184. *See also* education and training
Lebanon, 88
Legend Computers, 155
Lenovo, 155, 272

Leray, Thais, 107n
Lesotho, 182, 283
levy funds for in-service training, 187–88
LHC (Large Hadron Collider), 63
licensing, 108, 115, 116, 119b, 131n6
life expectancy, 33f
lifelong learning, 173, 174–75, 175t, 176
light bulbs, incandescent, 298, 299n8
linkages, 211–12, 229, 239–40b
Lipman, Geoffrey, 288
literacy rates, 170, 171, 174–75, 195n2
Little Red Schoolhouse Program, Philippines, 351b
livelihood partnerships model for pro-poor innovations, 347, 348t
local productive systems. *See* clusters

M
Macao, China, 193
macroreforms, scaling up to, 19, 20f
Madagascar, 171, 251, 304
Malaria Research and Training Center (MRTC), Bamako, Mali, 5b, 26, 340–41, 342b
Malawi, 275–76
Malaysia
   education and training in, 187, 189
   implementation of policy in, 245
   as new economic tiger, 42
   palm oil industry, 280, 281b
   small firms, support for, 85–86
   transition from developing to developed economy in, 43
Malaysian Agricultural Research and Development Institute, 281b
Maldives, 288
Mali
   education and training in, 171
   indigenous and traditional knowledge, using, 358b, 361b

International Centre for Research in Semi-Arid Tropics, 364*b*
MRTC, Bamako, 5*b*, 26, 340–41, 342*b*
Western cotton subsidies affecting, 111
Malta, 43
Manila Water Company, 346*b*, 347
Manufacturing Extension Partnership (MEP), U.S., 79–80, 100
manufacturing industries, 283–87, 284*b*, 285*b*
mapping clusters, 318–19
Marfouk, A., 191
market ideology, 54–55
marketing services, 12*b*, 75
Marshall, Alfred, 303
maturity of invention, 61–62
Mauritania, 171, 290
Mauritius, 251, 274*b*, 284, 330*n*1
Max Planck *Gesellschaft* model, 219
McCormick, Dorothy, 321*b*
McKinsey & Company, 209*b*, 347
MDGs (Millennium Development Goals), 336, 337*t*
medical tourism, 289–90
Men Sam An, 298*n*2
MEP (Manufacturing Extension Partnership), U.S., 79–80, 100
Mercedes-Benz, 287
metrology services for innovators, 12*b*, 75, 80–82
Mexico
  brain drain, 191
  business incubators in, 88
  business services in, 77–78
  clusters in, 94, 98
  community service initiatives, university-based, 341
  education and training in, 186, 187, 188*b*, 189
  guillotine approach in, 132*n*24
  implementation of policy in, 242, 244, 245, 246
  networks in, 102
  OECD innovation policy review of, 225
  pro-poor innovations in, 341, 346*b*, 347, 355
  productivity centers, 82
  R&D in, 158
  regional-level innovation policy reviews, 229
  supporting key industries in, 272, 297
  value chain, use of innovation policy to move up, 227
MFN (most favored nation) tariff regimes, 109
MG Rover, 101*b*
micro- and small enterprises (MSEs), 337–38
microcredit, 54, 347, 366–69, 368*b*
microlevel innovation surveys. *See under* evaluation
microreforms, 19, 20*f*, 259–63
Middle Ages, economic revolution of, 6–7, 7*f*, 33–35
Middle East and North Africa. *See also specific countries*
  brain drain, 191
  equity fundraising in emerging markets, 92–93, 92*t*
  informal sector in, 362
  procurement policy in, 126
MIGA (Multilateral Investment Guarantee Agency), 290
migration of skilled workers, 190–94, 192*f*, 194*f*, 250*b*, 257–58*b*, 309*b*
military programs, 63, 136
Millennium Development Goals (MDGs), 336, 337*t*
Mittal, 272
MNCs. *See* multinational corporations

mobile telephony, access to, 354, 356
models for innovation policy, 9–11, 11*f*
Moldova, 132*n*24
Monteverde Conservationist Association, Costa Rica, 292
Morocco, 92, 195*n*2, 304
Moscow University Science Park, 248*b*
mosquito nets, 366
most favored nation (MFN) tariff regimes, 109
Mozambique, 132*n*23
MRTC (Malaria Research and Training Center), Bamako, Mali, 5*b*, 26, 340–41, 342*b*
MSEs (micro- and small enterprises), 337–38
Multi-Fiber Agreement, 274*b*, 284
Multilateral Investment Guarantee Agency (MIGA), 290
multinational corporations (MNCs)
   involved in R&D, 135, 138, 140, 153–55
   regulatory framework and, 109
   in science parks, 310, 311
   SEZs attracting, 305, 306
Musau, Ben, 124
mutual insurance schemes, 338
mutual recognition agreements, 112
Mytelka, Lynn K., 318

N

Namibia, 185
Nano, 296
nano-bio-info-cogno complex, 35, 36*f*
National Agricultural Innovation Project, India, 339–40
National Association of Chambers of Commerce and Industry, France, 83
National Council for Science and Technology (CONACYT), Mexico, 77–78
National Federation of Clothing Professionals, Senegal, 98*b*
national implementation agendas, taxonomy of. *See under* implementation of policy
National Institute of Standards and Technology (NIST), U.S., 81
National Institutes of Health, 361*b*
national-level innovation policy reviews, 224–29, 226*b*
National Linkage Promotion program, Ireland, 239–40*b*, 246, 260, 297
National Qualification Frameworks (NQFs), 185
National Renewable Energy Law (2005), China, 273*b*
National Research and Development Corporation, United Kingdom, 56*b*
National Science Foundation, U.S., 56*b*, 219–20
Natura, 350
natural resources, 246–47
NCR Corporation, 355
Neotropic Foundation, 292
Netherlands
   angel investment in, 93
   business incubators in, 87
   clusters in, 104*n*10
   education and training in, 187
   knowledge vouchers in, 85*b*
   railways as network industries in, 121*b*
   Top Technology Institutes, 220
   urban environments, unemployment concentrated in, 326
network industries, 119–21, 121*b*
networks
   of champions, 264, 266–67
   grassroots innovation networks, 356–62

informal sector associations, 364–65
international R&D programs, 157–59
search networks, 267
supporting innovators, 99–100*b*, 99–102, 101*b*
Networks of Centers of Excellence, Canada, 220
new Asian economic tigers, 42, 244, 245, 251
new technology enterprises (NTEs), China, 261
New Zealand
    clusters in, 97
    competition policy in, 131*n*11
    natural resources in, 247
    NQFs, 185
    Oslo Manual innovation survey, 206–7
    program evaluation in, 219
    value chain, use of innovation policy to move up, 227
NGOs. *See* nongovernmental organizations
Nicaragua, 297, 341, 365
Niger, 171
Nigeria, 171, 172, 187
Nissan, 286
NIST (National Institute of Standards and Technology), U.S., 81
Nokia, 137, 247, 267
nongovernmental organizations (NGOs)
    pro-poor innovations, 22, 243, 338, 348, 349, 350, 352, 363, 364*b*, 369
    tourism activities in Costa Rica, 292
North Africa. *See* Middle East and North Africa
Norway
    CIS, 207
    clusters in, 104*n*10
    EIS SII, 203

implementation of policy in, 253
innovation policy review in, 228
natural resources in, 247
Oslo Manual innovation survey, 206–7
railways as network industries in, 121*b*
R&D in, 158
value chain, use of innovation policy to move up, 227
NQFs (National Qualification Frameworks), 185
NTEs (new technology enterprises), China, 261
Nueva Escuela Unitaria (NEU) program, Guatemala, 176*b*

O

obstacles to innovation, removing, 61–63
OECD. *See* Organisation for Economic Co-operation and Development
old Asian economic tigers, 42, 244, 251
Olyset Nets, 366
Oman, 43
one-stop shops
    business services, providing, 76–78, 103
    cross-border trade, 122
Organisation for Economic Co-operation and Development (OECD)
    on brain drain, 191, 192
    on competition policy, 131*n*12
    on education and training, 166, 170
    evolving understanding of innovation policy in, 9, 10–11*b*, 56–57*b*, 58–59
    Going for Growth studies, 215*b*
    Innovation Microdata Project, 215*b*
    innovation surveys, 45, 206–8, 210, 211

international policy learning, commitment to, 215*b*
national-level innovation policy reviews, 224–25, 226*b*, 228–29, 232*n*7
Oslo Manual, 206–8, 210, 211
public venture capital funds, 91
on R&D, 17, 136, 137, 138, 140–41, 153, 219
regional-level innovation policy reviews, 229–30
regulatory framework in, 13
on regulatory frameworks, 130
Science, Technology and Industry Scoreboard and Outlook, 209*b*
SEZs in, 304
small firms, importance of, 83–84
support for innovators in, 13
tariff regimes, 110, 130*n*3
urban environments, unemployment concentrated in, 326
Oryza longistaminata, 361*b*
OSEO, France, 65
Oslo Manual, 206–8, 210, 211

P

Pacific Region. *See* East Asia and Pacific
Pain, N., 217, 218
Pakistan, 95, 171
palm oil industry, Malaysia, 280, 281*b*
Panama, 362
Paraguay, 181, 187
Parkinson's Law, 62
Parkway Health, Singapore, 289
patents. *See* intellectual property rights
path dependence, 227–28
Patrimonio Hoy program, Mexico, 346*b*, 347
peanuts (groundnuts) research in Ghana, 363, 364*b*

pedagogical models, 17*b*, 175–83, 175*t*, 177–78*b*, 182*b*, 184. *See also* education and training
Penang Skills Development Center, Malaysia, 189
performance measurement. *See* evaluation
Perroux, François, 303
Peru, 94, 132*n*20, 195*n*2, 362
Pfizer, 137
pharmaceuticals industry
    Indian R&D in, 151
    IPR in, 113–14, 113*b*, 115*b*
Philippines
    brain drain, 191, 192
    clusters in, 95
    education and training in, 195*n*2, 349*b*
    indigenous and traditional knowledge, laws protecting, 362
    pro-poor innovations in, 346*b*, 349*b*, 351*b*, 356, 362
    regulatory reform in, 132*n*23
    SEZs in, 304
Piedmont Agency for Investment, Export and Tourism, Italy, 79
pilot programs, 132*n*20, 222, 265, 267, 268
PISA (Program for International Student Assessment), 170, 180
point-of-sale (POS) terminals, 354–55
Poland, 153, 241, 245
population growth, 32–34, 32*f*, 33*f*
Porsche, 287
Porter, Michael E., 104*n*10, 317
Portugal, 77, 362
POS (point-of-sale) terminals, 354–55
Pradhan, Sanjay, xvi
Prahalad, C. K., 345
precommercial (technology) procurement, 126–28, 128*f*, 129*b*

principles of innovation policy,
    7–11, 53–70
  broad perspective, importance of,
      9–11, 10f, 54, 55f
  building innovative sites, 63–64
  business environment, promoting,
      60–65, 60f
  capacity for innovation, factors
      affecting, 7, 8f
  cultural promotion of technical
      innovation, 61
  for developing countries, 68–69
  favorable conditions, creating, 58f
  financial incentives and
      instruments, 65–66
  government, role of, 7, 8–9
  groups, innovative, 64–65
  innovation systems, 8, 58–59, 59f
  institutions, role of, 65, 66f, 68
  models for, 9–11, 11f
  obstacles, removing, 61–63
  in OECD, 9, 10–11b, 56–57b, 58–59
  societal specificities, adapting
      innovation to, 67–68,
      67f, 69
  "whole of government"
      approach, 9–11, 54–60,
      58f, 59f
Prius, 25
private sector
  business services, providing, 76
  equity fundraising in emerging
      markets, 92t
  in-service training, involvement
      in, 189
  pro-poor innovations, 338, 344–52,
      345b, 346b, 348t, 349b, 351b
  R&D spending, 136–38, 141–42,
      142t, 143f, 150–57, 151–54t
  role in innovation, 3–4
  science parks, funding, 315
  self-discovery process, 238–39,
      239–40b

pro-poor innovations, 21–22, 335–70
  in agriculture, 338–40, 358b, 361b
  business models for, 347, 348t
  consumers, poor as, 344–45,
      345b, 346b
  CSR initiatives, 338, 350–52, 369n5
  defining, 335–36
  education and training, 349b, 351b,
      355, 363, 369
  employment opportunities,
      348, 351b
  financial services, access to, 354–55,
      356, 366–69, 368b
  financial support for, 343
  global networks, 338, 352–54
  government, role of
      e-government, 355
      private sector, encouraging,
          348–50, 349b
      public R&D, 340–44, 341b, 342b,
          363–66, 364b
  grassroots innovation networks,
      356–62
  ICT, using, 338, 354–56
  indigenous and traditional
      knowledge initiatives,
      356–62, 358b, 360–61b
  informal sector. *See* informal sector
  needs of the poor, understanding,
      336–38, 337t
  NGO involvement, 22, 243, 338, 348,
      349, 350, 352, 363, 364b, 369
  private sector involvement, 338,
      344–52, 345b, 346b, 348t,
      349b, 351b, 365–66
  producers, poor as, 347–50
  R&D
      in agriculture, 339–40
      global networks, 338, 352–54
      informal association networks
          fostering, 364–65
      private sector, government
          support for, 348–49

public and university initiatives, 340–44, 341*b*, 342*b*, 363–66, 364*b*
public-private partnerships, 365–66
role of, 336, 338
socially driven, 350–52, 351*b*
process innovation, 45–46, 45*t*
Proctor and Gamble, 349*b*
procurement policies, 122*b*, 125–30, 128*f*, 129*b*
product innovation, 45–46, 45*t*
productivity centers, 82
productivity of the poor, 347–50
productivity, total factor (TFP), 41–43, 42*f*
program evaluation, 213–24
  designing, 214–15*b*
  field experiments, 222–24
  financial incentives for R&D, 215–18
  national versus international, 215*b*
  public R&D, 218–22, 220*b*
  quantitative versus qualitative, 214*b*
Program for International Student Assessment (PISA), 170, 180
Project Shakti, 347
promotion and marketing services, 12*b*, 75
Public-Private Partnership for Handwashing, 349*b*
public research organizations, 220
public sector. *See* government, role of
public venture capital funds, 91

Q

Qatar n, 318*b*
quality control services for innovators, 12*b*, 75, 80–82
quotas, 110

R

railway systems, 119, 121*b*
Rajiv Gandhi National Drinking Water Mission, India, 341*b*
Ranbaxy, 151
RAND Science and Technology Capacity Index (STCI), 202
R&D. *See* research and development
regional issues
  fostering innovation in regions, 323–29, 325*b*, 328*b*, 329*b*
  implementation of policy at subnational level, 252–55
  innovation policy reviews, regional-level, 229–30
regulatory framework, 13–14, 107–34
  competition policy, 118–21
  cross-border trade, 121–23
  in developing countries, 68–69
  domestic business environment and, 116–24, 117*t*, 119*b*, 121*b*
  entry barriers, 117*t*, 118–21
  evaluating, 123–24
  guillotine approach to, 119*b*, 124, 132*n*24
  IPR, 14, 112–16, 113*b*, 115*b*, 131*n*6
  licensing, 108, 115, 116, 119*b*, 131*n*6
  network industries, 119–21, 121*b*
  obstacles, removing, 123–24
  procurement policies, 122*b*, 125–30, 128*f*, 129*b*
  technology transfer, 108–9
  trade and tax regimes, 13–14, 109–12
  transaction costs related to, 117*t*
Republic of Korea. *See* Korea, Republic of
research and development (R&D), 14–15, 135–63
  in agriculture, 339–40
  business incubators for, 87
  in developed versus developing economies, 139–40, 140*f*

direct government support
mechanisms, 15*t*
exporters and importers of, 138
financial incentives for, 215–18
GDP, as percentage of, 135–36, 136*f*, 140–42, 141*f*
global perspective on, 135–39
globalization of, 138–39
government, role of, 15*t*, 136–38, 137*t*, 141–43, 142*t*, 143*f*, 148–50, 149*b*, 153, 154*t*
    evaluating impact of public sector R&D, 218–22, 220*b*
    pro-poor innovation through public R&D, 340–44, 341*b*, 342*b*
historical growth of U.S. economy attributed to, 40
implementation of policy and, 244, 246, 261–62
in-service training, 186
incentives to innovate via, 145, 146*t*
innovation-related indicator, expenditure on R&D as, 209*b*
international networks, 157–59
IPR and, 115–16, 145–48
macroeconomic conditions affecting, 145
main actors, 136–38, 137*t*, 140–45, 141*f*, 142*t*, 151–52, 151–53*t*
MNCs, 135, 138, 140, 153–55
new areas of, 139
OECD, evolution of innovation policy in, 56*b*
patent numbers as indicator of, 142–45, 144*t*, 151, 152*t*
precommercial (technology) procurement, 126–28, 128*f*, 129*b*
private sector spending on, 136–38, 141–42, 142*t*, 143*f*, 150–57, 151–54*t*

pro-poor innovations. *See under* pro-poor innovations
program evaluation, 213, 215–22, 220*b*
small businesses, support for, 85
supporting innovators, 73, 89
tax incentives for, 153, 216, 217
TFP and, 41
at universities, 137–38, 137*t*, 155–57, 156*t*, 158*t*
Research and Technological Development (RTD) Framework Programme (FP), EU, 221
Research Assessment Exercise, United Kingdom, 219
Research Council of Norway, 228
Research Councils, 220
research institutes and groups conducting program evaluations, 219–20
research labs and location of science parks, 313
research programs, 221–22
research systems, 222
Reva, Anna, 335*n*
review. *See* evaluation
Revised Kyoto Convention on the Simplification and Harmonization of Customs Procedures (World Customs Organization), 131*n*16, 330*n*2
RICYT (Ibero-American Network on Science and Technology Indicators), 208
The Rise of the Creative Class (Florida), 325*b*
roadside motor mechanic shops in India, 363
Roberts, Edward B., 64
Rockefeller Foundation, 159
Romer, Paul, 40

Rural Extended Services and Care for Ultimate Emergency Relief program, Uganda, 355
Russian Federation
   implementation of policy in, 242, 244, 246–47, 248*b*, 249
   Oslo Manual innovation survey, 206–7
   pro-poor innovation in, 350
   R&D in, 136, 137, 140*f*
   regulatory reform in, 132*n*23
   supporting key industries in, 272, 285
Rwanda
   coffee industry, 273, 280–83, 295*t*, 296
   KIST, 340, 342*b*
   Technology Information Service, 366, 367*b*
   tourism in, 289, 290
   village phone program, 356

S
Safaricom, 271–72
Safeguard, 349*b*
SAIC, 286
salmon industry in Chile, 296
Samoa, 360–61*b*
sanitation, access to, 46–47, 48*t*
Saudi Arabia, 88, 252*t*
scalable, embedded distribution model for pro-poor innovations, 347, 348*t*
scaling up from micro- to macroreforms, 19, 20*f*
schools. *See* education and training; universities
Schumpeter, Joseph, 303
Science and Technology Capacity Index (STCI), RAND, 202
Science and Technology Policy Council, Finland, 57*b*, 65

science parks, 310–16
   consensus building and buy-in regarding, 316
   design, build-out, and services, 314–15, 314*b*
   developing, 329–30
   evolution of, 310–11, 312*f*
   financial support for, 315–16
   functional and physical components of, 311
   implementation of policy and, 248*b*, 260
   support for users, 315
   supporting innovators, 98–99
   urban environment, integration with, 311–13, 313*b*
science policy, innovation policy distinguished from, 9
Science, Technology, and Innovation Capacity Building Toolkit, World Bank, 24*n*1
scientific ideology, 54–55
Scotland, R&D in, 158
Scottish Employers Skill Survey, 170
Scottish Enterprise, 79
SEADCo, 308
search networks, 267
sector associations, 96–97, 97*b*, 98*b*
self-discovery process, 238–42, 239–40*b*
Senegal, 98*b*, 171
SERCOTEC (Technical Cooperation Service), Chile, 80, 100–101
Serena Hotels, 351*b*
services industries
   business services. *See* business services
   ICT, 293–95, 293*b*, 295*t*. *See also* information and communications technology
   tourism, 288–92, 291*t*
Seychelles, 330*n*1
SEZs. *See* special economic zones
Sfax science park, Tunisia, 313*b*

Shanghai R&D public service platform, China, 77, 78f
Sharia-compliant private venture capital bank, Bahrain-based, 92, 104n9
Shenzhen High-Tech Industrial and Software Parks, Guangdong province, China, 104n14, 309, 310b
Sichuan Tengzhong Heavy Industrial Machinery, 272
Sida (Swedish International Development Cooperation Authority), 127, 331n10
SII (Summary Innovation Index), EIS, 202–3
Silicon Valley, 64, 191, 192, 258b, 319, 321
Silk Road, 33
simple technologies with major impact on welfare, 46–47, 47t, 48t
Singapore
  business incubators in, 88, 89b
  education and training in, 186, 187
  ICT business services for technopreneurs, 83
  medical tourism in, 289
  as old economic tiger, 42
  supporting key industries in, 272, 297
  transition from developing to developed economy in, 43
  youth entrepreneurship programs in, 181
sites for innovation, building. *See* building innovative sites
Six Countries Program, 61, 69n1
SKS Microfinance, 368b
Slovak Republic, 132n23
small and medium-sized firms (SMEs)
  in-service training program for, 188b
  support for innovators, 83–86, 91–93, 101b

Small Business Administration, U.S., 56b
Small Business Research Initiative, U.K., 85
Smart Communications, Inc., 356
SME Development Plan, Malaysia, 85–86
SMEs. *See* small and medium-sized firms
Smith, Adam, 40
socially driven pro-poor innovations, 350–52, 351b
societal issues
  adapting innovation to societal specificities, 67–68, 67f, 69
  technical culture, promoting, 61
"soft" versus "hard" networks, 99
software industry. *See* information and communications technology
Solow, Robert, 40
South Africa
  automobile industry in, 285
  community service initiatives, university-based, 341
  education and training in, 174, 185, 195n2
  metrology, standards, testing, and quality control, 81
  OECD innovation policy review of, 225
  Oslo Manual innovation survey, 206–7
  pilot programs, 132n20
  pro-poor innovations in, 335, 341, 352
  R&D in, 153, 158
  tourism in, 289
  wine industry in, 54, 55f
South America. *See* Latin America and Caribbean
South Asia. *See also specific countries*
  agriculture in, 339
  education and training in, 171, 190

informal sector in, 335
MDGs in, 336, 337*t*
South Korea. *See* Korea, Republic of
Spain, 76–77, 187, 219, 249
Spark program, China, 262, 263*b*
special economic zones (SEZs), 304–10
   "catalyst" SEZs, 305–6, 305*f*
   defined, 304
   developing, 329
   education and training, 307–8, 308*t*
   high-technology investments, attracting, 308, 309*b*
   implementation of policy and, 21
   investment climate and, 310*b*
   "island" SEZs, 305*f*
   location of, 309–10
   management contracts, 308, 309*b*
   policies encouraging innovation through, 310, 311*t*
   reasons for developing, 305–7
   successful SEZs, pointers for building, 307–10, 308*t*, 309*b*, 310*b*, 311*t*
   supporting innovators, 104*n*14
   vertical spillovers or backward links, 306–7, 307*b*
spillover effects of SEZs, 306
Sri Lanka, 283
standards services for innovators, 12*b*, 75, 80–82
Starbucks, 280, 298*n*1
state. *See* government, role of
STCI (Science and Technology Capacity Index), RAND, 202
strategic incrementalism, 262–67, 264*f*
Sub-Saharan Africa. *See also specific countries*
   agriculture in, 340
   brain drain, 191
   clusters in, 95, 321*b*
   education and training in, 170, 171, 172–73, 173*t*, 181–82, 185, 189, 190
   implementation of policy in, 243, 249, 256
   informal sector in, 335, 362
   MDGs in, 336, 337*t*
   pro-poor innovations in, 335, 340
   road network in West Africa, effects of improving, 131*n*14
   supporting key industries in, 275–76
   trade preferences favoring, 111
subsidies, 110–11, 276
Sujala, 341*b*
Sumitomo, 366
Summary Innovation Index (SII), EIS, 202–3
supporting competitive and innovative industries, 20–21, 271–301. *See also specific industries, e.g. coffee industry, garment industry*
   agriculture. *See* agriculture
   education and training, importance of, 296–97
   examples of, 271–73
   financing, importance of, 297
   institutional framework, 298
   key dimensions of, 274–75, 295–98, 295*t*
   manufacturing, 283–87, 284*b*, 285*b*
   public policy, importance of, 297–98
   services
      business services. *See* business services
      ICT, 293–95, 293*b*, 295*t*. *See also* information and communications technology
      tourism, 288–92, 291*t*
   vulnerabilities, dealing with, 274*b*, 296
supporting innovators, 12–13, 73–105
   business incubators, 86–88, 86*b*, 88*b*, 89*b*, 104*n*6, 260, 320
   business services, providing. *See* business services

clusters, 93–99, 94b, 97b
financial support, 89–93, 90f, 92t
government, role of, 73–74
groups, innovative, 64
knowledge vouchers, 85b
networks, 99–100b, 99–102, 101b
R&D, 73, 89
small firms, 83–86, 91–93, 101b
surveys
    investment climate assessments and surveys, World Bank, 13, 130, 322, 330n4
    microlevel innovation surveys. *See under* evaluation
Sweden
    competence centers, 220
    Estonia and, 5b, 26
    implementation of policy in, 253
    precommercial (technology) procurement in, 129b
    railways as network industries in, 121b
    small firms, financial support for, 92
Swedish Energy Agency, 129b
Swedish International Development Cooperation Authority (Sida), 127, 331n10
Switzerland
    EIS SII, 203
    innovation policy review in, 228
    Oslo Manual innovation survey, 206–7
    railways as network industries in, 121b
    R&D in, 138
    reinvention of DDT in, 62
Syria, 88, 290

T
Tahiti, 288
TAI (Technology Achievement Index), UNDP, 201
Taiwan, China
    brain drain/diaspora, 257, 258b
    clusters in, 96
    developing countries, grouped with, 161n8
    FDI from diaspora populations in, 193
    implementation of policy in, 253, 257, 258b, 260
    as old economic tiger, 42
    R&D in, 142, 150, 151, 161n11
    SEZs in, 307, 309b, 310
    transition from developing to developed economy in, 43
Tanzania
    clusters in, 95
    education and training in, 186, 189
    indigenous and traditional knowledge, using, 358b
    pro-poor innovations in, 351b, 358b, 366
    tourism in, 289
    Western cotton subsidies affecting, 111
Tata, 272, 286
tax incentives for R&D, 153, 216, 217
tax/tariff regimes, 13–14, 109–12
Tbg, 91
TCM (traditional Chinese medicine), 357–58
teaching, new models of, 17b, 175–83, 175t, 177–78b, 182b, 184. *See also* education and training
Technical Cooperation Service (SERCOTEC), Chile, 80, 100–101
technical culture, promoting, 61
Technint, 245
TechnoAgro, 365
Technological Activity Index, UNICI, 201
technological innovation. *See* innovation policy

Technology Achievement Index (TAI), UNDP, 201
Technology Commercialization Handbook, World Bank, 24n1
technology extension services, 12b, 75, 79–80
technology incubators, 86b
Technology Information Service, Rwanda, 366, 367b
technology institutes, 96–97
technology parks. See science parks
technology policy, innovation policy distinguished from, 9
technology (precommercial) procurement, 126–28, 128f, 129b
technology transfer, 108–9
technopreneurs, 83, 150
TechnoServe, 365
TEKEL, Finland, 314b
Tekes, Finland, 57b, 65, 247
Telesecundaria, Mexico, 355
Territorial Development Agency of the Emilia Romagna Region (ERVET), Italy, 100
testing services for innovators, 12b, 75, 80–82
TFP (total factor productivity), 41–43, 42f
Thailand
 automobile industry in, 285
 clusters in, 318b
 implementation of policy in, 245
 indigenous and traditional knowledge, laws protecting, 362
 investment promotion services for innovators, 79
 medical tourism in, 289
 as new economic tiger, 42
 transition from developing to developed economy in, 43

Theus, Florian, 165n, 279b
tigers, Asian, 42, 244, 245, 251
TIMSS (Trends in International Mathematics and Science Study), 174
TNCs (transnational corporations). See multinational corporations
Top Technology Institutes, Netherlands, 220
Torch Programs, China, 87, 89b, 104n14, 261
total factor productivity (TFP), 41–43, 42f
tourism, 288–92, 291t
town and village enterprises (TVEs), China, 262, 263b
Toyota, 25, 137, 286
trade associations, 96–97, 97b, 98b
trade, cross-border, 121–23
trade preferences, 111
trade regimes, 109–12
Trade-Related Aspects of Intellectual Property Rights (TRIPS) agreement, 131n6
trademarks. See intellectual property rights
traditional Chinese medicine (TCM), 357–58
traditional knowledge. See indigenous and traditional knowledge
training. See education and training
transaction costs of regulatory framework, 117t
transnational corporations (TNCs). See multinational corporations
Treetap, 360b
Trends in International Mathematics and Science Study (TIMSS), 174
Trinidad and Tobago, 191
triple helix, 154

TRIPS (Trade-Related Aspects of Intellectual Property Rights) agreement, 131n6
Tropical Botanic Garden and Research Institute, 360b
Tropical Scientific Center, Costa Rica, 292
Tsipouri, Lena J., 125
Tunisia, 313b, 330n1
Turkey
   business incubators in, 104n13
   cities and regions, fostering innovation in, 327
   education and training in, 195n2
   EIS SII, 203
   gold jewelry production in, 283, 284b
   Oslo Manual innovation survey, 206–7
   science parks in, 98
TVEs (town and village enterprises), China, 262, 263b

U

Uganda
   education and training in, 171, 189
   Rural Extended Services and Care for Ultimate Emergency Relief program, 355
   supporting key industries in, 276
   village phone program, 356
   Western cotton subsidies affecting, 111
UIS (UNESCO Institute for Statistics), 210
Ukraine, 132n24, 237, 244
UN Habitat, 326
UN Industrial Development Organization, 331n10
UNCTAD (United Nations Conference on Trade and Development), 110, 201, 229
UNDP (United Nations Development Programme), 171, 201
UNESCO (United Nations Educational, Scientific, and Cultural Organization), 210, 215b, 289
UNICEF (United Nations Children's Fund), 349b
UNICI (United Nations Innovation Capability Index), 201
UNIDO (United Nations Industrial Development Organization), 94b
United Arab Emirates, 289, 324, 330n1
United Kingdom
   angel investment in, 93
   brain drain, 191
   business incubators in, 87
   business services in, 76, 77
   cities and regions, fostering innovation in, 324, 328
   education and training in, 170, 185, 187
   evolution of innovation policy in, 56b
   foresight approach in, 265, 266
   highway variable message signs, 127b
   IPR in, 131n7
   NQFs, 185
   one-stop shops in, 103
   public venture capital funds, 91
   R&D in, 138
   regional-level innovation policy reviews, 229
   Research Assessment Exercise, 219
   Research Councils, 220
   research systems, 222
   SEZs in, 310
   small firms, support for, 84–85, 91–92, 104n5
   urban environments, unemployment concentrated in, 326
   value chains, globalization of, 101b

United Nations
  CSR schemes, support for, 350
  HDI, 231*b*
United Nations Children's Fund
    (UNICEF), 349*b*
United Nations Conference on Trade
    and Development
    (UNCTAD), 110, 201, 229
United Nations Development
    Programme (UNDP),
    171, 201
United Nations Educational, Scientific,
    and Cultural Organization
    (UNESCO), 210, 215*b*, 289
United Nations Industrial
    Development Organization
    (UNIDO), 94*b*
United Nations Innovation Capability
    Index (UNICI), 201
United States
  angel investment in, 93
  ATP, 221–22
  automobile industry in, 285
  business services in, 76, 77
  clusters in, 318*b*
  competition policy in, 131*n*11
  EIS SII, 203
  evolution of innovation policy
      in, 56*b*
  Government Performance and
      Results Act and Program
      Assessment Rating Tool, 222
  historical economic growth in, 38–40
  implementation of policy in, 256
  indigenous and traditional
      knowledge, laws
      protecting, 362
  innovation surveys, 206–7, 232*n*4
  IPR in, 147
  manufacturing in, 283
  medical tourism from, 289–90
  metrology, standards, testing, and
      quality control, 81
  military programs in, 63
  networks in, 99–100
  precommercial procurement in, 127
  program evaluation in, 219–20
  R&D in, 136, 138, 139,
      143, 153, 154–55
  Silicon Valley, 64
  technology extension services
      in, 79–80
  tourism in, 289
  trade/tariff regimes, 110, 111
  urban environments,
      unemployment
      "concentrated in, 326
  youth entrepreneurship programs
      in, 181, 182*b*
universities. *See also* education
    and training
  community service initiatives, 341–44
  pro-poor innovations, role in,
      340–44, 341*b*, 342*b*,
      363–65, 364*b*
  R&D associated with, 137–38,
      137*t*, 155–57, 156*t*, 158*t*
  science parks and presence of, 313
urban environments
  fostering innovation in, 323–29,
      325*b*, 328*b*, 329*b*
  integration of science parks with,
      311–13, 313*b*
Urban Operational Centers for
    Economic Renewal
    (CUORE), Italy, 328*b*
Uruguay, 44
USAID (U.S. Agency for International
    Development), 281,
    297, 331*n*10
Utz, Anuja, 335*n*

V

vaccines
  access to, 46, 47*t*
  GAVI, 352, 353

value chains
   in critical mass agenda, 244–46
   globalization of, 80–81, 101b, 303
   use of innovation policy to move up, 227, 291t
van Pottelsberghe, B., 217
Van Welsum,. Désirée, 199n
Vancouver Agreement, 329b
variable message signs on U.K. highways, 127b
Vector Health International Ltd., 366
venture capital, 89–90, 90f, 91, 258b, 260
vertical spillovers or backward links from SEZs, 306–7, 307b
vested interests, as obstacle, 62
VET (vocational education and training), 183–86, 320
Vietnam
   ICT services in, 294–95, 295t
   implementation of policy in, 244, 251
   supporting key industries in, 273, 289, 294–95, 295t
   tourism in, 289
Vietnam Post and Telecommunication Corporation, 294
Vietnam Singapore investment park, 89b
Vinnova, 127, 331n10
virtuous cycle of innovation, creating, 242–43, 255–62
vocational education and training (VET), 183–86, 320
Volkswagen, 286

W

Wal-Mart, 25, 54, 363, 365
Walden International Investment Group, 258b
Walhalla High School, NC, entrepreneurship program, 182
walkie-talkies, 354, 355
water, access to, 46–47, 47t
Water and Sanitation Program, 349b
WBI (World Bank Institute), xv, 121
WEF. See World Economic Forum
Weichai Power, 287
Western Europe. See Europe/European Union
WGC (World Gold Council), 284b
White, Justin, 303n
WHO (World Health Organization), 366
"whole of government" approach, 9–11, 54–60, 58f, 59f
wind power in China, 272, 273b
wine industry
   California wine cluster, 317
   in South Africa, 54, 55f
Wockhardt, India, 289
women
   pro-poor innovations, 347
   SEZs employing, 307
World Bank
   clusters, 331n10
   country income classifications, 42
   Doing Business surveys, 13, 132n21, 322, 330n4
   on education and training, 171, 183, 184
   international policy learning, commitment to, 215b
   investment climate assessments and surveys, 13, 130, 322, 330n4
   KAM, 17, 200–201, 203, 204–5t, 205, 215b
   Public-Private Partnership for Handwashing, 349b
   Science, Technology, and Innovation Capacity Building Toolkit, 24n1
   support for key industries, 281, 297
   Technology Commercialization Handbook, 24n1

World Bank Institute (WBI), xv, 121
World Business/INSEAD Global
    Innovation Index (GII), 203,
    204–5t, 205
World Customs Organization,
    131n16, 330n2
World Development Report 2005
    (World Bank), 118
World Economic Forum (WEF)
    evaluation tools, 17
    GCI, 203, 204–5t, 205, 215, 232n2
World Gold Council (WGC), 284b
World Health Organization
    (WHO), 366
World Heritage sites, UNESCO, 289
World Resources Institute, 344
World Trade Organization (WTO), 110

Y

youth entrepreneurship
    programs, 180–81, 182b
Yunus, Muhammad, 54

Z

Zaï, 358b
Zambia, 276
Zetsche, Dieter, 287
Zhongguancun Science Park, Beijing,
    China, 104n14
Zimbabwe, 95, 276

ECO-AUDIT
Environmental Benefits Statement

The World Bank is committed to preserving endangered forests and natural resources. The Office of the Publisher has chosen to print ***Innovation Policy: A Guide for Developing Countries*** on recycled paper with 50 percent postconsumer fiber in accordance with the recommended standards for paper usage set by the Green Press Initiative, a nonprofit program supporting publishers in using fiber that is not sourced from endangered forests. For more information, visit www.greenpressinitiative.org.

Saved:
- 13 trees
- 4 million Btu of total energy
- 1,273 lb. of net greenhouse gases
- 6,131 gal. of waste water
- 372 lb. of solid waste

green press INITIATIVE